VOLUME ONE HUNDRED AND THIRTY TWO

PROGRESS IN MOLECULAR BIOLOGY AND TRANSLATIONAL SCIENCE

Trafficking of GPCRs

VOLUME ONE HUNDRED AND THIRTY TWO

PROGRESS IN MOLECULAR BIOLOGY AND TRANSLATIONAL SCIENCE

Trafficking of GPCRs

Edited by

GUANGYU WU

Department of Pharmacology and Toxicology, Medical College of Georgia, Georgia Regents University, Augusta, Georgia, USA

ELSEVIER

AMSTERDAM • BOSTON • HEIDELBERG • LONDON
NEW YORK • OXFORD • PARIS • SAN DIEGO
SAN FRANCISCO • SINGAPORE • SYDNEY • TOKYO

Academic Press is an imprint of Elsevier

Academic Press is an imprint of Elsevier
225 Wyman Street, Waltham, MA 02451, USA
525 B Street, Suite 1800, San Diego, CA 92101-4495, USA
125 London Wall, London, EC2Y 5AS, UK
The Boulevard, Langford Lane, Kidlington, Oxford OX5 1GB, UK

First edition 2015

ISBN: 978-0-12-802939-8
ISSN: 1877-1173

For information on all Academic Press publications visit our website at store.elsevier.com

CONTENTS

CONTRIBUTORS

Sana Al Awabdh
INSERM U894, and Centre de Psychiatrie et Neurosciences, Université Paris Descartes, Sorbonne Paris Cité, Paris, France

Annette G. Beck-Sickinger
Faculty of Biosciences, Pharmacy and Psychology, Institute of Biochemistry, Universität Leipzig, Leipzig, Germany

Shanna L. Bowman
Department of Biological Sciences, Carnegie Mellon University, Pittsburgh, Pennsylvania, USA

Christopher Cottingham
Department of Biology and Chemistry, Morehead State University, Morehead, Kentucky, USA

Michèle Darmon
INSERM U894, and Centre de Psychiatrie et Neurosciences, Université Paris Descartes, Sorbonne Paris Cité, Paris, France

Jason E. Davis
Department of Pharmacology and Toxicology, Medical College of Georgia, Georgia Regents University, Augusta, Georgia, USA

Denis J. Dupré
Department of Pharmacology, Dalhousie University, Halifax, Nova Scotia, Canada

Michel-Boris Emerit
INSERM U894, and Centre de Psychiatrie et Neurosciences, Université Paris Descartes, Sorbonne Paris Cité, Paris, France

Craig J. Ferryman
Department of Biology and Chemistry, Morehead State University, Morehead, Kentucky, USA

Catalin M. Filipeanu
Department of Pharmacology, College of Medicine, Howard University, Washington, District of Columbia, USA

Qin Fu
Department of Pharmacology, School of Basic Medicine, Tongji Medical College, Huazhong University of Science and Technology, Wuhan, PR China

Eugenia V. Gurevich
Department of Pharmacology, Vanderbilt University, Nashville, Tennessee, USA

Vsevolod V. Gurevich
Department of Pharmacology, Vanderbilt University, Nashville, Tennessee, USA

Yoshikazu Imanishi
Department of Pharmacology, School of Medicine, Case Western Reserve University, Cleveland, Ohio, USA

Justine E. Kennedy
Department of Molecular Pharmacology and Therapeutics, Loyola University Chicago, Health Sciences Division, Maywood, Illinois, USA

Wolfgang Klein
Leibniz-Institut für Molekulare Pharmakologie (FMP), Berlin, Germany

Adriano Marchese
Department of Molecular Pharmacology and Therapeutics, Loyola University Chicago, Health Sciences Division, Maywood, Illinois, USA

Justine Masson
INSERM U894, and Centre de Psychiatrie et Neurosciences, Université Paris Descartes, Sorbonne Paris Cité, Paris, France

Karin Mörl
Faculty of Biosciences, Pharmacy and Psychology, Institute of Biochemistry, Universität Leipzig, Leipzig, Germany

Ina Nemet
Department of Pharmacology, School of Medicine, Case Western Reserve University, Cleveland, Ohio, USA

Manojkumar A. Puthenveedu
Department of Biological Sciences, Carnegie Mellon University, Pittsburgh, Pennsylvania, USA

Kausik Ray
Scientific Review Branch, NIDCD, National Institutes of Health, Bethesda, MD, USA

Philip Ropelewski
Department of Pharmacology, School of Medicine, Case Western Reserve University, Cleveland, Ohio, USA

Claudia Rutz
Leibniz-Institut für Molekulare Pharmakologie (FMP), Berlin, Germany

Ralf Schülein
Leibniz-Institut für Molekulare Pharmakologie (FMP), Berlin, Germany

Qin Wang
Department of Cell, Developmental and Integrative Biology, University of Alabama at Birmingham, Birmingham, Alabama, USA

Jaime Wertman
Department of Microbiology and Immunology, Dalhousie University, Halifax, Nova Scotia, Canada

Guangyu Wu
Department of Pharmacology and Toxicology, Medical College of Georgia, Georgia Regents University, Augusta, Georgia, USA

Yang K. Xiang
Department of Pharmacology, University of California, Davis California, USA

Brent Young
Department of Pharmacology, Dalhousie University, Halifax, Nova Scotia, Canada

Maoxiang Zhang
Department of Pharmacology and Toxicology, Medical College of Georgia, Georgia Regents University, Augusta, Georgia, USA

PREFACE

G protein-coupled receptors (GPCRs) (also known as seven-transmembrane domain receptors or 7TMRs) constitute the largest family of cell surface receptors involved in signal regulation under diverse physiological and pathological conditions and are drug targets for many diseases. Extensive studies carried out over the past 2–3 decades have clearly demonstrated that the spatiotemporal regulation of GPCR intracellular trafficking, including the cell surface export, internalization, recycling, and degradation, is a crucial mechanism that controls receptor transport to the right place which in turn dictates the integrated responses of the cell to hormones and drugs at the right time. Adding to the complexity, each of these trafficking processes is mediated by multiple pathways and is highly regulated by many factors, such as structural determinants, specific motifs, interacting proteins, posttranslational modifications, and transport machineries, altogether coordinating receptor transport using very specialized routes. GPCR trafficking is rapidly evolving and has great potential to translate into new therapeutics.

The main purpose of this book is to review our current understanding of intracellular trafficking of some well-characterized GPCRs. In addition, this book will also highlight the roles of trafficking in regulating the functionality of the receptors and pinpoint current challenges and future directions in studying GPCR trafficking. The contributors are experts in this area with many years of experience. It is my hope that this book will be useful to graduate students, postdoctoral fellows, and researchers who are interested in general GPCR biology or intracellular trafficking of GPCRs.

I am grateful to each of the contributors for their valuable time and tremendous efforts to make this book possible. It is my great pleasure to work with them to put together a book on this very important topic in GPCR biology. I thank Dr. P. Michael Conn, the Chief Editor of the *Progress in Molecular Biology and Translational Science* series, for inviting me to edit this volume and always being supportive. I also would like to take this opportunity to thank my former mentor, Dr. Stephen M. Lanier, for leading me into the GPCR field.

GUANGYU WU

CHAPTER ONE

Arrestins: Critical Players in Trafficking of Many GPCRs☆

Vsevolod V. Gurevich[1], Eugenia V. Gurevich
Department of Pharmacology, Vanderbilt University, Nashville, Tennessee, USA
[1]Corresponding author: e-mail address: vsevolod.gurevich@vanderbilt.edu

Contents

Abstract

Arrestins specifically bind active phosphorylated G protein-coupled receptors (GPCRs). Receptor binding induces the release of the arrestin C-tail, which in non-visual arrestins contains high-affinity binding sites for clathrin and its adaptor AP2. Thus, serving as a physical link between the receptor and key components of the internalization machinery of the coated pit is the best-characterized function of non-visual arrestins in GPCR trafficking. However, arrestins also regulate GPCR trafficking less directly by orchestrating their ubiquitination and deubiquitination. Several reports suggest that arrestins play additional roles in receptor trafficking. Non-visual arrestins appear to be required for the recycling of internalized GPCRs, and the mechanisms of their function in this case remain to be elucidated. Moreover, visual and non-visual arrestins were shown to directly bind *N*-ethylmaleimide-sensitive factor, an important ATPase involved in vesicle trafficking, but neither molecular details nor the biological role of these interactions is clear. Considering how many different proteins arrestins appear to bind, we can confidently expect the elucidation of additional trafficking-related functions of these versatile signaling adaptors.

☆We use systematic names of arrestin proteins: arrestin-1 (historic names S-antigen, 48 kDa protein, visual or rod arrestin), arrestin-2 (β-arrestin or β-arrestin1), arrestin-3 (β-arrestin2 or hTHY-ARRX), and arrestin-4 (cone or X-arrestin; for unclear reasons, its gene is called "arrestin 3" in the HUGO database).

Progress in Molecular Biology and Translational Science, Volume 132
ISSN 1877-1173
http://dx.doi.org/10.1016/bs.pmbts.2015.02.010

ABBREVIATIONS

AIP4 atrophin-1-interacting protein 4
AP2 adaptor protein 2
β2AR β_2-adrenergic receptor
GPCR G protein-coupled receptor
GRK G protein-coupled receptor kinase
Nedd4 neural precursor cell expressed developmentally down-regulated protein 4

1. ARRESTINS AND GPCR TRAFFICKING

Preferential binding of arrestins to active phosphorylated receptors was discovered about 30 years ago.[1] The finding that arrestin binding suppresses receptor coupling to cognate G proteins was made soon after in the visual system.[2] The mechanism turned out to be remarkably simple: direct competition between arrestin and G protein for overlapping sites.[3,4] For some time, it appeared that the only function arrestins have is to bind active phosphorylated G protein-coupled receptors (GPCRs), precluding receptor interactions with G proteins by direct competition.[3,4] The first described non-GPCR binding partners of arrestins were trafficking proteins: clathrin in 1996[5] and clathrin adaptor AP2 a few years later.[6] These data demonstrated that arrestins play an essential role not only in GPCR desensitization[7] but also in receptor endocytosis,[8] via trafficking signals added by receptor-bound arrestins. The discovery that arrestins are ubiquitinated upon receptor binding and regulate ubiquitination of GPCRs[9] revealed yet another mechanism, whereby arrestins regulate receptor trafficking indirectly. Here, we discuss several known mechanisms of arrestin effects on GPCR trafficking and highlight observations that suggest that there are many other mechanisms that still remain to be elucidated.

2. NON-VISUAL ARRESTINS MEDIATE GPCR INTERNALIZATION VIA COATED PITS

Arrestins promote GPCR internalization by virtue of recruitment of clathrin and AP2 via fairly well-mapped binding sites in the C-tail of non-visual arrestins[5,6,10,11] (Fig. 1). Interestingly, the C-tail in the basal conformation of all arrestins is anchored to the N-domain,[12–16] whereas receptor binding triggers its release.[17–19] The expression of separated arrestin C-tail carrying these sites inhibits GPCR internalization, apparently by winning

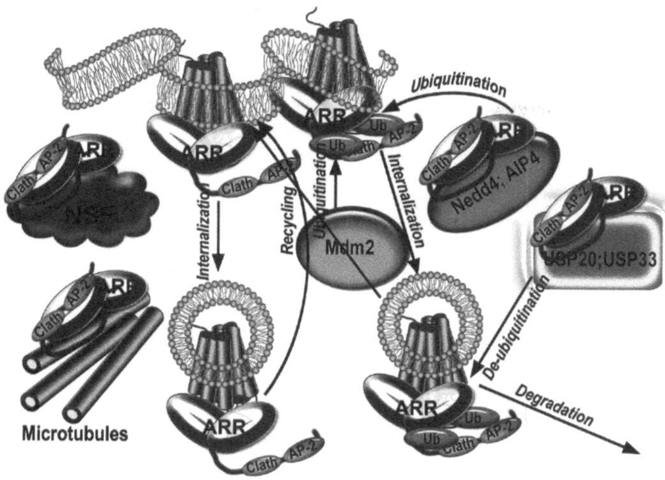

Figure 1 Arrestins play many roles in GPCR trafficking. Arrestins (ARR) bind active phosphorylated GPCRs (shown as a seven-helix bundle). Receptor binding induces the release of the arrestin C-tail, which carries binding sites for clathrin (Clath) and adaptor protein-2 (AP2). The interactions of these sites with clathrin and AP2 promote receptor internalization via coated pits. Arrestins also recruit ubiquitin ligases Mdfm2, Nedd4, and AIP4 to the complex, which favors ubiquitination of both non-visual arrestins and at least some GPCRs. Arrestins also recruit certain deubiquitination enzymes (USP20 and USP33 are shown), facilitating receptor deubiquitination. The role of arrestin interactions with microtubules, centrosome, and *N*-ethylmaleimide-sensitive factor (NSF) in trafficking of GPCRs and/or other proteins remains to be elucidated. (See the color plate.)

the competition with the arrestin–receptor complexes for clathrin and AP2.[20] This finding provided the first clear evidence of functional significance of shielding of the arrestin C-tail in the basal conformation and its release upon receptor binding. In free arrestins, the C-tail is anchored to the body of the molecule, which makes it inaccessible, preventing its competition with the receptor-bound arrestins for the components of internalization machinery (reviewed in Ref. 21).

Another known mechanism of arrestin recruitment to the coated pit is its direct binding to phosphoinositides, which was reported to be necessary for GPCR internalization.[22] Since resident coated pit protein AP2 is also recruited to this part of the membrane via phosphoinositide binding,[23] one might think that as soon as the arrestin–receptor complex is formed, it has no choice but to move to the coated pit. However, this does not appear to be the case. In muscarinic M2 receptor, which was among the first shown to bind arrestins,[24] two Ser/Thr clusters in the third cytoplasmic loop were identified as critical for arrestin binding and receptor desensitization.[25] Yet the elimination of these clusters, and even dominant-negative dynamin K44A mutant that blocks the internalization of β2AR in the same cells, did not prevent M2 endocytosis, suggesting that M2 receptor does not use coated pits and internalizes in an arrestin-independent manner.[25] Interestingly, overexpression of non-visual arrestins can redirect some M2 to coated pits,[25] suggesting that this receptor can use more than one route. Many other GPCRs were shown to have that choice. For example, chemokine receptor CCR5 uses both phosphorylation- and arrestin-dependent and -independent pathways.[26] Cysteinyl leukotriene type 1 receptor internalizes normally in mouse embryonic fibroblasts lacking both non-visual arrestins, yet arrestin expression facilitates its internalization,[27] apparently directing it to the arrestin-dependent pathway, which is usually not preferred, similar to M2 receptor.[25] Metabotropic glutamate receptor mGluR1a constitutively internalizes via arrestin-independent mechanism, whereas its agonist-dependent internalization appears to be mediated by arrestin-2.[28] Endogenous and overexpressed serotonin 5HT4 receptor internalizes via arrestin-dependent pathway, but the deletion of Ser/Thr cluster targeted by G protein-coupled receptor kinases (GRKs) redirects it to an alternative pathway and even facilitates its internalization.[29]

Thus, it appears that the ability of GPCRs to use more than one internalization pathway is a general rule, rather than an exception, likely representing one of the many backup mechanisms cells usually have. Many receptors have recognizable internalization motifs in their sequence, so arrestin binding simply adds new ones. The relative strength of these motifs, as well as the arrestin expression levels, likely determines the pathway(s) each receptor chooses in a particular cell. The dominant internalization pathway of a particular receptor is not necessarily the same in different cell types, or even at different functional states of the same cell (reviewed in Ref. 8). Variety, rather than uniformity, characterizes the world of GPCR signaling and trafficking.[30]

3. VISUAL ARRESTINS AND TRAFFICKING PROTEINS

In vertebrate rod photoreceptors, rhodopsin is localized on the discs, which are detached from the plasma membrane[31] and therefore are topologically equivalent to vesicles with internalized non-visual GPCRs. Thus, vertebrate rhodopsin is not supposed to be internalized. Indeed, arrestin-1, which is the prevalent arrestin isoform in both rods and cones,[32] does not have conventional clathrin- or AP2-binding elements in its C-tail.[33] However, sequence comparison of arrestin-1 and non-visual subtypes shows that in the region homologous to AP2-binding motif in arrestin-2 and -3, only one positive charge is missing.[34] Therefore, it is hardly surprising that arrestin-1 also binds AP2, albeit with ~30 times lower affinity.[34] Constitutively active rhodopsin–K296E is a naturally occurring mutant that causes autosomal dominant retinitis pigmentosa in humans, apparently due to constitutive phosphorylation and formation of a stable complex with arrestin-1.[35] The concentration of rhodopsin in the outer segment of rods reaches ~3 mM.[31] Rods also express roughly 8 arrestin molecules per 10 rhodopsins,[36–38] so the concentrations of both proteins and their complex formed in bright light are very high. It turns out that at these concentrations even low affinity matters: the presence of WT arrestin-1 facilitates rod death in animals expressing rhodopsin–K296E, with visible accumulation of AP2 in the outer segment, where it is not observed in normal mice.[34] In contrast, truncated arrestin-1 lacking the C-tail containing the low-affinity AP2-binding site protects photoreceptors in these animals and preserves their function.[34] Thus, in rod and cone photoreceptors, both of which express very high levels of arrestin-1,[32] even relatively low-affinity interactions, which would not matter in other cells, with submicromolar concentrations of both non-visual arrestins,[39,40] can become biologically relevant.

Interestingly, the localization of rhodopsin on invaginations of the plasma membrane in flies, in contrast to detached discs in vertebrate rods, is one of the many differences between vertebrate and invertebrate photoreceptors. Another difference directly follows from this localization: *Drosophila* rhodopsin is internalized, like "normal" vertebrate GPCRs, via clathrin- and AP2-mediated mechanism.[41] In fly photoreceptors, arrestin is evenly distributed, whereas in dark-adapted vertebrate rods, it is concentrated in the inner segment, with fairly small fraction in the outer segment, where rhodopsin resides.[36–38] However, in both types of photoreceptors upon illumination, arrestin translocates to rhodopsin-containing membranes.[36–38,42–45] Like non-visual arrestins, and in contrast to vertebrate visual arrestin,[22] visual arrestin in *Drosophila* has high-affinity phosphoinositide-binding site.[43] It was proposed that due to phosphoinositide binding, *Drosophila* arrestin translocates to rhodopsin on phosphoinositide-rich vesicles moved with the help of *Drosophila* myosin III (NINAC).[42] The participation of NINAC in metarhodopsin inactivation in *Drosophila* was independently confirmed,[46] but arrestin translocation was found to be largely driven by its binding to rhodopsin in flies,[44] just like in mice.[45] Thus, the internalization of invertebrate rhodopsin apparently follows the same rules as many non-visual GPCRs: active receptor recruits arrestin via direct binding,[47] which then links it to the key components of the coated pit.[5,6,41]

4. UBIQUITINATION AND DEUBIQUITINATION IN GPCR CYCLING AND SIGNALING

Monoubiquitination of many proteins regulates their trafficking and signaling, rather than proteasomal degradation.[48] Two GPCRs, β2AR[9] and chemokine receptor CXCR4,[49] were shown to be ubiquitinated in response to agonist activation. Arrestin ubiquitination upon receptor binding, as well as the role of arrestin in GPCR ubiquitination, was discovered a few years later than the interactions of non-visual arrestins with clathrin and AP2.[9] It appears that arrestin ubiquitination by Mdm2 prolongs the life of the arrestin–receptor complex.[50] As only receptor-bound arrestins facilitate ERK1/2 activation,[51,52] it is natural that arrestin ubiquitination increases ERK1/2 activation induced by GPCR stimulation.[53] Slow deubiquitination of the receptor-bound arrestin prolongs the dwell time of the complex inside the cell and slows down receptor recycling.[50] However, receptor or arrestin ubiquitination per se does not appear to be necessary for arrestin-dependent internalization: virtually complete suppression of agonist-induced ubiquitination of arrestin-2 does not appreciably affect endocytosis of β2AR.[54] Arrestin-2 recruits ubiquitin ligase AIP4 to ubiquitinate CXCR4, which affects endosomal sorting of this receptor.[55] Receptor-bound arrestin-3 recruits yet another ubiquitin ligase, Nedd4, which ubiquitinates β2AR, and this receptor modification is required for lysosomal targeting of internalized β2AR,[56] although arrestin domain-containing protein 3 was also suggested as the mediator of the interaction of Nedd4 with β2AR.[57,58] Finally, both non-visual arrestins bind a fourth ubiquitin ligase, parkin.[54] Interestingly, parkin binding enhances arrestin interactions with Mdm2, but paradoxically strongly reduces arrestin ubiquitination in response to receptor activation.[54] The possible role of parkin in receptor modification remains to be elucidated. To further complicate matters, arrestins were found to recruit deubiquitinating enzymes USP20 and USP33 to β2AR, which facilitate receptor recycling and resensitization.[59,60]

To summarize, it is clear that arrestins bind several ubiquitin ligases and recruit them at least to some GPCRs. Both arrestins and GPCRs are ubiquitinated upon receptor stimulation. Receptor ubiquitination appears to play a role in sorting and lysosomal targeting, whereas the ubiquitination of arrestins likely affects their affinity for receptors. However, arrestin-mediated recruitment of some deubiquitinating enzymes suggests that their role in GPCR trafficking is more complex and includes postendocytotic steps. Interestingly, the role of arrestins in recruiting deubiquitinases was shown on β2AR,[59,60] which appears to contradict the idea that arrestins bound to this particular receptor dissociate from it very quickly.[61] Thus, the biological functions of arrestin-assisted ubiquitination and deubiquitination of GPCRs and similar modifications of non-visual arrestins need to be further clarified. One should also keep in mind that the role of the same processes in trafficking of different GPCRs is not necessarily the same: the very fact that animals have so many members of this superfamily suggests that variety, rather than uniformity, is the key.[30]

5. FASTER CYCLING PREVENTS RECEPTOR DOWNREGULATION

With very few exceptions, the fate of internalized receptors is not predetermined: they can be recycled back to the plasma membrane and reused, or sent to lysosomes and destroyed.[7] The latter process leads to the reduction of overall receptor number, usually termed downregulation. We do not know how the choice between recycling and elimination is made, but it appears that the intensity and/or duration of signaling can tip the scales one way or another. In the process of internalization and recycling, most receptors transition through several functional states. First, in case of GPCRs that internalize via arrestin-dependent pathway, after phosphorylation by GRKs and arrestin binding receptors, move into coated vesicles and then to endosomes. The internal pH in endosomes is much lower than on the extracellular side of the membrane.[62] It is likely (but remains unproven) that acidification facilitates the dissociation of the ligand. The loss of the bound agonist and consequent transition into inactive state is the only conceivable mechanism of subsequent release of bound arrestins: both non-visual subtypes demonstrate lower binding to inactive phosphoreceptors,[63–65] even though the difference is not as dramatic as in the case of visual arrestin-1.[66,67] Arrestin dissociation is necessary to make receptor-attached phosphates accessible to phosphatases,[68] so it must precede receptor dephosphorylation. Since both non-visual arrestins require at least two phosphates for high-affinity binding,[63] dephosphorylation has to be a multistep process. It must be completed, as it appears that only fully dephosphorylated receptors are recycling competent.[69,70] One conceivable model is that only certain functional states of the receptor can be diverted to lysosomes and destroyed; and the other is that every state can be transported to lysosomes, so that the longer the time that a GPCR spends in the endosomal compartment, the higher the probability that it will be transported to lysosomes and destroyed.

Similar to visual arrestin-1, both non-visual arrestins can be made to bind active unphosphorylated GPCRs by mutations destabilizing the main phosphate sensor, the polar core, by mutations detaching the C-tail from the body of the molecule, or by C-tail deletions.[64,65,71] The effect of two different arrestin-2 mutants, one activated by polar core mutation and the other by the C-tail detachment, on cycling of β2AR was tested in cells.[72] Since these forms of arrestin-2 bind the same active receptor as GRKs, they actually compete with GRKs and suppress receptor phosphorylation both *in vitro*, in the system reconstituted from purified proteins, and in cells.[72] It turned out that in cells, these preactivated arrestin-2 mutants bind unphosphorylated β2AR and induce its internalization. Interestingly, unphosphorylated β2AR internalized in complex with these mutants recycles very rapidly, much faster than in the presence of WT arrestin-2 that only binds phosphorylated receptor.[72] Importantly, the expression of phosphorylation-independent arrestin-2 mutants protected the receptor from downregulation, so that, in sharp contrast to cells expressing WT arrestin-2, even after 24 h of agonist exposure virtually no β2AR was lost.[72] This was the first study of the effect of the nature of the arrestin–receptor complex on the fate of internalized receptor. It did not answer all questions. The results can be interpreted in the context of both models: (1) as an indication that rapid cycling reduces the chances of the receptor to be diverted to lysosomes, or (2) as a suggestion that only phosphorylated forms of the receptor are diverted to that compartment and destroyed. The use of nonphosphorylatable β2AR mutants in similar experiments is necessary to resolve this issue.

6. ARRESTINS IN RECEPTOR RECYCLING AND VESICLE TRAFFICKING: QUESTIONS WITHOUT ANSWERS

The mechanism whereby arrestin-2 and -3 participate in GPCR internalization is fairly well established: the C-tail of both non-visual arrestins is released upon receptor binding,[19] which increases the accessibility of clathrin and AP2-binding sites in this element.[10,73,74] In addition, arrestins appear to recruit ubiquitin ligases to GPCRs, and receptor ubiquitination plays a role in receptor sorting.[9,55,56] Yet it is still unclear how arrestins participate in other steps of GPCR trafficking. *N*-Formyl-peptide receptor binds arrestin-2 and -3 in an activation- and phosphorylation-dependent manner,[75,76] yet it was reported to internalize in the absence of both non-visual arrestins.[77] However, in mouse embryonic fibroblasts lacking both non-visual arrestins, internalized *N*-formyl-peptide receptor does not recycle.[77] The receptor travels to the perinuclear recycling compartment and gets stuck there, but its recycling can be rescued by the expression of either arrestin-2 or -3.[77] These data suggest that, as far as *N*-formyl-peptide receptor recycling is concerned, the two non-visual arrestins are functionally redundant. Yet we do not have many clues how exactly are arrestins involved in GPCR recycling. One conceivable scenario is that arrestins bind to this receptor after internalization and recruit deubiquitinating enzymes necessary for recycling, as was shown in the case of β2AR,[59,60] but this leaves open the question why arrestins do not bind it before endocytosis, similar to β2AR.[5,6,9] Existing evidence does not suggest any good answers to this question.

Another issue that needs experimental clarification is arrestin binding to the *N*-ethylmaleimide-sensitive factor (NSF), an ATPase involved in vesicle trafficking. Arrestin-2 binding to NSF was discovered 15 years ago,[78] but its functional significance in case of non-visual arrestins remains unclear. Interestingly, a few years ago, visual arrestin-1 was shown to interact with NSF in photoreceptors.[79] It appears that in rods, arrestin-1 is necessary to maintain proper NSF function and normal level of neurotransmitter release.[79] However, the molecular mechanism of this arrestin-1 effect remains to be elucidated.

7. CONCLUSIONS AND FUTURE DIRECTIONS

The role of non-visual arrestins in recruiting GPCRs to coated pits and facilitation of receptor internalization via this pathway is fairly well established. The case of ubiquitin modification of receptors and arrestins is less straightforward: arrestins seem to recruit enzymes responsible for ubiquitination and deubiquitination of GPCRs. These modifications play distinct roles in receptor trafficking, but the exact role of non-visual arrestins, which are also ubiquitinated in response to receptor stimulation, remains to be elucidated. The functions of non-visual arrestins in complex trafficking itineraries of individual GPCR subtypes might be different. How arrestins affect the recycling of internalized GPCRs, and how exactly arrestin binding regulates NSF function and vesicle trafficking, remains even less clear (Fig. 1). Cytoskeleton is intimately involved in trafficking of many proteins. Arrestins were shown to bind microtubules[80–82] and a very specialized structure containing polymerized tubulin, the centrosome.[83] However, the role of these interactions in the transport of receptors and/or other molecules within the cell still needs to be defined. Most likely, recent finding that non-visual arrestins recruit clathrin to microtubules targeting focal adhesions, thereby facilitating integrin internalization and focal adhesion disassembly,[84] is only the tip of the iceberg.

REFERENCES

1. Kuhn H, Hall SW, Wilden U. Light-induced binding of 48-kDa protein to photoreceptor membranes is highly enhanced by phosphorylation of rhodopsin. *FEBS Lett.* 1984;176:473–478.
2. Wilden U, Hall SW, Kühn H. Phosphodiesterase activation by photoexcited rhodopsin is quenched when rhodopsin is phosphorylated and binds the intrinsic 48-kDa protein of rod outer segments. *Proc Natl Acad Sci USA.* 1986;83:1174–1178.
3. Wilden U. Duration and amplitude of the light-induced cGMP hydrolysis in vertebrate photoreceptors are regulated by multiple phosphorylation of rhodopsin and by arrestin binding. *Biochemistry.* 1995;34:1446–1454.
4. Krupnick JG, Gurevich VV, Benovic JL. Mechanism of quenching of phototransduction. Binding competition between arrestin and transducin for phosphorhodopsin. *J Biol Chem.* 1997;272:18125–18131.
5. Goodman Jr OB, Krupnick JG, Santini F, et al. Beta-arrestin acts as a clathrin adaptor in endocytosis of the beta2-adrenergic receptor. *Nature.* 1996;383(6599):447–450.
6. Laporte SA, Oakley RH, Zhang J, et al. The 2-adrenergic receptor/arrestin complex recruits the clathrin adaptor AP-2 during endocytosis. *Proc Natl Acad Sci USA.* 1999;96:3712–3717.
7. Carman CV, Benovic JL. G-protein-coupled receptors: turn-ons and turn-offs. *Curr Opin Neurobiol.* 1998;8:335–344.
8. Gurevich VV, Gurevich EV. The structural basis of arrestin-mediated regulation of G protein-coupled receptors. *Pharmacol Ther.* 2006;110:465–502.
9. Shenoy SK, McDonald PH, Kohout TA, Lefkowitz RJ. Regulation of receptor fate by ubiquitination of activated beta 2-adrenergic receptor and beta-arrestin. *Science.* 2001;294:1307–1313.
10. Kim YM, Benovic JL. Differential roles of arrestin-2 interaction with clathrin and adaptor protein 2 in G protein-coupled receptor trafficking. *J Biol Chem.* 2002;277:30760–30768.
11. Kang DS, Kern RC, Puthenveedu MA, von Zastrow M, Williams JC, Benovic JL. Structure of an arrestin2-clathrin complex reveals a novel clathrin binding domain that modulates receptor trafficking. *J Biol Chem.* 2009;284:29860–29872.
12. Hirsch JA, Schubert C, Gurevich VV, Sigler PB. The 2.8 Å crystal structure of visual arrestin: a model for arrestin's regulation. *Cell.* 1999;97(2):257–269.
13. Han M, Gurevich VV, Vishnivetskiy SA, Sigler PB, Schubert C. Crystal structure of beta-arrestin at 1.9 Å: possible mechanism of receptor binding and membrane translocation. *Structure.* 2001;9(9):869–880.
14. Sutton RB, Vishnivetskiy SA, Robert J, et al. Crystal structure of cone arrestin at 2.3 Å: evolution of receptor specificity. *J Mol Biol.* 2005;354:1069–1080.
15. Zhan X, Gimenez LE, Gurevich VV, Spiller BW. Crystal structure of arrestin-3 reveals the basis of the difference in receptor binding between two non-visual arrestins. *J Mol Biol.* 2011;406:467–478.

16. Milano SK, Pace HC, Kim YM, Brenner C, Benovic JL. Scaffolding functions of arrestin-2 revealed by crystal structure and mutagenesis. *Biochemistry*. 2002;41(10):3321–3328.
17. Hanson SM, Francis DJ, Vishnivetskiy SA, et al. Differential interaction of spin-labeled arrestin with inactive and active phosphorhodopsin. *Proc Natl Acad Sci USA*. 2006;103:4900–4905.
18. Vishnivetskiy SA, Francis DJ, Van Eps N, et al. The role of arrestin alpha-helix I in receptor binding. *J Mol Biol*. 2010;395:42–54.
19. Zhuo Y, Vishnivetskiy SA, Zhan X, Gurevich VV, Klug CS. Identification of receptor binding-induced conformational changes in non-visual arrestins. *J Biol Chem*. 2014;289(30):20991–21002.
20. Orsini MJ, Benovic JL. Characterization of dominant negative arrestins that inhibit beta2-adrenergic receptor internalization by distinct mechanisms. *J Biol Chem*. 1998;273(51):34616–34622.
21. Gurevich VV, Gurevich EV. The new face of active receptor bound arrestin attracts new partners. *Structure*. 2003;11:1037–1042.
22. Gaidarov I, Krupnick JG, Falck JR, Benovic JL, Keen JH. Arrestin function in G protein-coupled receptor endocytosis requires phosphoinositide binding. *EMBO J*. 1999;18:871–881.
23. Gaidarov I, Keen JH. Phosphoinositide-AP-2 interactions required for targeting to plasma membrane clathrin-coated pits. *J Cell Biol*. 1999;146:755–764.
24. Gurevich VV, Richardson RM, Kim CM, Hosey MM, Benovic JL. Binding of wild type and chimeric arrestins to the m2 muscarinic cholinergic receptor. *J Biol Chem*. 1993;268(23):16879–16882.
25. Pals-Rylaarsdam R, Gurevich VV, Lee KB, Ptasienski J, Benovic JL, Hosey MM. Internalization of the m2 muscarinic acetylcholine receptor: arrestin-independent and -dependent pathways. *J Biol Chem*. 1997;272:23682–23689.
26. Kraft K, Olbrich H, Majoul I, Mack M, Proudfoot A, Oppermann M. Characterization of sequence determinants within the carboxyl-terminal domain of chemokine receptor CCR5 that regulate signaling and receptor internalization. *J Biol Chem*. 2001;276:34408–34418.
27. Naik S, Billington CK, Pascual RM, et al. Regulation of cysteinyl leukotriene type 1 receptor internalization and signaling. *J Biol Chem*. 2005;280:8722–8732.
28. Dale LB, Bhattacharya M, Seachrist JL, Anborgh PH, Ferguson SS. Agonist-stimulated and tonic internalization of metabotropic glutamate receptor 1a in human embryonic kidney 293 cells: agonist-stimulated endocytosis is beta-arrestin1 isoform-specific. *Mol Pharmacol*. 2001;60:1243–1253.
29. Barthet G, Gaven F, Framery B, et al. Uncoupling and endocytosis of 5-HT4 receptors: distinct molecular events with different GRK2 requirements. *J Biol Chem*. 2005;280:27924–27934.
30. Gurevich VV, Gurevich EV. Rich tapestry of G protein-coupled receptor signaling and regulatory mechanisms. *Mol Pharmacol*. 2008;74(2):312–316.
31. Pugh Jr EN, Lamb TD. Phototransduction in vertebrate rods and cones: molecular mechanisms of amplification, recovery and light adaptation. In: Stavenga DG, DeGrip WJ, Pugh Jr EN, eds. *Handbook of Biological Physics. Molecular Mechanisms in Visual Transduction*. Amsterdam: Elsevier; 2000:183–255.
32. Nikonov SS, Brown BM, Davis JA, et al. Mouse cones require an arrestin for normal inactivation of phototransduction. *Neuron*. 2008;59:462–474.
33. Gurevich EV, Gurevich VV. Arrestins are ubiquitous regulators of cellular signaling pathways. *Genome Biol*. 2006;7:236.
34. Moaven H, Koike Y, Jao CC, Gurevich VV, Langen R, Chen J. Visual arrestin interaction with clathrin adaptor AP-2 regulates photoreceptor survival in the vertebrate retina. *Proc Natl Acad Sci USA*. 2013;110(23):9463–9468.

35. Li T, Franson WK, Gordon JW, Berson EL, Dryja TP. Constitutive activation of phototransduction by K296E opsin is not a cause of photoreceptor degeneration. *Proc Natl Acad Sci USA*. 1995;92(8):3551–3555.
36. Hanson SM, Gurevich EV, Vishnivetskiy SA, Ahmed MR, Song X, Gurevich VV. Each rhodopsin molecule binds its own arrestin. *Proc Natl Acad Sci USA*. 2007;104:3125–3128.
37. Song X, Vishnivetskiy SA, Seo J, Chen J, Gurevich EV, Gurevich VV. Arrestin-1 expression in rods: balancing functional performance and photoreceptor health. *Neuroscience*. 2011;174:37–49.
38. Strissel KJ, Sokolov M, Trieu LH, Arshavsky VY. Arrestin translocation is induced at a critical threshold of visual signaling and is superstoichiometric to bleached rhodopsin. *J Neurosci*. 2006;26:1146–1153.
39. Gurevich EV, Benovic JL, Gurevich VV. Arrestin2 and arrestin3 are differentially expressed in the rat brain during postnatal development. *Neuroscience*. 2002;109:421–436.
40. Gurevich EV, Benovic JL, Gurevich VV. Arrestin2 expression selectively increases during neural differentiation. *J Neurochem*. 2004;91:1404–1416.
41. Orem NR, Xia L, Dolph PJ. An essential role for endocytosis of rhodopsin through interaction of visual arrestin with the AP-2 adaptor. *J Cell Sci*. 2006;119(Pt 15): 3141–3148.
42. Lee SJ, Montell C. Light-dependent translocation of visual arrestin regulated by the NINAC myosin III. *Neuron*. 2004;43:95–103.
43. Lee SJ, Xu H, Kang LW, Amzel LM, Montell C. Light adaptation through phosphoinositide-regulated translocation of Drosophila visual arrestin. *Neuron*. 2003;39:121–132.
44. Satoh AK, Xia H, Yan L, Liu CH, Hardie RC, Ready DF. Arrestin translocation is stoichiometric to rhodopsin isomerization and accelerated by phototransduction in Drosophila photoreceptors. *Neuron*. 2010;67(6):997–1008.
45. Nair KS, Hanson SM, Mendez A, et al. Light-dependent redistribution of arrestin in vertebrate rods is an energy-independent process governed by protein-protein interactions. *Neuron*. 2005;46:555–567.
46. Liu CH, Satoh AK, Postma M, Huang J, Ready DF, Hardie RC. Ca^{2+}-dependent metarhodopsin inactivation mediated by calmodulin and NINAC myosin III. *Neuron*. 2008;59(5):778–789.
47. Barak LS, Ferguson SS, Zhang J, Caron MG. A beta-arrestin/green fluorescent protein biosensor for detecting G protein-coupled receptor activation. *J Biol Chem*. 1997;272:27497–27500.
48. Mukhopadhyay D, Riezman H. Proteasome-independent functions of ubiquitin in endocytosis and signaling. *Science*. 2007;315:201–205.
49. Marchese A, Benovic JL. Agonist-promoted ubiquitination of the G protein-coupled receptor CXCR4 mediates lysosomal sorting. *J Biol Chem*. 2001;276: 45509–45512.
50. Shenoy SK, Lefkowitz RJ. Trafficking patterns of beta-arrestin and G protein-coupled receptors determined by the kinetics of beta-arrestin deubiquitination. *J Biol Chem*. 2003;278:14498–14506.
51. Coffa S, Breitman M, Hanson SM, et al. The effect of arrestin conformation on the recruitment of c-Raf1, MEK1, and ERK1/2 activation. *PLoS One*. 2011;6:e28723.
52. Luttrell LM, Roudabush FL, Choy EW, et al. Activation and targeting of extracellular signal-regulated kinases by beta-arrestin scaffolds. *Proc Natl Acad Sci USA*. 2001;98(5):2449–2454.
53. Shenoy SK, Barak LS, Xiao K, et al. Ubiquitination of beta-arrestin links seven-transmembrane receptor endocytosis and ERK activation. *J Biol Chem*. 2007;282:29549–29562.

54. Ahmed MR, Zhan X, Song X, Kook S, Gurevich VV, Gurevich EV. Ubiquitin ligase parkin promotes Mdm2-arrestin interaction but inhibits arrestin ubiquitination. *Biochemistry*. 2011;50:3749–3763.
55. Bhandari D, Trejo J, Benovic JL, Marchese A. Arrestin-2 interacts with the ubiquitin-protein isopeptide ligase atrophin-interacting protein 4 and mediates endosomal sorting of the chemokine receptor CXCR4. *J Biol Chem*. 2007;282:36971–36979.
56. Shenoy SK, Xiao K, Venkataramanan V, Snyder PM, Freedman NJ, Weissman AM. NEDD4 mediates agonist-dependent ubiquitination, lysosomal targeting and degradation of the beta 2 adrenergic receptor. *J Biol Chem*. 2008;283:22166–22176.
57. Nabhan JF, Pan H, Lu Q. Arrestin domain-containing protein 3 recruits the NEDD4 E3 ligase to mediate ubiquitination of the beta2-adrenergic receptor. *EMBO Rep*. 2010;11:605–611.
58. Han SO, Kommaddi RP, Shenoy SK. Distinct roles for β-arrestin2 and arrestin-domain-containing proteins in β2 adrenergic receptor trafficking. *EMBO Rep*. 2013;14(2):164–171.
59. Berthouze M, Venkataramanan V, Li Y, Shenoy SK. The deubiquitinases USP33 and USP20 coordinate beta2 adrenergic receptor recycling and resensitization. *EMBO J*. 2009;28(12):1684–1696.
60. Shenoy SK, Modi AS, Shukla AK, et al. Beta-arrestin-dependent signaling and trafficking of 7-transmembrane receptors is reciprocally regulated by the deubiquitinase USP33 and the E3 ligase Mdm2. *Proc Natl Acad Sci USA*. 2009;106:6650–6655.
61. Oakley RH, Laporte SA, Holt JA, Caron MG, Barak LS. Differential affinities of visual arrestin, barrestin1, and barrestin2 for G protein-coupled receptors delineate two major classes of receptors. *J Biol Chem*. 2000;275:17201–17210.
62. Van Dyke RW. Acidification of lysosomes and endosomes. *Subcell Biochem*. 1996;27:331–360.
63. Gurevich VV, Dion SB, Onorato JJ, et al. Arrestin interaction with G protein-coupled receptors. Direct binding studies of wild type and mutant arrestins with rhodopsin, β_2-adrenergic, and m2 muscarinic cholinergic receptors. *J Biol Chem*. 1995;270:720–731.
64. Celver J, Vishnivetskiy SA, Chavkin C, Gurevich VV. Conservation of the phosphate-sensitive elements in the arrestin family of proteins. *J Biol Chem*. 2002;277(11):9043–9048.
65. Kovoor A, Celver J, Abdryashitov RI, Chavkin C, Gurevich VV. Targeted construction of phosphorylation-independent β-arrestin mutants with constitutive activity in cells. *J Biol Chem*. 1999;274:6831–6834.
66. Gurevich VV, Benovic JL. Cell-free expression of visual arrestin. Truncation mutagenesis identifies multiple domains involved in rhodopsin interaction. *J Biol Chem*. 1992;267:21919–21923.
67. Gurevich VV, Benovic JL. Visual arrestin interaction with rhodopsin: sequential multisite binding ensures strict selectivity towards light-activated phosphorylated rhodopsin. *J Biol Chem*. 1993;268:11628–11638.
68. Palczewski K, McDowell H, Jakes S, Ingebritsen TS, Hargrave PA. Regulation of rhodopsin dephosphorylation by arrestin. *J Biol Chem*. 1989;264:15770–15773.
69. Hsieh C, Brown S, Derleth C, Mackie K. Internalization and recycling of the CB1 cannabinoid receptor. *J Neurochem*. 1999;73:493–501.
70. Morrison KJ, Moore RH, Carsrud ND, et al. Repetitive endocytosis and recycling of the beta 2-adrenergic receptor during agonist-induced steady state redistribution. *Mol Pharmacol*. 1996;50:692–699.
71. Gurevich VV, Pals-Rylaarsdam R, Benovic JL, Hosey MM, Onorato JJ. Agonist-receptor-arrestin, an alternative ternary complex with high agonist affinity. *J Biol Chem*. 1997;272:28849–28852.
72. Pan L, Gurevich EV, Gurevich VV. The nature of the arrestin × receptor complex determines the ultimate fate of the internalized receptor. *J Biol Chem*. 2003;278:11623–11632.

73. Xiao K, Shenoy SK, Nobles K, Lefkowitz RJ. Activation-dependent conformational changes in {beta}-arrestin 2. *J Biol Chem.* 2004;279(53):55744–55753.
74. Nobles KN, Guan Z, Xiao K, Oas TG, Lefkowitz RJ. The active conformation of beta-arrestin1: direct evidence for the phosphate sensor in the N-domain and conformational differences in the active states of beta-arrestins1 and -2. *J Biol Chem.* 2007;282(29):21370–21381.
75. Key TA, Foutz TD, Gurevich VV, Sklar LA, Prossnitz ER. N-Formyl peptide receptor phosphorylation domains differentially regulate arrestin and agonist affinity. *J Biol Chem.* 2003;278:4041–4047.
76. Bennett TA, Maestas DC, Prossnitz ER. Arrestin binding to the G protein-coupled N-formyl peptide receptor is regulated by the conserved "DRY" sequence. *J Biol Chem.* 2000;275:24590–24594.
77. Vines CM, Revankar CM, Maestas DC, et al. N-Formyl peptide receptors internalize but do not recycle in the absence of arrestins. *J Biol Chem.* 2003;278(43):41581–41584.
78. McDonald PH, Cote NL, Lin FT, Premont RT, Pitcher JA, Lefkowitz RJ. Identification of NSF as a beta-arrestin1-binding protein. Implications for beta2-adrenergic receptor regulation. *J Biol Chem.* 1999;274:10677–10680.
79. Huang SP, Brown BM, Craft CM. Visual arrestin 1 acts as a modulator for N-ethylmaleimide-sensitive factor in the photoreceptor synapse. *J Neurosci.* 2010;30:9381–9391.
80. Hanson SM, Cleghorn WM, Francis DJ, et al. Arrestin mobilizes signaling proteins to the cytoskeleton and redirects their activity. *J Mol Biol.* 2007;368(2):375–387.
81. Hanson SM, Francis DJ, Vishnivetskiy SA, Klug CS, Gurevich VV. Visual arrestin binding to microtubules involves a distinct conformational change. *J Biol Chem.* 2006;281:9765–9772.
82. Nair KS, Hanson SM, Kennedy MJ, Hurley JB, Gurevich VV, Slepak VZ. Direct binding of visual arrestin to microtubules determines the differential subcellular localization of its splice variants in rod photoreceptors. *J Biol Chem.* 2004;279:41240–41248.
83. Shankar H, Michal A, Kern RC, Kang DS, Gurevich VV, Benovic JL. Non-visual arrestins are constitutively associated with the centrosome and regulate centrosome function. *J Biol Chem.* 2010;285(11):8316–8329.
84. Cleghorn WM, Branch KM, Kook S, et al. Arrestins regulate cell spreading and motility via focal adhesion dynamics. *Mol Biol Cell.* 2015;26(4):622–635.

CHAPTER TWO

Regulation of GPCR Trafficking by Ubiquitin

Justine E. Kennedy, Adriano Marchese[1]
Department of Molecular Pharmacology and Therapeutics, Loyola University Chicago, Health Sciences Division, Maywood, Illinois, USA
[1]Corresponding author: e-mail address: amarchese@luc.edu

Contents

Abstract

G protein-coupled receptor (GPCR)-promoted signaling mediates cellular responses to a variety of stimuli involved in diverse physiological processes. In addition, GPCRs are also the largest class of target for many drugs used to treat a variety of diseases. Despite the role of GPCR signaling in health and disease, the molecular mechanisms governing GPCR signaling remain poorly understanding. Classically, GPCR signaling is tightly regulated by GPCR kinases and β-arrestins, which act in a concerted fashion to govern GPCR desensitization and also GPCR trafficking. Ubiquitination has now emerged as an important posttranslational modification that has multiple roles, either directly or indirectly, in governing GPCR trafficking. Recent studies have revealed a mechanistic link between GPCR phosphorylation, β-arrestins, and ubiquitination. Here, we review recent developments in our understanding of how ubiquitin regulates GPCR trafficking within the endocytic pathway.

Progress in Molecular Biology and Translational Science, Volume 132
ISSN 1877-1173
http://dx.doi.org/10.1016/bs.pmbts.2015.02.005

1. INTRODUCTION

G protein-coupled receptor (GPCR) signaling is classically known to be mediated through the associated guanine nucleotide (G)-binding proteins (G proteins) that are heterotrimers comprised of an α-subunit and a βγ heterodimer.[1] Agonist binding to GPCRs induces a conformational change in the associated G protein, thereby facilitating the exchange of GDP for GTP on the α-subunit and reversible disassociation of the βγ heterodimer.[1,2] Both the GTP-bound α-subunit and the released Gβγ heterodimer can signal to a diverse array of effector molecules involved in many signaling pathways leading to cellular responses.[1] Importantly, to ensure that the cellular responses are of the appropriate magnitude and duration, signaling is highly regulated to maintain normal cellular homeostasis.[3] The mechanisms that regulate GPCR signaling are complex and occur at every level of the signaling pathway, including at the level of the receptor itself. Two main families of proteins that regulate GPCRs directly include G protein-coupled receptor kinases (GRKs) and the multifaceted adaptor proteins referred to as β-arrestins.[3] β-Arrestins also regulate signaling by controlling receptor proximal degradation of classical second messengers cAMP and diacylglycerol (DAG) through interactions with phosphodiesterases or DAG enzymes, respectively.[4,5] In addition to their role as negative regulators of GPCR signaling, β-arrestins are now also commonly recognized as positive regulators or transducers of signaling.[6]

Direct phosphorylation of GPCRs is a common posttranslational modification that governs their signaling. Agonist activation usually results in rapid phosphorylation by GRKs on serine or threonine amino acid residues located within the intracellular domains of GPCRs.[3] GRK-mediated phosphorylation provides a binding surface for the adaptor proteins, β-arrestins,[7,8] which are recruited from the cytoplasm to the phosphorylated receptor at the plasma membrane.[9] This serves to uncouple the receptor from the associated G protein through a process that involves steric hindrance, thereby terminating or preventing further G protein signaling from the receptor.[8] This culminates in a process referred to as desensitization, a process in which even in the continued presence of stimulus the receptor is unable to signal.

In addition, activated GPCRs are typically removed from the cell surface via a complex process leading to their endocytosis or internalization into intracellular compartments known as endosomes.[10] Once in an endocytic compartment, GPCRs can be dephosphorylated by an endosomal-associated phosphatase and recycled to the cell surface whereby the GPCR again has access to the extracellular ligand leading to functional resensitization of receptor signaling.[11] Alternatively, GPCRs can be targeted to a terminal degradative compartment known as the lysosome, leading to degradation and a loss in the total cellular complement of a GPCR giving rise to a phenomenon known as downregulation culminating in long-term attenuation of signaling.[10,12] The mechanisms dictating whether a GPCR recycles or is targeted to lysosomes for degradation remain poorly understood, but recent advances have revealed a role for ubiquitin in this sorting decision.[13] Direct ubiquitination of GPCRs themselves in which ubiquitin acts in a *cis* manner or ubiquitination of adaptor proteins in which ubiquitin acts in a *trans* manner has been shown to regulate various steps of the itinerary that GPCRs follow along the endocytic pathway. Here, we focus on recent advances that have led to our current understanding of the mechanisms by which ubiquitin regulates GPCR trafficking.

2. UBIQUITINATION MACHINERY

Ubiquitin is a 76-amino acid protein that is generally covalently attached to protein substrates through the formation of an isopeptide bond between the C-terminal glycine (Gly76) residue of ubiquitin and the epsilon amino group of internal lysine residues on target substrates.[14,15] In certain circumstances, ubiquitin can also be attached to the free amino group at the N-terminus of a substrate[16,17] or other internal amino acid residues,[18] but whether this applies to GPCRs remains to be determined, to our knowledge.

Ubiquitin conjugation of proteins is carried out by an enzymatic cascade involving the sequential activity of three enzymes that are dedicated to ubiquitination reactions: E1, E2, and E3.[14,15] There is a single conserved E1 enzyme and approximately 40 identified E2 enzymes in the human genome.[19] In contrast, E3 ubiquitin ligases represent a diverse family of over 600 identified proteins in the mammalian genome.[20] A typical ubiquitination reaction can be divided into discrete steps. Ubiquitin is first activated at its C-terminus in an ATP-dependent manner by the E1-activating enzyme. This first step can be divided into two distinct events in which ubiquitin is initially activated via a ubiquitin-adenylate intermediate, which then reacts with a cysteine residue on the E1 to form an E1-ubiquitin intermediate. In the second step, ubiquitin is transferred from the E1 cysteine residue to a cysteine residue on the E2-conjugating enzyme. The E2 interacts with E3, which also binds to the substrate, and the E3 either directly or indirectly transfers ubiquitin to a nearby lysine residue on the substrate. Ubiquitination of protein substrates is typically transient and is reversed by deubiquitinating enzymes (DUBs).[21] DUBs have selective protease activity and mediate cleavage of the isopeptide bond between ubiquitin and its substrates. DUBs have been implicated in regulating the trafficking of GPCRs.[22–30]

Because E3s mediate the interaction with their substrates, they typically provide the specificity to an ubiquitination reaction. E3s that indirectly attach ubiquitin to substrate proteins essentially serve as a scaffold or bridging molecule for E2 and the substrate.[19] E3s that serve as scaffolds for ubiquitination reactions fall into two general families: RING domain or F box E3s.[19] RING domain ligases do not possess intrinsic catalytic activity; however, their ligase activity stems from the fact that the RING domain binds the E2 enzyme, while the substrate binds to another region in a manner that facilitates transfer of ubiquitin moieties from the ubiquitin-loaded E2 to the substrate.[19] The RING domain E3s form the largest family of E3s, and several RING domain E3s have been implicated in GPCR trafficking, via either ubiquitination of GPCRs or adaptor molecules.[31–34]

In contrast to RING domain E3s, HECT (homologous to E6-AP C-terminus) domain E3s are directly involved in ubiquitination reactions because they form a direct thioester intermediate with ubiquitin in which the ubiquitin-loaded E2 transfers ubiquitin to an active site cysteine residue on the E3 before the ubiquitin is transferred to a lysine residue on the substrate protein.[35] HECT domain E3s represent a smaller family (~30 members) within the large family of E3 ligases.[36] HECT domain E3 can be divided into three discrete groups.[36] The group known as the Nedd4-like HECT domain family of E3s has been implicated in GPCR trafficking.[37] The Nedd4 family comprises nine mammalian members: Nedd4, Nedd4-2, AIP4 (a.k.a. Itch), WWP1, WWP2, SMURF1, SMURF2, NEDL1, and NEDL2.[38] Nedd4-like E3s contain a N-terminal calcium-dependent phospholipid-binding domain, two to four tandemly linked WW domains, and a conserved C-terminal HECT domain[36]. The Nedd4-like E3 AIP4 uniquely contains a proline-rich region that can bind to SH3 domains.[39] AIP4 is the human ortholog of the mouse E3 ubiquitin ligase referred to as Itch.[40] WW domains are protein–protein interaction modules that contain two conserved tryptophan residues and they typically interact with PPXY or PPPY motifs, where X represents any amino acid.[41,42] Nedd4-like E3s typically interact directly with their substrates by interacting with PPXY motifs; however, not all substrates have these motifs and in such cases the interaction is indirect via an adaptor protein that has such a motif.[41–43] Nedd4-like E3 ubiquitin ligases have been shown to interact with GPCRs either directly through noncanonical WW-domain-mediated interactions[44] or indirectly through interactions involving adaptor proteins.[45–47] The HECT domain is located at the C-terminal end of Nedd4-like E3s and contains a highly conserved cysteine residue that directly accepts ubiquitin and therefore facilitates substrate ubiquitination directly.[35]

3. MECHANISMS OF GPCR UBIQUITINATION

GPCR ubiquitination can be regulated by agonist activation or it can occur in an agonist-independent manner. To the best of our knowledge, the first mammalian GPCRs shown to be ubiquitinated in an agonist-dependent manner were the β_2-adrenergic receptor (β_2AR) and the C-X-C receptor 4 (CXCR4) chemokine receptor.[32,48] Several GPCRs have since been shown to be ubiquitinated in an agonist-dependent manner.[49] GPCRs can also be ubiquitinated in an agonist-independent manner. For example, GPCRs such as GPR37[50] or the δ-opioid receptor (DOR)[51] can be ubiquitinated during biosynthesis, as a quality control measure to target misfolded receptors for ubiquitination and degradation by the proteasome. The trigger for ubiquitination during biosynthesis is likely detected as a conformational change in the misfolded GPCR, leading to its removal via endoplasmic reticulum protein degradation.[52] Interestingly, limiting the amount of ubiquitination during biosynthesis can enhance the cell surface levels of certain GPCRs. The DUB ubiquitin-specific protease 4 associates with the C-terminus of the A_{2A} adenosine receptor and possibility regulates its ubiquitination status, thereby facilitating its passage via the biosynthetic pathway to the plasma membrane.[30] GPCRs can also be constitutively ubiquitinated postsynthesis in a ligand-independent manner, but the trigger for this type of ubiquitination remains unknown, although it may be dependent upon the compartment to which the receptor localizes.[53] In this case, it appears that agonist activation can induce GPCR deubiquitination.[37,53] Surprisingly, given that there are many GPCRs, a relatively small number have been shown to be ubiquitinated.[37] The reason for this is not clear, but it is possible that not all GPCRs are regulated by ubiquitination.[53,54] Another possibility is that it may be due to technical difficulties in detecting GPCR ubiquitination.[55]

Lysine residues within any of the intracellular domains of GPCRs can be subject to ubiquitination. For example, the μ-opioid receptor (MOR) is mostly ubiquitinated on two lysine residues located within the first intracellular loop of the receptor, as determined by mutational analysis.[56] Although lysine residues are present on the other intracellular domains of MOR, and in particular the C-terminal tail, these lysine residues are not sufficient to support receptor ubiquitination.[56] The β_2AR appears to be ubiquitinated on lysine residues within the third intracellular loop, but in contrast to MOR, β_2AR is also ubiquitinated on lysine residues in the C-terminal tail.[57] Other GPCRs, such as CXCR4 and protease-activated receptor 1 (PAR1), seem to be mostly ubiquitinated on lysine residues located within the C-terminal tail, despite the fact that there are lysine residues located on other intracellular domains.[48,58] It remains unclear why certain lysine residues are subject to ubiquitination while others are not, but it is likely related to the structural constraints adopted by distinct ligand-induced receptor conformations that restrict E3 ligase access to certain intracellular domains and hence lysine residues. Further work will be required to understand this process in greater detail.

4. GPCR REGULATION BY E3 UBIQUITIN LIGASES

Agonist-dependent ubiquitination of GPCRs typically occurs at the plasma membrane and it typically requires receptor phosphorylation.[32,44,48] This is particularly well characterized for CXCR4. The E3 ubiquitin ligase AIP4 mediates agonist-dependent ubiquitination of CXCR4 at the plasma membrane.[59] The mechanism by which AIP4 recognizes and ubiquitinates CXCR4 was only recently elucidated.[44] A receptor mutant in which two consecutive serine residues (S324 and S325) are mutated to alanine residues is not ubiquitinated as efficiently as wild-type CXCR4.[44,48] These residues are rapidly phosphorylated by agonist activation at the plasma membrane, as assessed by confocal microscopy using a phospho-specific antibody directed against dually phosphorylated S324 and S325.[44] Phosphorylation of these residues is likely mediated by GRK6 and/or PKCδ.[60] Phosphorylation of these residues promotes the recruitment of the E3 ubiquitin ligase AIP4 to the plasma membrane following agonist stimulation, as assessed by TIRF microscopy.[44] Therefore, AIP4 binding to the phosphorylated receptor at the plasma membrane is required for ubiquitinating nearby lysine residues (Fig. 1A).

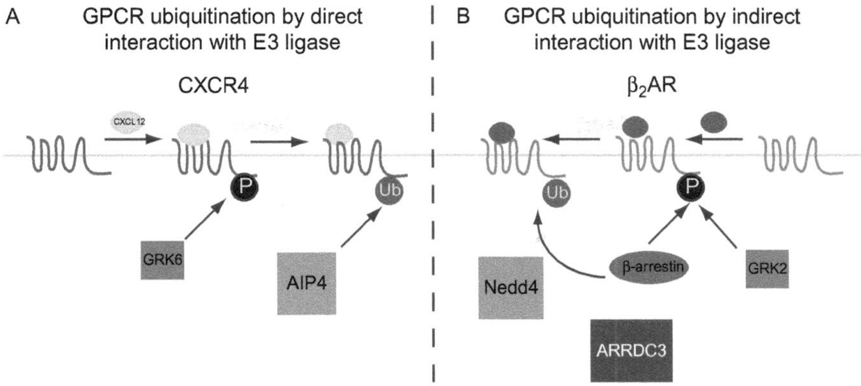

Figure 1 Mechanisms of GPCR ubiquitination at the plasma membrane by E3 ubiquitin ligases. E3 ubiquitin ligases belonging to the HECT domain Nedd4-like family can interact with GPCRs directly (A) or indirectly (B). (A) Upon binding to its cognate ligand CXCL12 (yellow oval), CXCR4 is phosphorylated by GRK6 on serine residues 324 and 325 at the plasma membrane. This is followed by recruitment of AIP4, a Nedd4-like E3 ubiquitin ligase, to the receptor, resulting in ubiquitination of nearby lysine residues. The WW domains of AIP4 mediate the interaction with CXCR4 via a noncanonical mechanism with phosphorylated serine residues 324 and 325. (B) Upon binding to the selective β-agonist isoproterenol (red oval) β_2AR is rapidly phosphorylated, likely by GRK2, leading to recruitment of β-arrestin-2. β-Arrestin-2 interacts with Nedd4, a Nedd4-like E3 ubiquitin ligase, and serves to bridge the interaction between β_2AR and Nedd4, thereby enabling ubiquitination of the receptor. The α-arrestin protein ARRDC3 has also been implicated in this event, but it may in fact act at a later step in the endocytic pathway. (See the color plate.)

The interaction between AIP4 and CXCR4 occurs directly via a noncanonical interaction.[44] As described above, AIP4 is a member of the Nedd4-like family of E3 ubiquitin ligases.[38] Similar to other members of the Nedd4-like family, AIP4 typically interacts with its substrates via the WW domains.[42] WW domains are protein–protein interaction modules approximately 40 amino acids residues, and as the name implies, they are defined by the presence of two conserved tryptophan residues.[61] The WW domains found in AIP4 interact with proline-rich sequences, such as PPXY or PPPY motifs.[41,42] However, the intracellular domains of CXCR4, or GPCRs in general, do not contain such motifs, suggesting that AIP4 interacts with CXCR4 via an indirect mechanism. However, biochemical interaction studies using purified proteins revealed that the C-terminal tail of CXCR4 interacts directly with full-length AIP4.[44] This is mediated by AIP4 WW domains I and II, but not III and IV, which appear to interact directly with the tandemly phosphorylated amino acid residues

S324 and S325,[44] thereby establishing a noncanonical mode of interaction of WW domains with phosphorylated serine and possibly threonine residues. This may not be restricted to CXCR4 as AIP4 may also mediate ubiquitination of DOR; however, the mechanism by which AIP4 recognizes and ubiquitinates DOR remains to be defined.[26] AIP4 is likely not going to be required for ubiquitination of all GPCRs, at least not for β_2AR,[46] MOR[47] nor the S1P receptor,[45] which have been shown to not require AIP4 for their ubiquitination. Interestingly, the S1P receptor may be regulated by phosphorylation and ubiquitination by WWP2 (a.k.a. AIP2), a member of the Nedd4-like family of E3 ligases, although the interaction between S1P and WW2 may not require prior phosphorylation.[45] Therefore, GPCR ubiquitination is likely going to be regulated by diverse mechanisms via distinct E3 ubiquitin ligases.

Although CXCR4 can interact directly with AIP4, other GPCRs appear to interact indirectly with E3 ubiquitin ligases via an adaptor protein, although phosphorylation of the receptor is also required. For example, β_2AR interacts with the E3 ubiquitin ligase Nedd4 via the adaptor protein β-arrestin-2[46] (Fig. 1B). Nedd4 is a member of the Nedd4-like family of E3 ubiquitin ligases.[38] Knockdown of Nedd4 by siRNA, but not other related E3s such as AIP4 or Nedd4-2, attenuates ubiquitination of β_2AR.[46] However, β_2AR ubiquitination is also attenuated in mouse embryonic fibroblasts (MEFs) isolated from β-arrestin-2 knockout mice,[32] suggesting a role of β-arrestin-2 in β_2AR ubiquitination.[46] Consistent with this idea, recruitment of Nedd4 to the β_2AR complex is attenuated in cells in which β-arrestin-2 is reduced by targeted siRNA.[46] These data suggest that it is likely that β-arrestin-2 serves as an adaptor for recruitment of Nedd4 to the activated β_2AR. This may be generalizable to other GPCRs such as MOR because ubiquitination of MOR is impaired in MEFs isolated from β-arrestin-1 knockout mice.[62] This is in contrast to CXCR4 in which ubiquitination of CXCR4 is not impaired in cells in which β-arrestin-1 and β-arrestin-2 have been silenced by siRNA.[63] This is likely because AIP4 can interact directly with CXCR4,[44] while other E3 ligases cannot and require an adaptor protein.[46] Why certain GPCRs such as CXCR4 can interact directly with Nedd4-like E3 ubiquitin ligases, while others require an adaptor protein remains to be explored.

In addition to β-arrestins, α-arrestins may also mediate recruitment of E3 ligases to GPCRs. α-Arrestins were recently discovered by bioinformatics analysis to be distantly related to β-arrestins.[64,65] There is very little amino acid sequence identity between β-arrestins and α-arrestins, but they share structural similarities.[64,66] Computational methods predict that α-arrestins share a so-called arrestin-fold with β-arrestins.[64,66] In contrast to β-arrestins, α-arrestins have a long C-terminal region, which has two or three PPXY motifs,[64] which as described above commonly interact with WW domains.[41] Indeed, α-arrestins have been shown to interact with WW domains of several Nedd4-like E3s via their PPXY motifs.[67,68] In this regard, the α-arrestin protein ARRDC3 has been shown to serve as an adaptor for β_2AR ubiquitination at the plasma membrane by Nedd4.[69] Knockdown of ARRDC3 by siRNA attenuates agonist-stimulated β_2AR ubiquitination and Nedd4 binds directly to the PPXY motifs of ARRDC3 through its WW domains.[69] α-Arrestins can also regulate the ubiquitination and trafficking of other GPCRs.[67] Therefore, in common with β-arrestins, α-arrestins seem to link the same E3 ligase to the same GPCR, suggesting a complex and possibly context-dependent regulation of GPCRs by ubiquitination. However, a recent study has suggested that ARRDC3 is not involved in β_2AR ubiquitination, but instead may be required at a later step of β_2AR trafficking.[70] Interestingly, β-arrestin-1 has been shown to regulate CXCR4 trafficking on endosomes,[71] suggesting that β-arrestins and α-arrestins may share a common function in GPCR trafficking at the level of the endosome.

Unlike α-arrestins, β-arrestins do not have PPXY motifs, although they can interact with Nedd4 and other members of the WW-domain containing Nedd4-like family of E3s, including AIP4 and Smurf2.[46,47,63] Despite a lack of PPXY motifs, the interaction between β-arrestin-1 and AIP4 is mediated by the WW domains of AIP4, suggesting a noncanonical WW-domain-mediated interaction.[63] In contrast, the interaction between β-arrestin-2 and Nedd4 is not mediated by the WW domains of Nedd4, suggesting that it occurs via another domain on Nedd4.[46] The C2 domain is a Ca^{2+}-dependent phospholipid-binding domain that can also mediate protein–protein interactions with Nedd4.[72] Whether the C2 domain of Nedd4 mediates the interaction with β-arrestin-2, to our knowledge, remains to be determined.

5. ROLE OF UBIQUITIN IN GPCR INTERNALIZATION

One function of GPCR ubiquitination is to promote GPCR internalization. In general, most GPCRs are internalized via a β-arrestin-dependent mechanism and likely do not require direct ubiquitination for internalization.[73] For example, a ubiquitin-deficient β_2AR mutant internalizes just as efficiently as the wild-type receptor[32] likely because internalization is efficiently promoted by β-arrestin-2.[74] However, not all GPCRs require β-arrestins for internalization,[75] indicating that there are additional mechanisms and adaptors to promote GPCR internalization. For example, internalization of the thrombin receptor, PAR1, is not impaired in MEFs isolated from embryos of β-arrestin knockout mice, indicating that PAR1 internalization occurs via a β-arrestin-independent mechanism.[75] Recently, AP2 was shown to act as an adaptor that in part mediates agonist-induced internalization of PAR1.[76] AP2 is a clathrin adaptor and can link certain types of receptors to clathrin-coated pits for clathrin-mediated endocytosis.[77] AP2 is a heterotetrameric protein complex comprised of α-, β_2-, μ_2-, and σ_2-adaptin subunits.[77] The β_2-adaptin subunit interacts with β-arrestins and this interaction is generally required for β-arrestin-mediated internalization of GPCRs.[73,78] However, the μ_2-adaptin subunit is essential for PAR1 internalization upon agonist activation.[76] The μ_2-adaptin subunit is believed to interact directly with phosphorylated serine and threonine residues within the C-terminal tail, thereby linking PAR1 to clathrin for internalization via clathrin-coated pits. In addition, PAR1 may also require the ubiquitin-binding adaptor protein epsin-1 for internalization.[76] PAR1 internalization via epsin requires ubiquitination of C-terminal lysine residues and an intact ubiquitin-binding domain (UBD) in epsin-1, suggesting that the ubiquitin moieties attached to PAR1 serve to link the receptor to epsin-1 via an ubiquitin–UBD-mediated interaction. This may not be unique to PAR1 because internalization of MOR also requires receptor ubiquitination and epsin-1, likely via a similar ubiquitin–UBD-mediated interaction.[47,56] Further work will be required to establish whether ubiquitin-mediated internalization can be generalizable to other GPCRs.

Ubiquitin also plays an indirect role in GPCR internalization via ubiquitination of β-arrestins. Agonist activation of β_2AR promotes ubiquitination of β-arrestin-2 by the RING domain E3 ubiquitin ligase Mdm2.[32] Ubiquitination of β-arrestins does not occur in Mdm2-null MEFs and agonist-promoted internalization of β_2AR is attenuated, although ubiquitination of β_2AR is not impaired.[32] While the precise role for ubiquitin remains to be elucidated, it appears that discrete ubiquitin moieties on β-arrestin may stabilize its interaction with GPCRs and clathrin, thereby facilitating its ability to promote GPCR internalization.[79] Interestingly, β-arrestins are modified by other types of posttranslational modifications that may have an impact on its role in GPCR internalization.[73] For example, agonist-dependent SUMOylation of β-arrestin-2 may facilitate its binding to AP2, thereby facilitating β_2AR internalization.[80] Similarly, S-nitrosylation of β-arrestin-2 may also enhance its interactions with clathrin and AP2 to facilitate β_2AR internalization.[81] SUMOylation and S-nitrosylation occur on distinct amino acid residues located on the C-terminal tail of β-arrestins, which becomes exposed once β-arrestins bind to activated and phosphorylated receptors.[80,81] This region also contains the binding sites for clathrin and the β_2-adaptin subunit of AP2.[73] Therefore, it is possible that SUMOylation and/or S-nitrosylation may somehow facilitate GPCR internalization by enabling access to the exposed tail to proteins of the internalization machinery. Further work will be required to delineate exactly if and how this occurs.

6. ROLE OF UBIQUITIN IN GPCR ENDOSOME TO LYSOSOME SORTING

In addition to a role in GPCR internalization, ubiquitin moieties attached to GPCRs promote GPCR sorting into the degradative pathway.[13] Sorting into this pathway typically occurs on early-to-late endosomes or maturing multivesicular bodies (MVBs).[13,82,83] MVBs are endocytic vesicular intermediates between early and late endosomes defined by the presence of many intraluminal vesicles (ILVs).[84,85] GPCRs destined for lysosomal degradation can be found on the limiting membrane of these structures and accumulate on the membranes of the ILVs[86–88] (Fig. 2). Because MVBs fuse with lysosomes where their contents are degraded,[84] targeting into the ILVs is likely required for complete degradation of GPCRs. The formation of ILVs and targeting of ubiquitinated GPCRs into ILVs occurs via the endosomal sorting complex required for transport (ESCRT) pathway, which comprises four protein complexes (ESCRT-0, -I, -II, and -III) and the AAA-ATPase-Vps4 complex.[89] In the canonical model of ESCRT sorting, the ESCRT complexes act in a coordinated fashion to recognize and concentrate ubiquitinated transmembrane proteins on membranes of the ILVs.[89] ESCRTs are multimeric protein complexes, and three ESCRTs complexes (-0, -I, and -II) exist as stable complexes with subunits that have UBDs. These ESCRTs interact directly with the ubiquitin moieties on cargo via their UBDs, thereby directing the cargo into the forming ILVs. This process is initiated by ESCRT-0, comprised of HRS and STAM1, which concentrates ubiquitinated cargo in microdomains of early endosomes, and with the help of ESCRT-I and -II sorts the ubiquitinated cargo into ILVs. This is followed by the recruitment of the individual monomeric

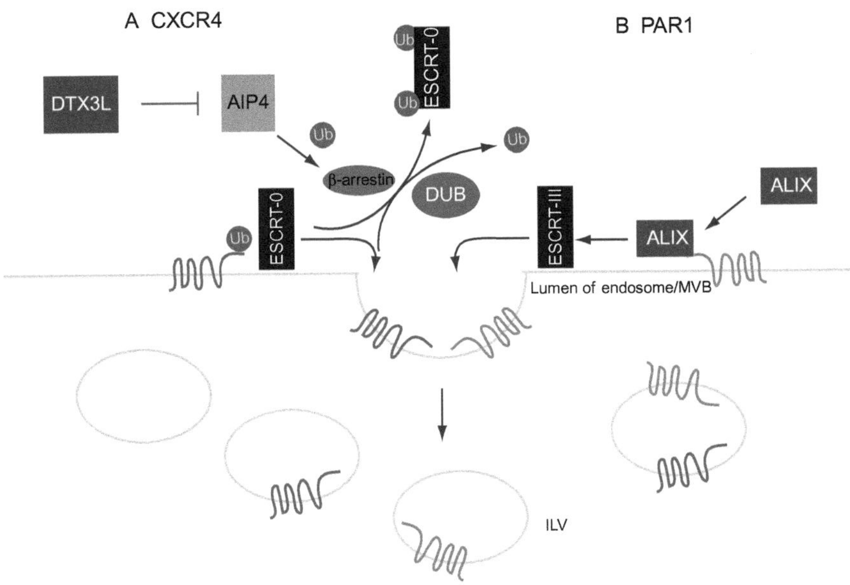

Figure 2 Mechanisms of GPCR sorting into intraluminal vesicles (ILVs) of endosomes or multivesicular bodies (MVBs). GPCRs targeted for lysosomal degradation can be found on the limiting membrane of endosomes or MVBs and on the membrane of ILVs. A mature MVB fuses with lysosomes where degradation of its contents occurs. The ILVs are formed by a complex process that involves the ESCRT (endosomal sorting complex required for transport) pathway. ESCRTs also deliver ubiquitinated GPCRs or other ubiquitinated transmembrane proteins into the invaginating regions of the limiting membrane that eventually give rise to the ILVs. (A) For GPCRs such as CXCR4, the attached ubiquitin moieties interact with ubiquitin-binding ESCRTs such as ESCRT-0. ESCRT-I, -II, and –III are also required and act in a coordinated manner to deliver CXCR4 into ILVs, but for clarity are not shown in the figure. ESCRT-0 recognizes and recruits ubiquitinated CXCR4 into the pathway. This step is regulated by ubiquitination of ESCRT-0. β-Arrestin-1 likely serves as an adaptor for ESCRT-0 ubiquitination by the Nedd4-like E3 ubiquitin ligase AIP4. Ubiquitination of ESCRT-0 is believed to inhibit its sorting activity, possibly to allow ubiquitinated CXCR4 to interact with the other ubiquitin-binding ESCRTs. AIP4 ligase activity is inhibited by the RING-finger E3 ubiquitin ligase DTX3L through a mechanism that is not completely understood, but this is thought to reduce ESCRT-0 ubiquitination and facilitate CXCR4 sorting into ILVs. (B) For some GPCRs, for which PAR1 serves as an example, delivery into ILVs does not require receptor ubiquitination and nor are the ubiquitin-binding ESCRTs (ESCRT-0, -I, and –II) required. Instead, the adaptor protein ALIX is required to target PAR1 into ILVs. ALIX interacts with a $YP(X)_3L$ motif present in the second intracellular loop of PAR1 and it also interacts with ESCRT-III, thereby delivering PAR1 into ILVs. (See the color plate.)

subunits that comprise ESCRT-III which assemble on the limiting membrane of endosomes and act at the scission step to form ILVs-containing receptors destined for degradation on their membranes. The AAA-ATPase Vps4 complex acts to disassemble ESCRT-III subunits, thereby allowing for recycling of the ESCRT components to initiate additional rounds of ESCRT-mediated sorting.

The role of ESCRTs in targeting GPCRs into the degradative pathway is particularly well characterized for CXCR4[90] (Fig. 2). Although CXCR4 is ubiquitinated at the plasma membrane, it is internalized in a ubiquitin-independent manner, but the ubiquitin moiety is required for sorting CXCR4 into the degradative pathway.[48] CXCR4 ubiquitination-deficient mutants when transiently expressed in HEK293 cells show attenuated agonist-induced degradation.[44,48] Internalized CXCR4 is found within specialized microdomains of endosomes together with ESCRT-0 where sorting into ILVs likely begins.[59] Treating cells with siRNA directed against ESCRT-0 (i.e., against HRS subunit) or a dominant-negative Vps4 attenuates CXCR4 degradation.[59] This established for the first time, to our knowledge, a role of ESCRTs in lysosomal trafficking of mammalian GPCRs.[59]

The ubiquitin moieties attached to CXCR4 likely initially interact with UBDs present on subunits of ESCRT-0 for entry into the ESCRT pathway (Fig. 2A). The other UBD-containing ESCRTs, ESCRT-I and -II, have also been implicated in CXCR4 trafficking to lysosomes.[91,92] CXCR4 sorting to lysosomes is very efficient because only a small fraction of CXCR4 escapes and recycles to the plasma membrane.[71] In other cell types, such as T cells, CXCR4 is also efficiently degraded following exposure to agonist[48] and only a small portion recycles back to the plasma membrane via Rab11-positive recycling endosomes.[93] Interestingly, CXCR4 lysosomal sorting may be compromised in breast cancer cells leading to greater recycling and may explain in part why CXCR4 levels are upregulated in a subset of breast tumors.[87,94,95] Other GPCRs such as DOR, PAR2, and β_2AR also sort to lysosomes via the ESCRT pathway, suggesting that this pathway may be generalizable to most GPCRs.[86]

Although most GPCRs likely require direct ubiquitination for entry into ILVs, for some GPCRs direct ubiquitination may not be required. For example, ubiquitination of PAR1 is not required for its sorting into ILVs of MVBs as a ubiquitination-deficient mutant can be easily detected on the membrane of ILVs[86] (Fig. 2B). Accordingly, UBD-containing ESCRTs, such as ESCRT-0 and ESCRT-I, and likely ESCRT-II, are not required for PAR1 lysosomal degradation.[96] However, ESCRT-III is required, as siRNA targeting ESCRT-III subunits blocks agonist-induced degradation of PAR1.[86] The connectivity of PAR1 to ESCRT-III is bridged by the adaptor protein ALIX.[86] ALIX binds to ESCRT-III and to a tyrosine-based ($YPX_{(3)}L$) motif found on the second intracellular loop of PAR1. Mutation of the tyrosine residue within this motif disrupts binding to ALIX and attenuates PAR1 degradation. ALIX is not involved in lysosomal degradation of PAR2, a GPCR that is dependent upon ubiquitin-binding ESCRTs for delivery to lysosomes.[86,97] The $YPX_{(3)}L$ motif is not unique to PAR1 and is found on the intracellular loops of several GPCRs, suggesting that these GPCRs may be sorted for lysosomal degradation via an ALIX/ESCRT-III-dependent mechanism.[86] Therefore, as shown schematically in Fig. 2, direct ubiquitination of GPCRs may be required for some (e.g., CXCR4) but not all GPCRs (e.g., PAR1) for sorting in ILVs.

Ubiquitin also regulates GPCR sorting into the degradative pathway in an indirect manner. The E3 ubiquitin ligase AIP4 that mediates CXCR4 ubiquitination at the plasma membrane is also found on microdomains of early endosomes where internalized CXCR4 and ESCRT-0 are present.[59] CXCR4 activation induces ubiquitination of ESCRT-0 subunits HRS and STAM, which may be mediated by endosomally localized AIP4. The role of ESCRT-0 ubiquitination on CXCR4 sorting remains to be clearly defined, but the attached ubiquitin moieties on HRS and STAM1 may be linked to terminating ESCRT-0 sorting activity.[71] This is based on the unexpected finding in CXCR4 degradation experiments that suggest that ESCRT-0 subunits HRS and STAM have opposing roles in targeting CXCR4 for lysosomal degradation.[59,71] While siRNA-mediated knockdown of HRS inhibits CXCR4 degradation,[59] STAM1 knockdown has the opposite effect and accelerates CXCR4 degradation.[71] This may be explained by the fact that STAM1 interacts with β-arrestin-1 to regulate ubiquitination of HRS.[71] Expression of a dominant-negative β-arrestin-1 that disrupts the β-arrestin-1/STAM1 interaction attenuates HRS ubiquitination while accelerating or enhancing CXCR4 degradation.[71] Because β-arrestin-1 also interacts with AIP4, it is likely that β-arrestin-1 serves as an adaptor that binds to STAM1 to promote AIP4-mediated ubiquitination of HRS and likely STAM1.[63] To counter this, AIP4-mediated ubiquitination of HRS and STAM1 is negatively regulated by the RING domain E3 ubiquitin ligase DTX3L.[33] Ultimately, it appears that the ubiquitination status of ESCRT-0 is tightly linked to the amount of CXCR4 that is sorted into ILVs eventually degraded in lysosomes (see model in Fig. 2A). Ubiquitination of ESCRT-0 may induce intra- or intermolecular interactions between the ubiquitin moieties and its own UBDs, possibly inhibiting its sorting activity by preventing it from interacting with ubiquitin moieties on cargo.[33] Although this remains to be tested directly, it does represent a unique process by which GPCRs can regulate their own sorting efficiency by regulating the activity of the sorting machinery.

Proteins can also interact with the ESCRT machinery to modulate GPCR lysosomal targeting. A multidomain protein termed G protein-coupled receptor interacting scaffold protein (GISP) was identified in a yeast two-hybrid screen of a rat brain cDNA library to interact with the C-terminal domain of the $GABA_{B1}$ subunit, but not the $GABA_{B2}$ subunit, of the $GABA_B$ receptor.[98] Overexpression of GISP increased the surface expression of the $GABA_{B1}$ subunit, possibly by promoting or enhancing its surface expression via the biosynthetic pathway.[98] In another yeast two-hybrid screen, Tsg101, a subunit of ESCRT-I, was identified to be an interacting partner of GISP.[99] The interaction between GISP and Tsg101 was confirmed in the adult rat brain and HEK293 cells, as assessed by coimmunoprecipitation studies and GISP and Tsg101 were shown to colocalize in neurons by fluorescence microscopy. Overexpression of GISP delays lysosomal degradation of the $GABA_{B2}$ subunit in HEK293 cells, while a mutant of GISP that is unable to bind to Tsg101 does not prevent degradation of this subunit.[99] Therefore, GISP binding to Tsg101 can inhibit the sorting activity of ESCRT-I, possibly by preventing ESCRT-I from interacting with ubiquitinated receptors or other elements of the ESCRT pathway, thereby leading to increased $GABA_{B2}$ levels in cells and possibly increased GABAergic responsiveness.[98] This highlights the fact that proteins like GISP can regulate ESCRT function and underscores the notion that together with ubiquitination the ESCRT machinery is susceptible to multiple modes of regulation. Ultimately, this is important because changes in the sorting efficiency or the sorting itinerary of GPCRs can impact signaling and therefore impact physiological processes.

7. ROLE OF DEUBIQUITINATION IN GPCR LYSOSOMAL SORTING

Although ubiquitin attachment to GPCRs is required for their degradation, the removal of ubiquitin seems to be equally important. Removal of ubiquitin can occur before or after ubiquitinated GPCRs enter the

ESCRT pathway.[59] Two DUBs, known as USP8 (ubiquitin-specific protease 8; a.k.a. UBPY) and AMSH (associated molecule with the SH3 domain of STAM) are found on endosomes and are linked to the ESCRT machinery.[100] USP8 and AMSH have also been linked to GPCR deubiquitination. For example, both USP8 and AMSH have been shown to regulate the ubiquitination status of PAR2.[24] USP8 or AMSH knockdown by siRNA results in somewhat greater ubiquitination of PAR2, suggesting that these DUBs regulate PAR2 deubiquitination.[24] Deubiquitination may be necessary to efficiently target PAR2 to lysosomes because USP8 or AMSH knockdown moderately attenuated PAR2 degradation.[24] In the case of CXCR4, knockdown of USP8 by siRNA attenuates agonist-induced degradation of CXCR4,[23] however, knockdown of AMSH does not.[71] In contrast to PAR2, USP8 (or AMSH) does not impact CXCR4 ubiquitination but instead modulates the ubiquitin status of ESCRT-0 that is ubiquitinated by the E3 ligase AIP4,[23] reinforcing the idea that ubiquitination of the transport machinery represents an important regulatory event in GPCR trafficking. Although AMSH or USP8 does not seem to regulate the ubiquitination status of CXCR4, CXCR4 may be deubiquitinated by another DUB.[25] Interestingly, USP8 may regulate the deubiquitination and trafficking of Frizzled[27] and Smoothened,[28] other members of the GPCR/7-transmembrane domain superfamily of receptors. Therefore, ubiquitin removal from either the GPCR or sorting machinery is essential to sort GPCRs into ILVs for eventual degradation in lysosomes.

Deubiquitination of GPCRs may also represent an important regulatory event dictating their recycling and resensitization of receptor signaling. Two highly related DUBs, USP20 and USP30, have been shown to regulate the ubiquitination status of β_2AR possibly by interacting directly with the third intracellular loop.[29] The interactions may occur on endosomal compartments where deubiquitination of β_2AR may promote its recycling and resensitization of receptor signaling.[29] Deubiquitination of Smoothened by USP8 may regulate trafficking of Smoothened from an intracellular compartment to the cell surface, thereby enhancing cellular responsiveness to the ligand hedgehog.[28] Interestingly, β_2AR recycling may also be governed by ubiquitination of Rab11, a GTPase involved in recycling.[101] In any event, deubiquitination of GPCRs may serve as a generalizable mechanism to regulate cell surface expression of GPCRs and hence hormonal responsiveness. Dysregulation of deubiquitination could potentially be a contributing factor to certain diseases. For example, decreased ubiquitination of CXCR4 contributes to CXCR4 recycling and increased responsiveness of tumor cells to the CXCR4 agonist CXCL12, which is a contributing factor to metastasis.[95] While to our knowledge defective deubiquitination of CXCR4 has not been shown, it does highlight the fact that the ubiquitination status of GPCRs is under tight control and any perturbations could be deleterious.

8. EFFECT OF BIASED AGONISM ON GPCR TRAFFICKING: ROLE OF UBIQUITIN

Given that biased agonism dictates β-arrestin versus G protein-dependent signaling,[102] and since as discussed above, β-arrestins can serve as adaptors for E3 ubiquitin ligases, it is therefore likely that GPCR ubiquitination may also be influenced by ligand bias. This is likely best demonstrated by MOR, with which two distinct MOR agonists, morphine and DAMGO, have differential effects on promoting β-arrestin-mediated MOR ubiquitination.[62] Remarkably, DAMGO is able to promote MOR ubiquitination, while morphine is not. DAMGO-promoted MOR ubiquitination is attenuated in β-arrestin-1-null MEFs, indicating a role of β-arrestin-1 in MOR ubiquitination. The difference in the MOR ubiquitination profile induced by DAMGO and morphine may be linked to the fact that DAMGO is better able to induce β-arrestin recruitment to the receptor than morphine. It is likely that DAMGO and morphine induce distinct receptor conformations that are differentially recognized by GRKs and β-arrestin-1. Interestingly, although DAMGO also promotes robust β-arrestin-2 recruitment, β-arrestin-2 is not required for DAMGO-induced MOR ubiquitination.[62] In contrast, DADLE, another opioid agonist, induces rapid ubiquitination of MOR that may require β-arrestin-2.[47] In this case, β-arrestin-2 may serve as an adaptor for Smurf2, a member of the Nedd4-family of E3 ubiquitin ligases.[47] The ligase that promotes DAMGO-dependent ubiquitination of MOR remains to be determined, as far as we know. But this highlights the fact that distinct ligands can regulate the recruitment of E3 ligases likely by regulating differential recruitment of β-arrestins to activated GPCRs.

Ligand-biased ubiquitination of GPCRs may be linked to ligand-biased interactions between β-arrestins and E3 ubiquitin ligases. The β-blocker carvedilol is a biased agonist for β-arrestin-dependent signaling but is an antagonist for G protein-dependent signaling.[103] Recently, carvedilol was shown to regulate ubiquitination of β_2AR by the E3 ubiquitin ligase MARCH2 (membrane-associated RING-CH2).[70] Carvedilol, but not the β-blocker propranolol, induces rapid ubiquitination of β_2AR, similar to the β-agonist isoproterenol.[70] Accordingly, carvedilol induces lysosomal trafficking of endogenous β_2AR in vascular smooth muscle cells or overexpressed in HEK293 cells, very similar to what is observed with the β-agonist isoproterenol.[70] However, in contrast to isoproterenol, β-arrestins and Nedd4 are not required for carvedilol-mediated ubiquitination of β_2AR, suggesting that carvedilol-promoted β_2AR ubiquitination is mediated by another E3 ubiquitin ligase. A mass spectrometry-based approach to identify binding partners of β_2AR after carvedilol treatment compared with isoproterenol treatment revealed that the E3 ubiquitin ligase MARCH2 prefers to bind to β_2AR in cells treated with carvedilol compared with isoproterenol.[70] Carvedilol promotes recruitment of a GFP-tagged MARCH2 to the plasma membrane leading to a stable interaction with β_2AR. Interestingly, this stable interaction may hinder β_2AR deubiquitination and thereby limits receptor recycling and facilitates lysosomal degradation of β_2AR leading to attenuation of signaling. The fact that carvedilol mediates receptor ubiquitination via MARCH2 and limits receptor recycling, and receptor signaling may explain some of the beneficial effects that carvedilol has in treating heart failure.[70]

9. CONCLUSION

Ubiquitin has emerged as an important posttranslational modification that regulates GPCR trafficking. Recent developments have led to a greater understanding of the mechanisms governing the role of ubiquitin in GPCR internalization and endosomal sorting. It is apparent that GPCRs are differentially regulated by ubiquitin, either directly or indirectly, and it is now clear that ligand bias can differentially dictate GPCR ubiquitination and also possibly regulate deubiquitination. However, a large gap exists in our mechanistic understanding of these processes. Future studies aimed at identifying the underlying mechanisms governing the ubiquitination status of GPCRs will be essential in understanding the role that ubiquitin has in GPCR trafficking. Given that GPCRs are the targets for many drugs used to treat a variety of diseases, it is likely that elucidating these mechanisms will translate into a better understanding of GPCR signaling in human health and disease.

ACKNOWLEDGMENTS

Supported by NIH Grant GM106727 to A.M. and American Heart Association Predoctoral Fellowship 13PRE14280030 to J.E.K.

REFERENCES

1. Oldham WM, Hamm HE. Heterotrimeric G protein activation by G-protein-coupled receptors. *Nat Rev Mol Cell Biol.* 2008;9(1):60–71.
2. Pierce KL, Premont RT, Lefkowitz RJ. Seven-transmembrane receptors. *Nat Rev Mol Cell Biol.* 2002;3(9):639–650.
3. Moore CA, Milano SK, Benovic JL. Regulation of receptor trafficking by GRKs and arrestins. *Annu Rev Physiol.* 2007;69:451–482.
4. Perry SJ, Baillie GS, Kohout TA, et al. Targeting of cyclic AMP degradation to beta 2-adrenergic receptors by beta-arrestins. *Science.* 2002;298(5594):834–836.
5. Nelson CD, Perry SJ, Regier DS, Prescott SM, Topham MK, Lefkowitz RJ. Targeting of diacylglycerol degradation to M1 muscarinic receptors by beta-arrestins. *Science.* 2007;315(5812):663–666.
6. Shenoy SK, Lefkowitz RJ. Beta-arrestin-mediated receptor trafficking and signal transduction. *Trends Pharmacol Sci.* 2011;32(9):521–533.
7. Gurevich VV, Dion SB, Onorato JJ, et al. Arrestin interactions with G protein-coupled receptors. Direct binding studies of wild type and mutant arrestins with rhodopsin, beta 2-adrenergic, and m2 muscarinic cholinergic receptors. *J Biol Chem.* 1995;270(2):720–731.
8. Lohse MJ, Benovic JL, Codina J, Caron MG, Lefkowitz RJ. Beta-arrestin: a protein that regulates beta-adrenergic receptor function. *Science.* 1990;248(4962):1547–1550.
9. Barak LS, Ferguson SS, Zhang J, Caron MG. A beta-arrestin/green fluorescent protein biosensor for detecting G protein-coupled receptor activation. *J Biol Chem.* 1997;272(44):27497–27500.
10. Marchese A, Chen C, Kim YM, Benovic JL. The ins and outs of G protein-coupled receptor trafficking. *Trends Biochem Sci.* 2003;28(7):369–376.
11. Pitcher JA, Payne ES, Csortos C, DePaoli-Roach AA, Lefkowitz RJ. The G-protein-coupled receptor phosphatase: a protein phosphatase type 2A with a distinct subcellular distribution and substrate specificity. *Proc Natl Acad Sci USA.* 1995;92(18): 8343–8347.
12. Tsao P, von Zastrow M. Downregulation of G protein-coupled receptors. *Curr Opin Neurobiol.* 2000;10(3):365–369.
13. Marchese A, Trejo J. Ubiquitin-dependent regulation of G protein-coupled receptor trafficking and signaling. *Cell Signal.* 2013;25(3):707–716.
14. Hershko A, Ciechanover A. The ubiquitin system. *Annu Rev Biochem.* 1998;67:425–479.
15. Pickart CM. Mechanisms underlying ubiquitination. *Annu Rev Biochem.* 2001;70:503–533.
16. Varshavsky A. The ubiquitin system, an immense realm. *Annu Rev Biochem.* 2012;81:167–176.
17. Kirisako T, Kamei K, Murata S, et al. A ubiquitin ligase complex assembles linear polyubiquitin chains. *EMBO J.* 2006;25(20):4877.
18. Wang X, Herr RA, Hansen TH. Ubiquitination of substrates by esterification. *Traffic.* 2012;13(1):19–24.
19. Metzger MB, Hristova VA, Weissman AM. HECT and RING finger families of E3 ubiquitin ligases at a glance. *J Cell Sci.* 2012;125(pt 3):531–537.
20. Li W, Bengtson MH, Ulbrich A, et al. Genome-wide and functional annotation of human E3 ubiquitin ligases identifies MULAN, a mitochondrial E3 that regulates the organelle's dynamics and signaling. *PLoS One.* 2008;3(1):e1487.
21. Reyes-Turcu FE, Ventii KH, Wilkinson KD. Regulation and cellular roles of ubiquitin-specific deubiquitinating enzymes. *Annu Rev Biochem.* 2009;78:363–397.
22. Sierra MI, Wright MH, Nash PD. AMSH interacts with ESCRT-0 to regulate the stability and trafficking of CXCR4. *J Biol Chem.* 2010;285(18):13990–14004.

23. Berlin I, Higginbotham KM, Dise RS, Sierra MI, Nash PD. The deubiquitinating enzyme USP8 promotes trafficking and degradation of the chemokine receptor 4 at the sorting endosome. *J Biol Chem*. 2010;285(48):37895–37908.
24. Hasdemir B, Murphy JE, Cottrell GS, Bunnett NW. Endosomal deubiquitinating enzymes control ubiquitination and down-regulation of protease-activated receptor 2. *J Biol Chem*. 2009;284(41):28453–28466.
25. Mines MA, Goodwin JS, Limbird LE, Cui FF, Fan GH. Deubiquitination of CXCR4 by USP14 is critical for both CXCL12-induced CXCR4 degradation and chemotaxis but not ERK activation. *J Biol Chem*. 2009;284(9):5742–5752.
26. Hislop JN, Henry AG, Marchese A, von Zastrow M. Ubiquitination regulates proteolytic processing of G protein-coupled receptors after their sorting to lysosomes. *J Biol Chem*. 2009;284(29):19361–19370.
27. Mukai A, Yamamoto-Hino M, Awano W, Watanabe W, Komada M, Goto S. Balanced ubiquitylation and deubiquitylation of Frizzled regulate cellular responsiveness to Wg/Wnt. *EMBO J*. 2010;29(13):2114–2125.
28. Xia R, Jia H, Fan J, Liu Y, Jia J. USP8 promotes smoothened signaling by preventing its ubiquitination and changing its subcellular localization. *PLoS Biol*. 2012;10(1): e1001238.
29. Berthouze M, Venkataramanan V, Li Y, Shenoy SK. The deubiquitinases USP33 and USP20 coordinate beta2 adrenergic receptor recycling and resensitization. *EMBO J*. 2009;28(12):1684–1696.
30. Milojevic T. The ubiquitin-specific protease USP4 regulates the cell surface level of the A_{2A} receptor. *Mol Pharmacol*. 2006;69(4):1083–1094.
31. Jacob C, Cottrell GS, Gehringer D, Schmidlin F, Grady EF, Bunnett NW. c-Cbl mediates ubiquitination, degradation, and down-regulation of human protease-activated receptor 2. *J Biol Chem*. 2005;280(16):16076–16087.
32. Shenoy SK, McDonald PH, Kohout TA, Lefkowitz RJ. Regulation of receptor fate by ubiquitination of activated beta 2-adrenergic receptor and beta-arrestin. *Science*. 2001;294(5545):1307–1313.
33. Holleman J, Marchese A. The ubiquitin ligase Deltex-3l regulates endosomal sorting of the G protein-coupled receptor CXCR4. *Mol Biol Cell*. 2014;25(12):1892–1904.
34. Moriyoshi K, Iijima K, Fujii H, Ito H, Cho Y, Nakanishi S. Seven in absentia homolog 1A mediates ubiquitination and degradation of group 1 metabotropic glutamate receptors. *Proc Natl Acad Sci USA*. 2004;101(23):8614–8619.
35. Huibregtse JM, Scheffner M, Beaudenon S, Howley PM. A family of proteins structurally and functionally related to the E6-AP ubiquitin-protein ligase. *Proc Natl Acad Sci USA*. 1995;92(7):2563–2567.
36. Rotin D, Kumar S. Physiological functions of the HECT family of ubiquitin ligases. *Nat Rev Mol Cell Biol*. 2009;10(6):398–409.
37. Dores MR, Trejo J. Ubiquitination of G protein-coupled receptors: functional implications and drug discovery. *Mol Pharmacol*. 2012;82(4):563–570.
38. Ingham RJ, Gish G, Pawson T. The Nedd4 family of E3 ubiquitin ligases: functional diversity within a common modular architecture. *Oncogene*. 2004;23(11):1972–1984.
39. Malik R, Soh UJ, Trejo J, Marchese A. Novel roles for the E3 ubiquitin ligase atrophin-interacting protein 4 and signal transduction adaptor molecule 1 in G protein-coupled receptor signaling. *J Biol Chem*. 2012;287(12):9013–9027.
40. Perry WL, Hustad CM, Swing DA, O'Sullivan TN, Jenkins NA, Copeland NG. The itchy locus encodes a novel ubiquitin protein ligase that is disrupted in a18H mice. *Nat Genet*. 1998;18(2):143–146.
41. Chen HI, Sudol M. The WW domain of Yes-associated protein binds a proline-rich ligand that differs from the consensus established for Src homology 3-binding modules. *Proc Natl Acad Sci USA*. 1995;92(17):7819–7823.

42. Ingham RJ, Colwill K, Howard C, et al. WW domains provide a platform for the assembly of multiprotein networks. *Mol Cell Biol.* 2005;25(16):7092–7106.
43. Shearwin-Whyatt L, Dalton HE, Foot N, Kumar S. Regulation of functional diversity within the Nedd4 family by accessory and adaptor proteins. *BioEssays.* 2006;28(6):617–628.
44. Bhandari D, Robia SL, Marchese A. The E3 ubiquitin ligase atrophin interacting protein 4 binds directly to the chemokine receptor CXCR4 via a novel WW domain-mediated interaction. *Mol Biol Cell.* 2009;20(5):1324–1339.
45. Oo ML, Chang S-H, Thangada S, et al. Engagement of S1P1-degradative mechanisms leads to vascular leak in mice. *J Clin Investig.* 2011;121(6):2290–2300.
46. Shenoy SK, Xiao K, Venkataramanan V, Snyder PM, Freedman NJ, Weissman AM. Nedd4 mediates agonist-dependent ubiquitination, lysosomal targeting, and degradation of the beta2-adrenergic receptor. *J Biol Chem.* 2008;283(32):22166–22176.
47. Henry AG, Hislop JN, Grove J, Thorn K, Marsh M, von Zastrow M. Regulation of endocytic clathrin dynamics by cargo ubiquitination. *Dev Cell.* 2012;23(3):519–532.
48. Marchese A, Benovic JL. Agonist-promoted ubiquitination of the G protein-coupled receptor CXCR4 mediates lysosomal sorting. *J Biol Chem.* 2001;276(49): 45509–45512.
49. Dores MR, Trejo J. Ubiquitination of GPCRs, functional implications and drug discovery. *Mol Pharmacol.* 2012;82:563–570.
50. Imai Y, Soda M, Inoue H, Hattori N, Mizuno Y, Takahashi R. An unfolded putative transmembrane polypeptide, which can lead to endoplasmic reticulum stress, is a substrate of Parkin. *Cell.* 2001;105(7):891–902.
51. Petaja-Repo UE, Hogue M, Laperriere A, Bhalla S, Walker P, Bouvier M. Newly synthesized human delta opioid receptors retained in the endoplasmic reticulum are retrotranslocated to the cytosol, deglycosylated, ubiquitinated, and degraded by the proteasome. *J Biol Chem.* 2001;276(6):4416–4423.
52. Petaja-Repo UE, Hogue M, Bhalla S, Laperriere A, Morello JP, Bouvier M. Ligands act as pharmacological chaperones and increase the efficiency of delta opioid receptor maturation. *EMBO J.* 2002;21(7):1628–1637.
53. Canals M, Scholten DJ, de Munnik S, Han MK, Smit MJ, Leurs R. Ubiquitination of CXCR7 controls receptor trafficking. *PLoS One.* 2012;7(3):e34192.
54. Baugher PJ, Richmond A. The carboxyl-terminal PDZ ligand motif of chemokine receptor CXCR2 modulates post-endocytic sorting and cellular chemotaxis. *J Biol Chem.* 2008;283(45):30868–30878.
55. Caballero A, Marchese A. Ubiquitination of GPCRs. *Methods Mol Biol.* 2011;746:251–259.
56. Hislop JN, Henry AG, von Zastrow M. Ubiquitination in the first cytoplasmic loop of mu-opioid receptors reveals a hierarchical mechanism of lysosomal down-regulation. *J Biol Chem.* 2011;286(46):40193–40204.
57. Xiao K, Shenoy SK. Beta2-adrenergic receptor lysosomal trafficking is regulated by ubiquitination of lysyl residues in two distinct receptor domains. *J Biol Chem.* 2011;286(14):12785–12795.
58. Wolfe BL, Marchese A, Trejo J. Ubiquitination differentially regulates clathrin-dependent internalization of protease-activated receptor-1. *J Cell Biol.* 2007;177(5):905–916.
59. Marchese A, Raiborg C, Santini F, Keen JH, Stenmark H, Benovic JL. The E3 ubiquitin ligase AIP4 mediates ubiquitination and sorting of the G protein-coupled receptor CXCR4. *Dev Cell.* 2003;5(5):709–722.
60. Busillo JM, Armando S, Sengupta R, Meucci O, Bouvier M, Benovic JL. Site-specific phosphorylation of CXCR4 is dynamically regulated by multiple kinases and results in differential modulation of CXCR4 signaling. *J Biol Chem.* 2010;285(10):7805–7817.

61. Sudol M, Chen HI, Bougeret C, Einbond A, Bork P. Characterization of a novel protein-binding module—the WW domain. *FEBS Lett.* 1995;369(1):67–71.
62. Groer CE, Schmid CL, Jaeger AM, Bohn LM. Agonist-directed interactions with specific beta-arrestins determine mu-opioid receptor trafficking, ubiquitination, and dephosphorylation. *J Biol Chem.* 2011;286(36):31731–31741.
63. Bhandari D, Trejo J, Benovic JL, Marchese A. Arrestin-2 interacts with the ubiquitin-protein isopeptide ligase atrophin-interacting protein 4 and mediates endosomal sorting of the chemokine receptor CXCR4. *J Biol Chem.* 2007;282(51):36971–36979.
64. Alvarez CE. On the origins of arrestin and rhodopsin. *BMC Evol Biol.* 2008;8:222.
65. Lin CH, MacGurn JA, Chu T, Stefan CJ, Emr SD. Arrestin-related ubiquitin-ligase adaptors regulate endocytosis and protein turnover at the cell surface. *Cell.* 2008;135(4):714–725.
66. Aubry L, Guetta D, Klein G. The arrestin fold: variations on a theme. *Curr Genomics.* 2009;10(2):133–142.
67. Shea FF, Rowell JL, Li Y, Chang TH, Alvarez CE. Mammalian alpha arrestins link activated seven transmembrane receptors to Nedd4 family e3 ubiquitin ligases and interact with beta arrestins. *PLoS One.* 2012;7(12):e50557.
68. Rauch S, Martin-Serrano J. Multiple interactions between the ESCRT machinery and arrestin-related proteins: implications for PPXY-dependent budding. *J Virol.* 2011;85(7):3546–3556.
69. Nabhan JF, Pan H, Lu Q. Arrestin domain-containing protein 3 recruits the NEDD4 E3 ligase to mediate ubiquitination of the beta2-adrenergic receptor. *EMBO Rep.* 2010;11(8):605–611.
70. Han SO, Kommaddi RP, Shenoy SK. Distinct roles for beta-arrestin2 and arrestin-domain-containing proteins in beta2 adrenergic receptor trafficking. *EMBO Rep.* 2013;14(2):164–171.
71. Malik R, Marchese A. Arrestin-2 interacts with the endosomal sorting complex required for transport machinery to modulate endosomal sorting of CXCR4. *Mol Biol Cell.* 2010;21(14):2529–2541.
72. Vecchione A, Marchese A, Henry P, Rotin D, Morrione A. The Grb10/Nedd4 complex regulates ligand-induced ubiquitination and stability of the insulin-like growth factor I receptor. *Mol Cell Biol.* 2003;23(9):3363–3372.
73. Kang DS, Tian X, Benovic JL. Role of beta-arrestins and arrestin domain-containing proteins in G protein-coupled receptor trafficking. *Curr Opin Cell Biol.* 2014;27:63–71.
74. Ahn S, Nelson CD, Garrison TR, Miller WE, Lefkowitz RJ. Desensitization, internalization, and signaling functions of beta-arrestins demonstrated by RNA interference. *Proc Natl Acad Sci USA.* 2003;100(4):1740–1744.
75. Paing MM, Stutts AB, Kohout TA, Lefkowitz RJ, Trejo J. Beta-arrestins regulate protease-activated receptor-1 desensitization but not internalization or down-regulation. *J Biol Chem.* 2002;277(2):1292–1300.
76. Chen B, Dores MR, Grimsey N, Canto I, Barker BL, Trejo J. Adaptor protein complex-2 (AP-2) and epsin-1 mediate protease-activated receptor-1 internalization via phosphorylation- and ubiquitination-dependent sorting signals. *J Biol Chem.* 2011;286(47):40760–40770.
77. Traub LM. Tickets to ride: selecting cargo for clathrin-regulated internalization. *Nat Rev Mol Cell Biol.* 2009;10(9):583–596.
78. Kim YM, Benovic JL. Differential roles of arrestin-2 interaction with clathrin and adaptor protein 2 in G protein-coupled receptor trafficking. *J Biol Chem.* 2002;277(34):30760–30768.
79. Shenoy SK, Barak LS, Xiao K, et al. Ubiquitination of beta-arrestin links seven-transmembrane receptor endocytosis and ERK activation. *J Biol Chem.* 2007;282(40):29549–29562.

80. Wyatt D, Malik R, Vesecky AC, Marchese A. Small ubiquitin-like modifier modification of arrestin-3 regulates receptor trafficking. *J Biol Chem*. 2011;286(5):3884–3893.
81. Ozawa K, Whalen EJ, Nelson CD, et al. S-nitrosylation of β-arrestin regulates β-adrenergic receptor trafficking. *Mol Cell*. 2008;31(3):395–405.
82. Hislop JN, von Zastrow M. Role of ubiquitination in endocytic trafficking of G-protein-coupled receptors. *Traffic*. 2011;12(2):137–148.
83. Dores MR, Trejo J. Atypical regulation of G protein-coupled receptor intracellular trafficking by ubiquitination. *Curr Opin Cell Biol*. 2014;27:44–50.
84. Piper RC, Katzmann DJ. Biogenesis and function of multivesicular bodies. *Annu Rev Cell Dev Biol*. 2007;23:519–547.
85. Gruenberg J, Stenmark H. The biogenesis of multivesicular endosomes. *Nat Rev Mol Cell Biol*. 2004;5(4):317–323.
86. Dores MR, Chen B, Lin H, et al. ALIX binds a YPX(3)L motif of the GPCR PAR1 and mediates ubiquitin-independent ESCRT-III/MVB sorting. *J Cell Biol*. 2012;197(3):407–419.
87. Slagsvold T, Marchese A, Brech A, Stenmark H. CISK attenuates degradation of the chemokine receptor CXCR4 via the ubiquitin ligase AIP4. *EMBO J*. 2006;25(16):3738–3749.
88. Henry AG, White IJ, Marsh M, von Zastrow M, Hislop JN. The role of ubiquitination in lysosomal trafficking of delta-opioid receptors. *Traffic*. 2011;12(2):170–184.
89. Henne WM, Buchkovich NJ, Emr SD. The ESCRT pathway. *Dev Cell*. 2011;21(1):77–91.
90. Marchese A. Endocytic trafficking of chemokine receptors. *Curr Opin Cell Biol*. 2014;27:72–77.
91. Valiathan RR, Resh MD. Differential control of CXCR4 and CD4 downregulation by HIV-1 Gag. *Virol J*. 2008;5:23.
92. Malerod L, Stuffers S, Brech A, Stenmark H. Vps22/EAP30 in ESCRT-II mediates endosomal sorting of growth factor and chemokine receptors destined for lysosomal degradation. *Traffic*. 2007;8(11):1617–1629.
93. Kumar A, Kremer KN, Dominguez D, Tadi M, Hedin KE. Galpha13 and Rho mediate endosomal trafficking of CXCR4 into Rab11+ vesicles upon stromal cell-derived factor-1 stimulation. *J Immunol*. 2011;186(2):951–958.
94. Benovic JL, Marchese A. A new key in breast cancer metastasis. *Cancer Cell*. 2004;6(5):429–430.
95. Li YM, Pan Y, Wei Y, et al. Upregulation of CXCR4 is essential for HER2-mediated tumor metastasis. *Cancer Cell*. 2004;6(5):459–469.
96. Gullapalli A, Wolfe BL, Griffin CT, Magnuson T, Trejo J. An essential role for SNX1 in lysosomal sorting of protease-activated receptor-1: evidence for retromer-, Hrs-, and Tsg101-independent functions of sorting nexins. *Mol Biol Cell*. 2006;17(3):1228–1238.
97. Hasdemir B, Bunnett NW, Cottrell GS. Hepatocyte growth factor-regulated tyrosine kinase substrate (HRS) mediates post-endocytic trafficking of protease-activated receptor 2 and calcitonin receptor-like receptor. *J Biol Chem*. 2007;282(40):29646–29657.
98. Kantamneni S, Correa SA, Hodgkinson GK, et al. GISP: a novel brain-specific protein that promotes surface expression and function of GABA(B) receptors. *J Neurochem*. 2007;100(4):1003–1017.
99. Kantamneni S, Holman D, Wilkinson KA, et al. GISP binding to TSG101 increases GABA receptor stability by down-regulating ESCRT-mediated lysosomal degradation. *J Neurochem*. 2008;107(1):86–95.
100. Clague MJ, Coulson JM, Urbe S. Cellular functions of the DUBs. *J Cell Sci*. 2012;125(pt 2):277–286.

101. Lachance V, Degrandmaison J, Marois S, et al. Ubiquitylation and activation of a Rab GTPase is promoted by a beta(2)AR-HACE1 complex. *J Cell Sci.* 2014;127(pt 1): 111–123.
102. Reiter E, Ahn S, Shukla AK, Lefkowitz RJ. Molecular mechanism of β-arrestin-biased agonism at seven-transmembrane receptors. *Annu Rev Pharmacol Toxicol.* 2012;52(1):179–197.
103. Wisler JW, DeWire SM, Whalen EJ, et al. A unique mechanism of beta-blocker action: carvedilol stimulates beta-arrestin signaling. *Proc Natl Acad Sci USA.* 2007;104(42):16657–16662.

CHAPTER THREE

Rhodopsin Trafficking and Mistrafficking: Signals, Molecular Components, and Mechanisms

Ina Nemet, Philip Ropelewski, Yoshikazu Imanishi[1]
Department of Pharmacology, School of Medicine, Case Western Reserve University, Cleveland, Ohio, USA
[1]Corresponding author: e-mail address: yxi19@case.edu

Contents

Abstract

Rhodopsin is a seven-transmembrane G protein-coupled receptor (GPCR) and is the main component of the photoreceptor outer segment (OS), a ciliary compartment essential for vision. Because the OSs are incapable of protein synthesis, rhodopsin must first be synthesized in the inner segments (ISs) and subsequently trafficked across the connecting cilia to the OSs where it participates in the phototransduction cascade. Rapid turnover of the OS necessitates a high rate of synthesis and efficient trafficking of rhodopsin to the cilia. This cilia-targeting mechanism is shared among other ciliary-localized GPCRs. In this review, we will discuss the process of rhodopsin trafficking from the IS to the OS beginning with the trafficking signals present on the protein. Starting from the endoplasmic reticulum and the Golgi apparatus within the IS, we will

Progress in Molecular Biology and Translational Science, Volume 132
ISSN 1877-1173
http://dx.doi.org/10.1016/bs.pmbts.2015.02.007

cover the molecular components assisting the biogenesis and the proper sorting. We will also review the confirmed binding and interacting partners that help target rhodopsin toward the connecting cilium as well as the cilia-localized components which direct proteins into the proper compartments of the OS. While rhodopsin is the most critical and abundant component of the photoreceptor OS, mutations in the rhodopsin gene commonly lead to its mislocalization within the photoreceptors. In addition to covering the trafficking patterns of rhodopsin, we will also review some of the most common rhodopsin mutants which cause mistrafficking and subsequent death of photoreceptors. Toward the goal of understanding the pathogenesis, three major mechanisms of aberrant trafficking as well as putative mechanisms of photoreceptor degeneration will be discussed.

1. INTRODUCTION

Rhodopsin is a prototypical G protein-coupled receptor (GPCR) mediating perception of light. In vertebrate photoreceptor neurons, rhodopsin is localized to a part of sensory cilium called the outer segment (OS). Visual sensation is initiated in the OS when rhodopsin absorbs a photon and activates a heterotrimeric G protein, transducin. Transducin then activates phosphodiesterase which hydrolyzes cGMP. The reduction in the cGMP concentration is sensed by cGMP-gated channel located on the plasma membrane, leading to hyperpolarization of the membrane potential. Photoactivated rhodopsin eventually becomes deactivated through phosphorylation by rhodopsin kinase and subsequent binding of arrestin, which sterically interferes with the binding of transducin (for reviews see Refs. 1–3). To accomplish these orchestrated events for maximum effectiveness and specificity, each phototransduction component must locate to the appropriate membrane region of the OSs. By concentrating rhodopsin and downstream components in the OSs, interaction among correct signaling partner proteins can be assured.

In addition to rhodopsin, cilia targeting is a characteristic of many GPCR family members including neuropeptide Y family receptors,[4] vasoactive intestinal receptor 2,[5] somatostatin receptor 3 (SSTR3),[6–8] melanin-concentrating hormone receptor 1,[7,8] serotonin receptor 6 (5HT6),[9] dopamine receptor 1,[10] and smoothened in the hedgehog morphogen signaling pathway.[11] Recent studies suggest that some mechanisms of cilia targeting are shared among different GPCRs, and the disruption of these mechanisms is the likely cause of pleiotropic ciliopathies such as Bardet–Biedl syndrome (BBS). Thus, through the studies of vertebrate rhodopsin, the best-studied ciliary GPCR, we will be able to gain insight into the trafficking mechanisms of other GPCRs as well. In this chapter, we will focus on the trafficking signals and molecular components mediating the trafficking of vertebrate rhodopsin. For the studies of invertebrate rhodopsin, we recommend an excellent review published recently.[12]

2. TRAFFICKING SIGNALS OF RHODOPSIN

Early studies of rhodopsin trafficking signals were focused on rhodopsin mutations that lead to its mislocalization.[13–15] Truncation mutant that lacks the terminal five amino acids of rhodopsin, also known as Q344Ter mutant (Fig. 1), exhibits dramatically attenuated trafficking of rhodopsin to the OS.[16,17] Likewise, mutation of the penultimate amino acid, proline to serine, leads to similar OS trafficking defect.[18] Those studies suggested that an essential trafficking signal exists on the C-terminal (Ct)-tail region of rhodopsin, which was later shown to be sufficient for rhodopsin's OS localization.[19] Further, mechanistic studies were conducted to understand the role of the Ct-tail region using retinal cell-free translation system.[20]

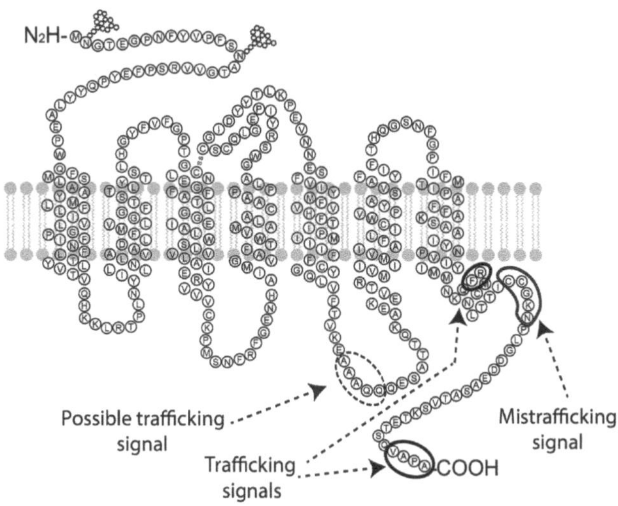

Figure 1 Rhodopsin trafficking and mistrafficking signals. VAPA (VXPX) trafficking motif is located at the very end of the C-terminus of rhodopsin. Defects in this signal lead to rhodopsin mislocalization. Another putative trafficking signal is FR, which has not been validated *in vivo*. AAAQQ is a possible trafficking signal because it shares homology with the trafficking signal responsible for ciliary targeting of other GPCRs (SSTR3 and 5HTR6). CCGKN is a mistrafficking signal. Disruption of the VXPX motif leads to unmasking of this mistrafficking signal and causes rhodopsin mislocalization. (See the color plate.)

In this system, monoclonal antibody blocking the Ct-tail leads to accumulation of newly synthesized rhodopsin in *trans*-Golgi network, strongly suggesting that Ct-tail is critical for proper exit of rhodopsin from the Golgi apparatus. Since the *trans*-Golgi network regulates protein sorting, these studies are consistent with the Ct-tail region playing a role in the OS targeting of rhodopsin. More specifically, the VXPX sequence motif is preserved across rhodopsins and cone visual pigments of different species.[21] VXPX motif also appears to play a critical role in cilia targeting not only for rhodopsin but also for other proteins such as retinol dehydrogenase in vertebrate photoreceptors,[22] polycystin 1 and 2 in renal cilia,[23,24] and rodent cGMP-gated channel in olfactory sensory cilia.[25]

While the VXPX motif appears to regulate the targeting of rhodopsin, questions still remain as to why such a simple motif can mediate the cilia targeting. While the VXPX motif is located at the very end of rhodopsin, the location of VXPX is not critical for its function. For example, the VXPX signal of polycystin 2 is localized in the N-terminus region,[24] whereas the VXPX signal of polycystin 1 at the Ct-tail is followed by additional amino acids.[23] Furthermore, rhodopsin fused to fluorescent proteins (FPs) can properly traffic to the OS[26]; thus, the VXPX motif can locate in the middle of a polypeptide. As the VXPX motif is defined only by two amino acids, valine and proline, such a motif may exist at random every 400 amino-acid sequence. The question then becomes how such a frequently occurring motif can define the specific localization of proteins. Recent studies on rhodopsin and other GPCRs indicate that there are multiple trafficking signals that can direct cilia targeting. Thus, it is unlikely that the VXPX motif is the only signal important for the cilia targeting of rhodopsin.

In addition to the VXPX signal, other trafficking signals may also play roles in the OS targeting of rhodopsin. Those signals include the "FR" motif within the eighth cytoplasmic alpha helix which is essential for the cilia targeting of the GPCR smoothened.[11] The FR motif is also critical for the cilia targeting of rhodopsin when heterologously expressed in Inner Medullary Collecting Duct (IMCD) cells,[27] though the role of FR in OS targeting has not been tested *in vivo*. In addition to the FR signal, the AX[S/A]XQ motif was identified in the third intracellular loop[7] based on the homology between rhodopsin and other cilia-targeted GPCRs SSTR3 and 5HTR6. This motif is sufficient to direct the ciliary localization of nonciliary localized GPCRs such as SSTR5 and HTR7.[7] Furthermore, the signal and additional amino-acid residues were sufficient to target another nonciliary membrane protein CD8α to the primary cilia.[28] Those results support that the third cytoplasmic loop plays a critical role in the cilia targeting of GPCRs, and it is possibly sufficient for ciliary targeting in general. The role of the AX[S/A]XQ motif has not been tested for OS trafficking of rhodopsin. It is not clear why rhodopsin requires so many putative signals for cilia targeting. Multiple signals may be required for accomplishing extremely high concentration (~4 mM) of rhodopsin within the OS.[29] To the best of our knowledge, this is the highest concentration of GPCR observed throughout any biological system and leads to a formation of paracrystalline arrays in the disk membranes.[30] Though quantitative assessment of OS targeting efficacy has been difficult in the past, FPs can be fused to rhodopsin and accurately calibrated to known concentrations of soluble FPs.[26,31] This approach was effectively used to compare the trafficking efficacy of a rhodopsin-FP fusion either containing or lacking the VXPX motif. Consistent with the redundancy in the cilia-targeting signals, rhodopsin lacking the VXPX motif can still reach the rod OSs.[26] However, the concentration attainable in the absence of the VXPX motif was lower than in its presence.[26] Thus, coordination of the VXPX motif and the other cilia-targeting signals allows the high concentration of rhodopsin in the OSs.

Another supportive trafficking signal was identified in rhodopsin based on the studies using Madin-Darby Canine Kidney (MDCK) epithelial cell line. When expressed heterologously in this cell line, rhodopsin preferably localizes to the apical microvilli.[32] The location of the apical sorting signal does not overlap with any of the ciliary-targeting signals described above. Instead, the region essential for this apical targeting closely coincides with the trafficking signal which directed mislocalization of rhodopsin in *Xenopus laevis* rod photoreceptors ($_{322}$CCGKN$_{326}$)[26] (Fig. 1). Rhodopsin lacking the last 22 amino acids (which includes the VXPX motif) was effectively trafficked to the apical plasma membrane region of the MDCK cells[32] and also trafficked to the inner segment (IS) plasma membrane of the rod photoreceptor cells.[26] On the contrary, rhodopsin lacking the last 32 amino acids (and hence lacking $_{322}$CCGKN$_{326}$ sequence) was not effectively trafficked to the apical membrane of MDCK cells[32] and also was not effectively trafficked to the IS plasma membrane of the rod photoreceptor cells.[26] The role of the apical sorting signal for the OS targeting of rhodopsin is unclear, since rhodopsin lacking the last 32 amino acids can be still delivered to the OS. Nevertheless, the apical sorting signal contributes to the IS plasma membrane mistrafficking of rhodopsin. It is possible that this apical sorting signal works together with the rest of the trafficking signals for rhodopsin mislocalization. In the absence of the VXPX signal, these combined signals become unmasked and contribute to the mislocalization of rhodopsin.

3. MOLECULAR COMPONENTS AND MECHANISMS FOR SPECIFIC OS TARGETING OF RHODOPSIN

3.1 Biogenesis of Rhodopsin in Endoplasmic Reticulum and Subsequent Maturation in Golgi Apparatus

Rhodopsin is a seven-transmembrane protein and is synthesized and folded in the endoplasmic reticulum (ER; Fig. 2). During its biogenesis, rhodopsin becomes glycosylated at two asparagine residues: N2 and N15. Removal of both glycans by tunicamycin leads to failure of rhodopsin targeting to the OS[33] and to malformation of the OS. A possible explanation for this phenotype, however, is that the OS malformation after tunicamycin treatment was due to glycosylation deficiency in OS proteins other than rhodopsin. Consistent with this idea, glycosylation-deficient rhodopsin can pass through the ER quality control mechanism and reach the plasma membrane when expressed in cultured cells.[34] Glycosylated proteins are recognized by chaperone proteins such as calnexin and calreticulin. Again consistent with the dispensable role of glycosylation for trafficking, rhodopsin does not

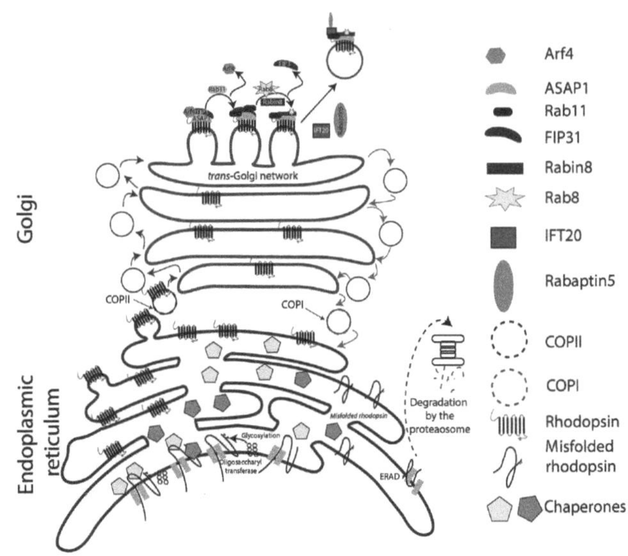

Figure 2 Biogenesis of rhodopsin in the endoplasmic reticulum and formation of rhodopsin transport carriers in the Golgi. After folding in the ER, rhodopsin is transported to the Golgi where rhodopsin interacts with Arf4, FIP3, ASAP1, Rab8, and Rabbin8. Then, rhodopsin transport carriers bud off from the *trans*-Golgi network. IFT20 and Rabaptin5 probably bind to the carrier, but the timing of the binding is unclear. (See the color plate.)

require calnexin for its biosynthesis as tested in mouse embryonic fibroblast lacking a functional calnexin.[35] The role of calreticulin for rhodopsin folding is currently unclear, although glycosylation independent of rhodopsin trafficking seems to suggest that calreticulin may not play a critical role either. Instead, BiP, a Hsp70 molecular chaperone located in the lumen of the ER, interacts with unfolded rhodopsin and plays a critical role in the prevention of WT and mutant rhodopsin aggregation within ER.[36–39] BiP recognizes unfolded proteins regardless of glycosylation status. Taken together, rhodopsin glycosylation plays a minor role in its biogenesis and trafficking. A recent study, however, suggests that mutations in glycosylation sites lead to light-dependent instability and mislocalization[40] which will be discussed in a later section (see Section 4.2).

In addition to glycosylation, rhodopsin also receives palmitoylation at the cysteine residues C322 and C323. Enzymes responsible for rhodopsin palmitoylation have been characterized biochemically,[41] but they have not been identified genetically. Palmitoylation may not play a direct role in the OS targeting of rhodopsin. Disruption of palmitoylation sites does not lead to mislocalization of rhodopsin.[42] Nevertheless, palmitoylations appear to play a role in the stability of opsin: an unliganded form of rhodopsin. Palmitoylation-deficient opsin causes rapid rod photoreceptor degeneration and opsin mislocalization.[43] Thus, palmitoylation may be required for opsin trafficking under intense light conditions when chromophore is of scarce supply and palmitoylation-deficient rhodopsins become unliganded and misfolded in the ER.

Exit of rhodopsin from the ER is likely mediated by the COPII coat as knockdown of COPII component sec13 in zebrafish leads to impaired rhodopsin trafficking.[44] While the mechanism of COPII assembly is unclear for vertebrate rhodopsin, COPII assembly can be specifically mediated by cytosolic phosphatidic acid phospholipase A1 (PAPLA1) in drosophila, which binds to both rhodopsin and COPII.[45] Other GPCRs appear to use the same mechanism for COPII assembly and ER exit. For this assembly, phospholipase activity of PAPLA1 is not essential.[45] As vertebrates also contain homologs of drosophila PAPLA1, it would be intriguing to test the roles of these homologs in the trafficking of vertebrate rhodopsin.

3.2 Sorting of Rhodopsin at the Golgi Apparatus and Post-Golgi Trafficking

Unlike phototransduction components of the OS which are abundant, the components involved in trafficking have evaded discovery due to their comparatively low quantities and transient nature of their interactions. Since studies suggested the Ct-tail played a critical role in rhodopsin trafficking, many efforts were focused on uncovering the partners which binds to the Ct-tail. A biochemical study indicated that the Ct of rhodopsin binds to Arf4, a small GTPase.[46] Arf4 binding requires the presence of the VXPX motif at the Ct-tail and localizes to the *trans*-Golgi network where rhodopsin transport carriers (post-Golgi vesicles carrying rhodopsin, named by Dr. Deretic) bud off (Fig. 2).[46] Expression of Arf4 I46D mutant, which is constitutively active, led to aberrant rhodopsin trafficking,[27] suggesting Arf4 plays a critical role in rhodopsin trafficking. Arf4 facilitates the recruitment of other trafficking components, such as ASAP1, Rab11, Rabin8, and Rab8, to the rhodopsin transport carriers[27] (Fig. 2). Thus, Arf4 binding is a critical step in rhodopsin trafficking; the binding allows ASAP1 interaction and the formation of Rab11–Rabin8–Rab8 complex which directs the targeting of rhodopsin toward the cilium.[27] Some components of Arf4 pathway, such as Rab11, may bind to rhodopsin independent of Arf4.[27,47] In addition to the VXPX motif, the FR motif is also responsible for the recruitment of these components as ASAP1, which acts as a scaffold for these proteins, directly binds to FR.[27]

Rhodopsin trafficking mediated by the Rabin8/Rab8 complex may play a critical role not only in OS trafficking but also in OS morphogenesis. Rabin8 and Rab8 are known to play critical roles in ciliogenesis in cultured cells. For ciliogenesis, Rabin8 and Rab8 cooperate with the products of genes responsible for BBS. Several of these BBS gene products form a complex called BBSome.[48] BBS is characterized by multitude of symptoms including retina degeneration, and BBS model mice demonstrate compromised rhodopsin trafficking leading to its mislocalization or shorter rod OSs.[49–51] In addition to indirect interaction via Rab8/Rabin8, rhodopsin family GPCRs can interact with BBSome via its cilia-targeting signal in the third cytoplasmic loop. The third cytoplasmic loop of a rhodopsin family member SSTR3 includes the AX[S/A]XQ motif which directly interacts with BBS proteins.[28] A complex of BBS proteins, BBSome, appears to form structures resembling COPI, COPII, and clathrin coats.[28] Since the BBSome is found within the cilia,[28] it may mediate the trafficking of GPCRs including rhodopsin through the cilia. Consistent with this notion BBSome dysfunction causes accumulation of vesicles near the base of the cilia.[52] Based on the studies of *Caenorhabditis elegans*, several BBS proteins move inside the cilia at the rates comparable to those of intraflagellar trafficking (IFT) particles,[53] which are carried by kinesin and dynein motors. Roles of IFT particles in the ciliary trafficking of rhodopsin are discussed in Section 3.4.

The Rab8 pathway is linked to several critical components whose defects lead to photoreceptor degeneration and rhodopsin mistrafficking. For example, a component important for rhodopsin trafficking through the cilia, IFT20, is anchored to the Golgi apparatus by the protein GMAP210/Trip11.[54,55] Additionally, IFT20 indirectly associates with Rab8 through Elipsa, which is a coiled-coil protein. Elipsa localizes to cilia and binds to IFT20, and also binds to Rabaptin5. Rabaptin5 can then interact with Rab8, which appears critical for rhodopsin targeting to cilia.[56] In mouse rods, defects in IFT20 cause rhodopsin accumulation in Golgi.[57] Therefore, IFT20 may also play an important role in the exit of rhodopsin from Golgi apparatus and delivery of rhodopsin transport carriers to the base of the cilia. This Golgi function of IFT20 is regulated by a small G protein Arl3 and its GTPase activator RP2. Loss of RP2 or dysregulation of Arl3 leads to dispersal of Golgi-localized IFT20. The role of RP2 and Arl3 in rhodopsin trafficking is suggested from the studies of mouse models. RP2-deficient mice demonstrate reduced rhodopsin contents in the OS,[58] while Arl3-deficient mice exhibit severe defects in rhodopsin trafficking to the OS.[59] Thus, components associated with Rab8 are involved in trafficking of rhodopsin in general.

3.3 Vectorial Targeting of Rhodopsin from Golgi to the Connecting Cilium: Possible Involvement of Microtubules

In general, vectorial targeting of cargoes requires cytoskeleton and associated motor proteins. Within the photoreceptor IS, plus-ends of microtubules are proximal to the Golgi apparatus, whereas minus-ends are proximal to the basal body.[60] Thus, the minus-end-directed motor, dynein, has been estimated to play a role in the trafficking of rhodopsin toward the basal body[61] (Fig. 3). There is also a hypothesis opposing the role of microtubules in rhodopsin trafficking. The hypothesis was derived from a study in which disruption of microtubules did not lead to apparent compromise in the disk membrane formation monitored by lucifer yellow incorporation.[62] The observation, however, was not derived from direct monitoring of rhodopsin trafficking and thus needs to be interpreted with caution.

Cytoplasmic dynein 1 and 2 are motors which carry cargo from the plus- to the minus-end of microtubules. Dynein 1 and 2 deficiencies lead to OS disorganization and shortening in zebrafish.[63,64] As rhodopsin is the major component of the OS, these OS defects may have stemmed from aberrant

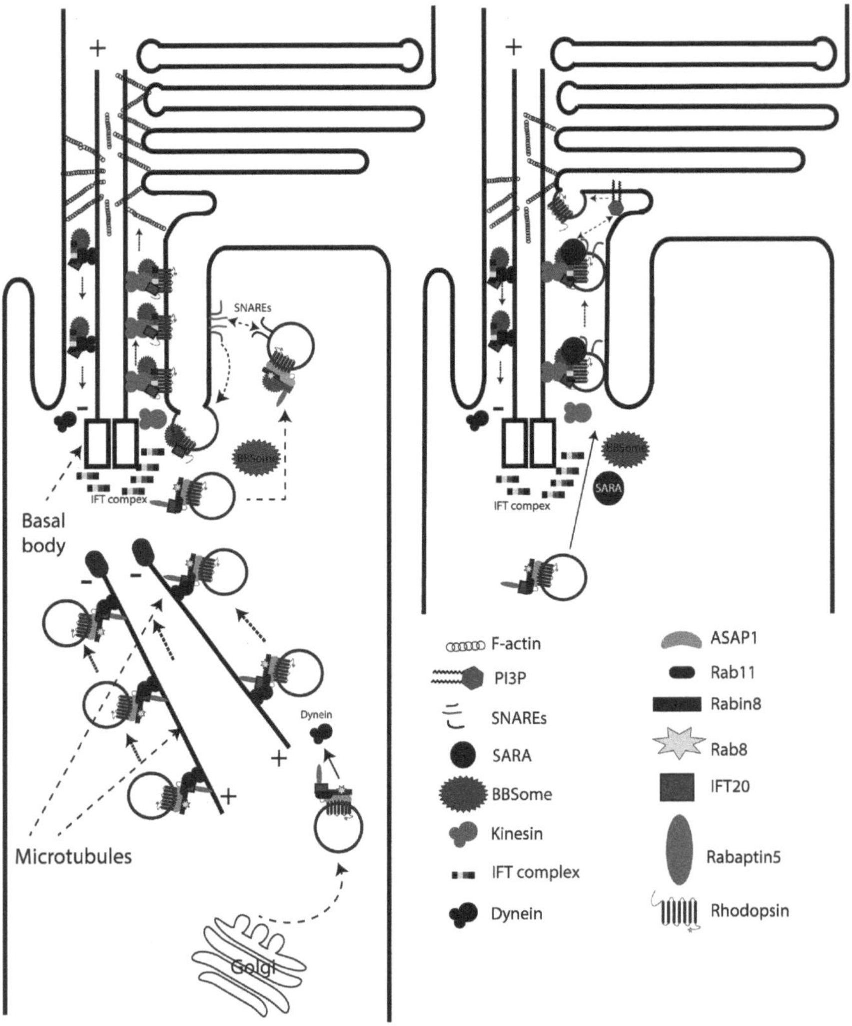

Figure 3 Vectorial trafficking of rhodopsin from the Golgi to the distal end of the connecting cilium. Dynein and microtubules may mediate the vectorial trafficking of rhodopsin transport carriers from the Golgi to the basal body. The BBSome, IFT particles, and kinesin may mediate the subsequent trafficking of rhodopsin to the region of new disk synthesis. Two possible mechanisms for rhodopsin delivery are proposed: the first is the fusion of rhodopsin transport carriers with the IS periciliary plasma membrane and then trafficking of rhodopsin-containing rafts to the nascent disks (left panel); the second is the trafficking of rhodopsin transport carriers directly to the nascent disks (right panel). (See the color plate.)

rhodopsin trafficking. TcTex-1, one of the dynein light-chain subunits, was documented to bind to rhodopsin Ct-tail region. In MDCK cells, TcTex-1 facilitates the apical membrane sorting of rhodopsin.[65] The rod OS is located in the apical region of the cell, and thus, apical sorting may facilitate the OS enrichment. While the roles of TcTex-1 have been characterized *in vitro* and in cultured cells, its role has not been investigated in rod photoreceptor cells. Thus, it is still unclear if this TcTex-1 is involved in rhodopsin trafficking *in vivo*. Additional rhodopsin-binding factors may facilitate dynein-mediated trafficking. As described above, Arf4 recruits additional factors such as Rab11 and ASAP1 on the Ct-tail region of rhodopsin. Furthermore, ASAP1 recruits an effector of Rab11, FIP3.[66,67] Rab11–FIP3 complex is known to bind the components of cytoplasmic dynein 1.[68,69] Taken together, microtubule–dynein system is a promising mediator of post-Golgi rhodopsin trafficking toward the base of the connecting cilia (Fig. 3).

3.4 Trafficking of Rhodopsin Toward the Distal End of the Connecting Cilium

To enter the OS from the IS, rhodopsin must travel through the connecting cilium's transition zone. This process has never been directly visualized, and as such, it is not clear exactly how rhodopsin enters the OS. Nevertheless, studies of the components involved in IFT have been instructive to our understanding of IS to OS rhodopsin trafficking (Fig. 3). These components assemble into stable complexes, IFT-A and IFT-B particles, which were originally identified and characterized in *Chlamydomonas*.[70] The IFT particles are localized to the ciliary portion of rod photoreceptors[71] and are involved in dynamic interaction between cargo and Kinesin II.[70,72,73] Deficiency in the IFT-B particle component, IFT88, leads to abnormal trafficking of rhodopsin.[74] IFT88 forms the IFT-B particle along with other components such as IFT57, IFT52, IFT27, IFT25, and IFT20.[72,75] Mutations or knockdown of these IFT-B components leads to defects in the OS maintenance in zebrafish.[76] IFT-A particle also plays a role in rhodopsin trafficking. For example, deficiency of an IFT-A component IFT140 leads to rhodopsin mislocalization to the plasma membrane.[77] These studies suggest that IFT-A and IFT-B play critical roles in rhodopsin trafficking.

IFT-A and -B particles interact with lebercilin,[78] which is localized to the connecting cilia. Lebercilin gene mutation causes Leber congenital amaurosis, which is one of the most devastating forms of blindness in humans. The IFT components also interact with related B9d2, which is homologous to Meckel–Gruber syndrome 1 gene product.[79]

Meckel–Gruber syndrome is a ciliopathy characterized by abnormal development of the eye, including the retina.[80] B9d2 in turn interacts with nephrocystin 2 (also called inversin),[79] which is associated with another ciliopathy, nephronophthisis.[81] B9d2 and nephrocystin2, along with nephrocystin 5, are critical for opsin trafficking, as morpholino knockdown of these proteins leads to compromised opsin trafficking.[79] The same morpholino knockdowns did not alter the trafficking of peripherin/rds, suggesting that these genes are involved in unique trafficking of opsins. It is currently unclear which sequence motif of rhodopsin is recognized by IFT particles and which protein component recognizes rhodopsin and anchors it to the IFT complex.

Plus-ends of microtubules are directed toward the OS in connecting cilia. Kinesins are the motors that direct cargo trafficking toward the plus-end of microtubules. There are two types of kinesins which associate with the IFT machinery: KIF3 and KIF17 complexes.[82] The KIF3 complex consists of KIF3A, and KIF3B or KIF3C motor subunit, as wells as a regulatory subunit Kap3. The function of KIF3A and IFT is associated with Rab8 pathway through RPGR. RPGR, a product of the gene primarily responsible for X-linked retinitis pigmentosa, interacts with KIF3A, IFT88, and Rab8.[83] Defects in RPGR and its interacting protein RPGRIP lead to severe photoreceptor degeneration in humans,[84,85] and rhodopsin mistrafficking in mice.[86,87] Past studies of the KIF3 complex also focused on mouse models in which KIF3A was conditionally knocked out. The degree of rhodopsin mislocalization varied among the studies, and in general, rhodopsin was still trafficked to the OS[88–91] while some studies also detected rhodopsin mislocalization in the IS plasma membrane.[88,91] It is possible that rapid degradation/elimination of mistrafficked rhodopsin[26] leads to less pronounced mislocalization in some of the studies using KIF3A knockout mice.[90,91] Nevertheless, rhodopsin mistrafficking is significant enough to cause rapid rod photoreceptor degeneration.[91] Consistent with this idea, overwhelming the degradation mechanism leads to aberrant homeostasis of proteins and subsequent photoreceptor degeneration.[92] In KIF3A-deficient rods, rhodopsin mislocalization contributes to the degeneration of rod photoreceptors. KIF3A facilitates the transport of rhodopsin along the ciliary plasma membrane of immortalized human retinal pigment epithelial cells (hTert-RPE1), consistent with the proposed role of KIF3A in the trafficking of rhodopsin through the connecting cilia of photoreceptors.[93] The role of KIF3B in mouse rods is unclear because a conditional knockout has not been studied so far. Nevertheless, a study utilizing a zebrafish mutant suggests that

KIF3B also contributes to opsin transport.[94] KIF3C is not involved in the trafficking of rhodopsin, as KIF3C homozygote knockout mouse does not show rhodopsin mislocalization.[95] The OS targeting of rhodopsin was not completely abrogated in any of these KIF3 mutants, suggesting that other motors likely contribute to the IS to OS trafficking of rhodopsin.

With regard to the IS to OS trafficking of rhodopsin, the roles of KIF17 kinesin complex and myosin VIIa were also intensely studied. KIF17 complex is a homodimer. Morpholino knockdown of *kif17* in zebrafish leads to severely compromised OS and impaired rhodopsin targeting to the OS.[96] However, in zebrafish *kif17* mutant with null or near-null phenotype, rhodopsin mislocalization and OS malformation were not observed.[94] Thus, the role of KIF17 in photoreceptor sensory cilia has been controversial and requires further investigation. The actin-based motor myosin may play a supportive role in the trafficking of rhodopsin. An unconventional myosin, myosin VIIa, is expressed in photoreceptor and retinal pigment epithelial cells.[97,98] Within the photoreceptors, myosin VIIa is concentrated in the connecting cilia.[98] Photoreceptors deficient in myosin VIIA accumulate rhodopsin on the plasma membrane portion of the connecting cilia,[99] suggesting that rhodopsin is cargo of myosin VIIa. For this trafficking, the giant spectrin βV is a candidate adaptor linking rhodopsin to myosin VIIa.[100] This function of myosin VIIa would be only supportive, at least in mouse, because myosin VIIa mutation does not lead to photoreceptor degeneration or morphological defects of the OS. Furthermore, this myosin VIIa function is potentially actin-independent. Disruption of actin by small molecules does not lead to an overt trafficking defect of rhodopsin in the connecting cilia, but rather leads to a disk membrane morphogenesis defect which is associated with later steps of rhodopsin trafficking[101] (discussed in Section 3.5).

The motors discussed thus far mediate the anterograde trafficking of cargoes. Retrograde trafficking, mediated by dynein motor, is critical to sustain this anterograde trafficking. This is because IFT particles, kinesins, and associated components need to be recycled back to the base of the connecting cilia for continuous trafficking of rhodopsin. Based on studies using *C. elegans*, the BBSome links IFT-B to IFT-A, which then binds to cytoplasmic dynein for retrograde trafficking.[102] In photoreceptors, there are two types of cytoplasmic dyneins: dynein 1 and 2. Components of both the dyneins localize to the connecting cilia of bovine photoreceptors.[103] A defect in cytoplasmic dynein heavy chain 1 (dync1h1) leads to severe OS malformation. Furthermore, antisense oligonucleotide-mediated knockdown of dync1h1 caused OS abnormality without any observable IS trafficking defects.[64] Dynein 2 plays a critical role in the recycling of IFT particles, as morpholino-mediated knockdown of dynein 2 components leads to accumulation of IFT88 in the distal part of connecting cilia.[63] Thus, retrograde trafficking mediated by dyneins appears important for recycling of IFT proteins in photoreceptor cells. Furthermore, in our recent studies of *Xenopus* rods, retrograde trafficking of rhodopsin occurs during early stage of OS development.[104] Such retrograde trafficking would be directly mediated by dynein.

3.5 Trafficking of Rhodopsin Within the OS

After reaching the base of the OS, the majority of rhodopsin is incorporated into the disk membranes, whereas a small fraction is also incorporated into the OS plasma membrane. Although the processes of disk morphogenesis and OS plasma membrane renewal have been poorly understood, recent studies have shed light on the underlying mechanisms.[105,106] To understand the trafficking of rhodopsin within the OS, the membrane architecture of the rod OS shall be defined in detail. The OS consists of two major compartments: disk and plasma membrane. Our recent study indicates that the OS plasma membrane is further divided into two domains: phototransductive plasma membrane domain and disk morphogenic plasma membrane domain, which are separated by a diffusion barrier (Fig. 4).[105] The former domain is responsible for phototransduction and contains cGMP-gated channels, whereas the latter domain is the site of the disk membrane morphogenesis and devoid of cGMP-gated channels. The diffusion barrier prevents the incorporation of cGMP-gated channels into disk membranes, while simultaneously preventing rhodopsin in the disk morphogenic domain from leaking into the phototransductive plasma membrane. Accordingly, disk membranes have an approximately fourfold higher concentration of rhodopsin than the phototransductive plasma membrane.[105]

The mechanism of disk morphogenesis has been under active debate for more than three decades. Discussion of the disk morphogenesis mechanism has largely revolved around two models: the evagination/rim formation model and endosome model. The "evagination/rim model" postulates that new disk membranes are generated by initial plasma membrane evagination within the disk morphogenic domain.[107] Rhodopsin is trafficked to the evaginations which are open to the extracellular space and then sealed into disks as rim and plasma membrane form (Fig. 4). The second is the "endosome model" in which nascent disk membranes are intracellular

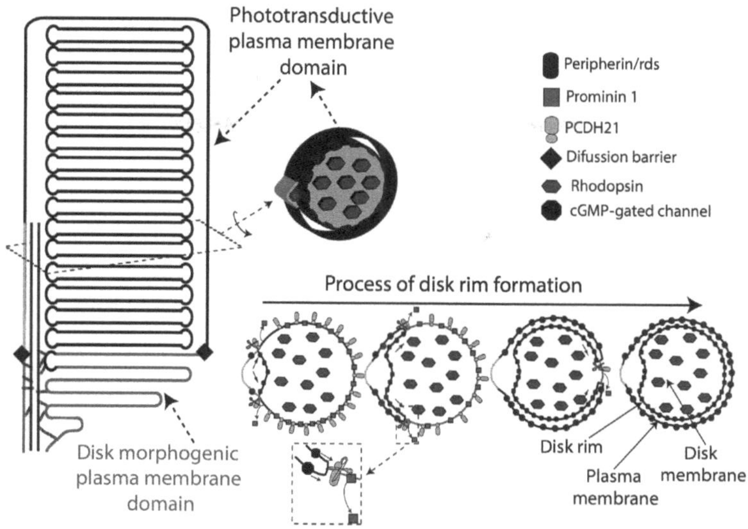

Figure 4 Sequestration of rhodopsin into disk membranes. The OS consists of two major compartments: disk and plasma membrane. OS plasma membrane is further divided into two domains: phototransductive plasma membrane domain and disk morphogenic plasma membrane domain. After delivery of rhodopsin to the disk morphogenic plasma membrane, rhodopsin is incorporated into the mature disks in the process of disk rim formation. Rim formation includes the entrance of peripherin/rds to the evaginations, the proteolytic cleavage of PCDH21, and the removal of prominin1 by an unknown mechanism. (See the color plate.)

structures formed via fusion of multiple endosomes containing rhodopsin.[108–110] The disk morphogenic domain is proposed to be a potential source of endosomes.[110]

The controversy regarding rhodopsin trafficking to new disks stems from variable electron microscopy (EM) studies of mouse rod photoreceptors in which evaginations were either present[111] or absent.[109,110] Nevertheless, evaginations (open disks) were consistently observed in rods of rhesus monkeys[107] and frogs.[112] Discrepancies observed in mouse rods are mostly attributed to different fixation conditions. In addition to fixation condition discrepancies, different sample preparation methods also affect the structure surrounding disk morphogenic domain. For example, closed nascent disks were also observed for dissociated and unfixed rod OS imaged by cryo-EM,[52] however, we recently found that OS dissociation itself causes the structural change surrounding the disk morphogenic domain.[106] Other than fixation or tissue preparation conditions, one of the major factors that may affect the appearance of nascent disks is the rate of rim/plasma membrane formation in the evagination model. As nascent disks of mice are the smallest among the species, rapid closure of rim/plasma membrane may lead to the majority of rods exhibiting closed nascent disk structures.

Evagination/rim formation model is supported from the studies of F-actin. F-actin filaments were observed at the site of disk morphogenesis as a cluster of short filaments with various orientations that bridge ciliary microtubule and plasma membrane.[113] F-actins interact with prominin-1 and photoreceptor-specific cadherin (PCDH21), both of which are located in nascent disks and appear to play critical roles in disk morphogenesis.[114–117] When F-actin filaments are disrupted with cytochalasin D treatment, evaginations do not seal and instead overgrow.[101] This overgrowth is due to the failure of sealing the evaginations by the phototransductive plasma membrane domain which contains cGMP-gated channel.[105,106] Although these findings are consistent with the evagination model, it was not clear if evaginations were formed *de novo* or were already present prior to disruption of F-actin function. To resolve this issue, we distinguished rhodopsins synthesized before and after cytochalasin D treatment and demonstrated that the overgrowths were formed from evaginations that contained rhodopsin prior to the actin disruption.[106] Thus, the overgrowths were not formed *de novo*, but formed from the preexisted evaginations.

The disk membrane is surrounded by the rim region which has a hairpin-shaped membrane curvature (Fig. 4). Peripherin/rds is localized to the rim region of the disk and is capable of facilitating high degrees of membrane curvature. Mice lacking peripherin/rds are incapable of forming OSs and their photoreceptors undergo degeneration.[118–120] Peripherin/rds haploinsufficiency, on the other hand, results in severe OS abnormalities including the appearance of abnormally large disks, swirled disk membranes, and disk misalignment.[121] Peripherin/rds is a tetraspanin membrane protein that forms homo- and heterotetramers with another OS tetraspanin protein, Rom1. Consistent with the evagination/rim formation model, peripherin/rds is not present in the immature nascent disks except the ciliary side of the evaginations where rim formation initiates[122] (Fig. 4).

Rim formation appears to involve biochemical and biophysical transformations of the membrane components. For rim formation, peripherin/rds needs to be incorporated into nascent disks. Recent evidence suggests that peripherin/rds carrier vesicles are trafficked through a secretory pathway distinct from that used by rhodopsin.[123] After trafficking of the carrier vesicles, incorporation of peripherin/rds to rim would be facilitated by its fusogenic activity[124] which may allow the docking/fusion of peripherin/rds carrier vesicles to the growing rim region. Prior to rim formation, the circumference of nascent disk is occupied with prominin-1 and PCDH21. During rim formation, peripherin/rds displaces prominin-1 and PCDH21. PCDH21 undergoes irreversible proteolysis during this process.[125] It was suggested that proteolysis of PCDH21 could be a driving force for disk morphogenesis for correct stacking of evaginations or "zippering up" of the rim between adjacent pairs of two evaginations and separation of disks from the phototransductive plasma membrane.[125] This "zipping up" process is also affected by defects in a component of phototransductive plasma membrane, cGMP-gated channel. Mutations in the gene encoding cGMP-gated channel beta subunit and GARP proteins lead to malformation and enlargement of the disk membrane[126]: a phenotype similar to that observed in prominin-1-deficient rod photoreceptors.[116] Therefore, the formation of disks is dependent on the formation of the phototransductive plasma membrane.

There is also evidence that vesicle fusion is required for disk membrane morphogenesis. SARA and syntaxin3, which are required for vesicle docking and fusion, are also associated with vesicular structures proximal to axonemal microtubules. The presence of a vesicle fusion mechanism has been apparent from the proteomic analysis of bovine rod OSs which identified several Rab and SNARE proteins involved in vesicular fusion including NSF, syntaxin 3, munc 18, and VAMP 2/3.[127] Syntaxin 3 and NSF are localized at the base of the OS, and VAMP2/3 and munc 18 throughout the OS.[127] Vesicle fusion also plays a critical role in the membrane homeostasis of the OS as SARA downregulation as well as SARA–Pi3P interaction interference leads to the accumulation of vesicular structures at the base of the OS. This observation was used to support the endosome model, as multiple vesicles form the nascent disks in this model and inhibition of their fusion with the disk membrane may lead to their accumulation. In the evagination/rim formation model, disks do not originate directly from the fusion of multiple vesicles, but the growth of evaginations or the process of rim/plasma membrane formation is possibly dependent on the vesicle fusion, and inhibition of this fusion mechanism may still lead to the accumulation of multiple vesicles. Such fusion events may, for example, be associated with the delivery of rhodopsin transport carriers to the evaginations, delivery of cGMP-gated channel and peripherin/rds to their respective compartments, or even disk membrane shedding. To further understand the role of vesicle fusion, the contents of the axonemal vesicular structures should be determined, as well as their movement visualized in real time to test if they are involved in disk morphogenesis.

4. MISLOCALIZATION OF RHODOPSIN MUTANTS

Mutations in the rhodopsin gene cause the degenerative disease retinitis pigmentosa (RP) that leads to progressive loss of photoreceptor cells and blindness. The first mutation associated with RP, P23H, was reported in 1990 and was found to cause autosomal dominant RP (ADRP).[128] Since then, over 100 unique point mutations in rhodopsin have been described; these mutations account for approximately 30–40% of all ADRP cases (https://sph.uth.edu/retnet/disease.htm). Initially, these mutants were classified based on cell culture studies which are substantially different from the highly differentiated, polarized, and ciliated photoreceptor neurons. These studies did not provide accurate insight into the individual molecular defects caused by the mutants. Mislocalization and mistrafficking of rhodopsin appear to be causative to the photoreceptor degeneration associated with the majority of rhodopsin mutations. Here, we summarize three major mechanisms leading to mistrafficking of rhodopsin and consequences of the mislocalization.

4.1 Mislocalization Due to Defects of Trafficking Signals

As rhodopsin requires specific targeting signals for OS localization, mutations causing signal defects lead to rhodopsin mislocalization. Mutations that cause defects in the major signal, VXPX, have been categorized as Class I. Defects caused by Class I mutations became apparent from the studies of transgenic animals where Class I mutants were expressed in rod photoreceptors.[16,18,129,130] These studies revealed that mutations in VXPX lead to rhodopsin mislocalization and severe photoreceptor degeneration. Among Class I mutations, Q344Ter has been extensively studied because this mutant completely lacks the putative cilia-targeting signal VXPX. Q344Ter mislocalizes to the IS, outer nuclear layer (ONL), and synaptic terminus, but it also correctly reaches the OS. This OS localization is observed even in the absence of endogenous WT rhodopsin.[17] As rhodopsin is required for the formation of the OS, this observation suggests that Q344Ter can contribute to OS formation. In addition to the IS plasma membrane, Q344Ter expressed in frog rods also mislocalizes in the calyceal processes (CPs)[26] which are vertically-oriented protrusions of the IS plasma membrane that surround the basal region of the OS. The dysfunction of CPs may be associated with Usher syndrome type I,[131] although their specific functions are unknown. Multiple CPs are elaborated in primates and frogs, but only one CP develops in mice. Notably, mice do not exhibit Usher syndrome type I.[131] Thus, studying rhodopsin mislocalization in models anatomically closer to human would be favorable as local toxicity at the CPs could contribute to the photoreceptor degeneration. Interestingly, mislocalized rhodopsin is actively removed from rod photoreceptors. Multiple rhodopsin-laden submicrometer-sized vesicles have been observed in the interphotoreceptor matrix of mutant retinas,[17,18,26] suggesting that the secretion of extracellular vesicles is a mechanism of eliminating the mislocalized rhodopsin.[26] Additional lysosome-mediated mechanisms may also lead to degradation of mislocalized rhodopsin.[26] As the toxicity of Class I rhodopsin appears to be dose-dependent, degradation of Class I mutant rhodopsin would be a neuroprotective mechanism. Thus, understanding of the pathways controlling the degradation/removal of Class I mutant rhodopsin will be crucial for the development of therapies for RP.

The mechanism(s) by which mislocalized rhodopsin leads to the photoreceptor cell death is under active debate. One hypothesis is the ectopic activation of G protein signaling cascade which leads to apoptotic events; however, G protein-independent mechanisms have been also proposed. Based on an *in vitro* study of dissociated salamander retina culture, it was suggested that photoactivation of mislocalized rhodopsin activates transducin (or other G proteins) outside the OSs.[132] Transducin, in the absence of its normal target, phosphodiesterase, appears to activate adenylate cyclase which leads to an increase in the intracellular levels of cAMP followed by neuritic sprouting via PKA–CREB pathway[133] and eventual apoptotic cell death via caspase-3 activation.[132] These observations are in line with a mouse study in which Q344Ter rhodopsin is capable of activating transducin.[17] In support of this hypothesis, in two zebrafish models of rhodopsin mislocalization (IFT88 mutant and transgenic Q344Ter), morpholino downregulation of transducin and inhibition of adenylyl cyclase significantly increased the number of surviving rod photoreceptor cells.[134] Taken together, cAMP-dependent proapoptotic pathway is a likely cause of photoreceptor degeneration caused by Class I mutant rhodopsin. There is also evidence that mislocalized rhodopsin is toxic even in the absence of G protein activation. While light exposure exacerbated the degeneration caused by Q344ter rhodopsin in a mouse model, less pronounced degeneration occurs for mice reared in darkness, where G protein activation does not occur.[17] Light independency was also proven by a transgenic *X. laevis* model where a second mutation K296R was introduced to the Q344Ter mutant, to block 11-*cis*-retinal binding and G protein activation.[135] These combined mutations led to a similar level of degeneration and neurite sprouting as caused by Q344Ter single mutation.[129] This finding is consistent with observation in $Q344Ter^{rho+/-}$ mice bred into transducin knockout background where light-induced retinal degeneration was only partially reduced, suggesting that another pathway causes photoreceptor degeneration. This pathway, however, is not dependent on aberrant arrestin binding[17] as discussed below (see Section 4.3).

4.2 Mislocalization Due to Misfolding

Rhodopsin mislocalization in the ER may occur by mutations leading to its misfolding. Those mutations have been traditionally categorized as Class II. Class II mutants were initially characterized using cell culture systems. In cultured cells, Class II mutants exhibited lower yields than WT and remained primarily in the ER.[14,15] In these initial studies, Class II mutant rhodopsins were synthesized in HEK293 cells in the absence of 11-*cis*-retinal, which is usually continuously supplied in intact rod photoreceptors. As 11-*cis*-retinal can act as a chaperone for mutant opsin, its absence might have led to increased misfolding of Class II mutant opsins in cultured cells. After synthesis, Class II mutant opsins do not bind to 11-*cis*-retinal effectively or at all, suggesting that misfolding of Class II opsins is a root cause of photoreceptor degeneration. One of the most common rhodopsin mutants in North America and a representative of misfolded rhodopsin is P23H which accounts for approximately 15% of all cases of ADRP.[15,128] Misfolded proteins are usually not allowed to leave the ER; instead, they become subjected to ER-associated degradation (ERAD). If proteins are not efficiently degraded by ERAD, then they accumulate in the ER and/or may form aggresomes as observed in HEK293 or COS-7 cells.[14,15,34,136,137] Mislocalization of the P23H mutant was also observed in rod photoreceptors of different transgenic animal models. In P23H transgenic mice, rhodopsin accumulates in the ONL and within the synaptic terminal when expressed in WT background.[138] In a *X. laevis* model, P23H rhodopsin is also retained in the ER.[139]

The animal models expressing P23H rhodopsin exhibit retinal degeneration, but the mechanisms that cause the degeneration remain an open question. Multiple animal and cell culture models undoubtedly indicated an elevated ER stress and an activation of unfolded protein response (UPR). UPR activation induced the expression of chaperones and stimulated ERAD.[140,141] Failure of ERAD causes the UPR to switch from protective to proapoptotic as indicated by increased level of a proapoptotic protein CHOP. CHOP mRNA level was significantly higher in P23H transgenic rats than in WT rats.[140] However, in P23H knock-in mouse, UPR activation led to a very efficient removal of the mutant rhodopsin, but did not lead to the induction of proapoptotic pathways[141] which is consistent with the finding that ablation of CHOP or other regulators of UPR-mediated apoptosis did not alter the rate of photoreceptor loss in P23H transgene mice.[142] Unlike cultured mammalian cells, photoreceptor cells are capable of efficiently eliminating P23H rhodopsin and therefore do not form aggresomes. Thus, the studies of P23H knock-in mice suggest that a mechanism other than ER stress should contribute to photoreceptor degeneration. An example of an ER stress-independent mechanism is proteasome overload observed in P23H transgenic mice.[92] A small quantity of P23H that escapes ERAD localizes to the OS.[139,143] This quantity of P23H that reaches the OS may also cause toxicity via OS malformation which is possibly a mechanism for photoreceptor degeneration. Basal disks were highly disorganized in transgenic animals expressing rhodopsin with three missense mutations including P23H,[144] whereas disks in P23H knock-in mouse were disorganized and sagittally oriented.[143] These observations suggest that abnormal disk morphogenesis is a cause of retinal degeneration.

Rhodopsin mutations T4K and T17M lead to loss of one glycan conjugation and cause sector RP in humans.[145–147] Similar to T4K patients, a dog T4R model exhibits distinct topographic pattern of retina degeneration.[148] In a human T17M patient, sector RP was characterized by retina degeneration and rhodopsin mislocalization in the superior equatorial region, while no defect was detected in the rest of the retina.[145] Studies on animal models suggest that sector RP is caused by light exposure because T4R dog and transgenic T17M mouse models exhibited photoreceptor degeneration exacerbated by light.[149,150] Recent studies using *X. laevis* model suggest that these glycosylation mutants destabilize the activated structure of rhodopsin which leads to rhodopsin misfolding.[40] This light dependency is inconsistent with the categorization of T17M mutant as a Class II mutant which cannot correctly fold after synthesis, because T17M needs to be able to fold to be photoactive. In mouse models, T17M rhodopsin localizes to the ER and induces upregulation of genes associated with the UPR, autophagy, and apoptosis.[151] Thus, protein misfolding may have enhanced the ER stress in the mouse model. Among species, there were notable variabilities of T17M rhodopsin localization. For example, T17M localized to the ER of mouse,[151] whereas it localized to the OS of *X. laevis*.[40,152] Differences in the localizations of those two models could be a result of lower stability of T17M rhodopsin at higher temperature in mice (37 °C) than in frogs (16–18 °C); however, temperature cannot account for the difference observed between humans and mice. In humans, mislocalization of T17M only occurs in the region afflicted by sector RP. This mislocalization could be due to light-dependent misfolding of rhodopsin in the ER, or a secondary consequence of rod photoreceptor degeneration. Thus for glycosylation-deficient mutants, it remains to be elucidated whether the light-dependent rhodopsin toxicity occurs in the OS or the ER.

4.3 Mislocalization Due to Aberrant Arrestin Binding

Vertebrate rhodopsins bind to visual arrestin (also called arrestin-14). For other GPCRs, β-arrestins (also called arrestin-2 and -3) bind and quench the receptors.[153] β-Arrestin not only sterically inhibits G protein binding but also leads to receptor desensitization by promoting endocytosis/internalization of the receptors[154,155] which effectively prevents the activation of receptors by extracellular ligands. β-Arrestins act as adaptors for clathrin and AP-2 which are required for endocytosis.[154,155] In the past, visual arrestins were considered incapable of mediating endocytosis because of poor binding to these endocytic components.[154,156,157] However, recent studies suggest that visual arrestin indeed can recruit these components under limited pathological conditions caused by rhodopsin mutants with enhanced binding to arrestin.[158] These mutants mislocalize to endosomes where rhodopsin activates proapoptotic pathways.

Constitutively active mutants of rhodopsin demonstrate enhanced binding to arrestin (for review of constitutively active mutants, see Ref. 159). Mutations K296E and K296M prevent 11-*cis*-retinal binding to K296, but at the same time allow activation of transducin constitutively in the absence of the chromophore and the light.[160] Based on those observations, it was assumed that photoreceptor death in K296E mutants is caused by permanent activation of the phototransduction cascade in a manner similar to the death caused by the constant light exposure. However, in transgenic K296E mice, mutant rhodopsin was phosphorylated and bound to arrestin which renders rhodopsin incapable of activating transducin *in vivo*.[161] The K296E/arrestin complex activates a transducin-independent process that eventually leads to photoreceptor cell death.[162] Furthermore, the capability of the K296E/arrestin complex to bind to endocytic adaptor protein (AP-2) appears to be an important event leading to cell death. Removal of the AP-2 binding site from arrestin prevented E296E-induced retinal degeneration and rescued photoreceptor function.[158]

Another example of constitutive activity geared toward aberrant arrestin binding is R135 rhodopsin mutants (e.g., R135A and R135L).[163,164] These mutants are characterized by hyperphosphorylation in the absence of light activation of rhodopsin.[163,164] The hyperphosphorylation of rhodopsin is followed by arrestin binding.[163,164] When ectopically expressed in mouse rods, R135L mislocalizes to the IS with arrestin distributed on the IS plasma membrane and through the cell body.[164] As discussed for K296E mutant, a stable complex of R135L and arrestin is potentially the underlying cause of photoreceptor degeneration. The pathways activated by arrestin/rhodopsin complex remain to be characterized.

5. FUTURE PERSPECTIVES

As reviewed in this chapter, tremendous progress has been made toward the understanding of rhodopsin trafficking in the past three decades, leading to identification of multiple trafficking signals that coordinate together for OS targeting. While many molecular components were identified *in vitro*, there are gaps in understanding their exact roles *in vivo* in relation to rhodopsin trafficking. In this regard, technical advances were made in detecting molecular interactions *in situ*, such as fluorescence complementation approach[165] and the proximity ligation assay.[27] Detection of *in vivo* interactions is insightful since the location of the interaction will provide the information about the step of rhodopsin trafficking involved. Multicolor super-resolution fluorescence microscopy[166–169] will also provide insight into the proximity of those interacting proteins with rhodopsin in the secretory pathway. Considering a large number of proteins associated with rhodopsin trafficking, inhibiting or modulating these molecular interactions, *in situ*, is very challenging. Technical advances in gene editing, such as those mediated by CRISPR-Cas9 or TALENs,[170] will accelerate the research by disrupting components of rhodopsin trafficking in living, developed animals in a medium-to-high-throughput manner.

Techniques for visualizing the trafficking of rhodopsin are still very limited, and it is unclear how many different protein components play roles in the specific steps of rhodopsin trafficking *in vivo*. To comprehensively understand the process of rhodopsin trafficking, we must establish imaging technologies to visualize rhodopsin from the site of its synthesis to the rod OS with high spatiotemporal resolution. Such imaging technology, ideally, should provide quantitative information about rhodopsin trafficking as well. One of the promising approaches is the use of rhodopsin-FP fusion.[171,172] By calibrating rhodopsin-FP fusions with known quantities of soluble FPs, this approach allows accurate determination of the quantity and concentration of rhodopsin-FP in subcellular compartments of rod photoreceptors.[26,31] Furthermore, several methods were developed to increase the contrast for recently synthesized rhodopsin-FP. For example, by expressing rhodopsin-FP under the regulation of specific promoters which are inducible by small molecule or heat shock,[79,173] it is possible to visualize and track the movement of newly synthesized rhodopsin-FP. A major shortcoming of inducible promoters, however, is the relatively long time (hours to days) required for induction, and thus this approach may pose limitations in visualizing rapid motion (on the timescale of seconds to minutes) of newly synthesized rhodopsin. In this regard, it is favorable to have a technique allowing rapid increase in the contrast for newly synthesized protein. The photoconversion technique, developed in our laboratory, takes advantage of a photoconvertible FP Dendra2. Dendra2 FP turns from green to red upon photoconversion by violet or blue light in an irreversible fashion. Thus, by photoconverting rhodopsin fused to Dendra2, we were able to instantaneously increase the contrast for newly synthesized rhodopsin-Dendra2 *in vivo*. This contrast-enhancing method, combined with video-rate confocal microscopy,[174] two-photon microscopy,[175] or super-resolution microscopy techniques,[166–169] will allow visualization of newly synthesized rhodopsin trafficking events in unprecedented detail.

ACKNOWLEDGMENT

This work was supported by the U.S. National Institutes of Health grants EY020826, EY011373, and EY007157.

REFERENCES

1. Burns ME, Arshavsky VY. Beyond counting photons: trials and trends in vertebrate visual transduction. *Neuron*. 2005;48(3):387–401.
2. Ridge KD, Abdulaev NG, Sousa M, Palczewski K. Phototransduction: crystal clear. *Trends Biochem Sci*. 2003;28(9):479–487.
3. Yau KW, Hardie RC. Phototransduction motifs and variations. *Cell*. 2009;139(2):246–264.
4. Loktev AV, Jackson PK. Neuropeptide Y family receptors traffic via the Bardet-Biedl syndrome pathway to signal in neuronal primary cilia. *Cell Rep*. 2013;5(5):1316–1329.
5. Soetedjo L, Glover DA, Jin H. Targeting of vasoactive intestinal peptide receptor 2, VPAC2, a secretin family G-protein coupled receptor, to primary cilia. *Biol Open*. 2013;2(7):686–694.

6. Handel M, Schulz S, Stanarius A, et al. Selective targeting of somatostatin receptor 3 to neuronal cilia. *Neuroscience*. 1999;89(3):909–926.
7. Berbari NF, Johnson AD, Lewis JS, Askwith CC, Mykytyn K. Identification of ciliary localization sequences within the third intracellular loop of G protein-coupled receptors. *Mol Biol Cell*. 2008;19(4):1540–1547.
8. Berbari NF, Lewis JS, Bishop GA, Askwith CC, Mykytyn K. Bardet-Biedl syndrome proteins are required for the localization of G protein-coupled receptors to primary cilia. *Proc Natl Acad Sci USA*. 2008;105(11):4242–4246.
9. Brailov I, Bancila M, Brisorgueil MJ, Miquel MC, Hamon M, Verge D. Localization of 5-HT(6) receptors at the plasma membrane of neuronal cilia in the rat brain. *Brain Res*. 2000;872(1–2):271–275.
10. Zhang Q, Nishimura D, Vogel T, et al. BBS7 is required for BBSome formation and its absence in mice results in Bardet-Biedl syndrome phenotypes and selective abnormalities in membrane protein trafficking. *J Cell Sci*. 2013;126(Pt 11):2372–2380.
11. Corbit KC, Aanstad P, Singla V, Norman AR, Stainier DY, Reiter JF. Vertebrate Smoothened functions at the primary cilium. *Nature*. 2005;437(7061):1018–1021.
12. Xiong B, Bellen HJ. Rhodopsin homeostasis and retinal degeneration: lessons from the fly. *Trends Neurosci*. 2013;36(11):652–660.
13. Sung CH, Schneider BG, Agarwal N, Papermaster DS, Nathans J. Functional heterogeneity of mutant rhodopsins responsible for autosomal dominant retinitis pigmentosa. *Proc Natl Acad Sci USA*. 1991;88(19):8840–8844.
14. Sung CH, Davenport CM, Nathans J. Rhodopsin mutations responsible for autosomal dominant retinitis pigmentosa. Clustering of functional classes along the polypeptide chain. *J Biol Chem*. 1993;268(35):26645–26649.
15. Sung CH, Davenport CM, Hennessey JC, et al. Rhodopsin mutations in autosomal dominant retinitis pigmentosa. *Proc Natl Acad Sci USA*. 1991;88(15):6481–6485.
16. Sung CH, Makino C, Baylor D, Nathans J. A rhodopsin gene mutation responsible for autosomal dominant retinitis pigmentosa results in a protein that is defective in localization to the photoreceptor outer segment. *J Neurosci*. 1994;14(10):5818–5833.
17. Concepcion F, Chen J. Q344ter mutation causes mislocalization of rhodopsin molecules that are catalytically active: a mouse model of Q344ter-induced retinal degeneration. *PLoS One*. 2010;5(6):e10904.
18. Li T, Snyder WK, Olsson JE, Dryja TP. Transgenic mice carrying the dominant rhodopsin mutation P347S: evidence for defective vectorial transport of rhodopsin to the outer segments. *Proc Natl Acad Sci USA*. 1996;93(24):14176–14181.
19. Tam BM, Moritz OL, Hurd LB, Papermaster DS. Identification of an outer segment targeting signal in the COOH terminus of rhodopsin using transgenic Xenopus laevis. *J Cell Biol*. 2000;151(7):1369–1380.
20. Deretic D, Puleo-Scheppke B, Trippe C. Cytoplasmic domain of rhodopsin is essential for post-Golgi vesicle formation in a retinal cell-free system. *J Biol Chem*. 1996;271(4):2279–2286.
21. Yokoyama S. Molecular evolution of vertebrate visual pigments. *Prog Retin Eye Res*. 2000;19(4):385–419.
22. Luo W, Marsh-Armstrong N, Rattner A, Nathans J. An outer segment localization signal at the C terminus of the photoreceptor-specific retinol dehydrogenase. *J Neurosci*. 2004;24(11):2623–2632.
23. Ward HH, Brown-Glaberman U, Wang J, et al. A conserved signal and GTPase complex are required for the ciliary transport of polycystin-1. *Mol Biol Cell*. 2011;22(18):3289–3305.
24. Geng L, Okuhara D, Yu Z, et al. Polycystin-2 traffics to cilia independently of polycystin-1 by using an N-terminal RVxP motif. *J Cell Sci*. 2006;119(Pt 7): 1383–1395.

25. Jenkins PM, Hurd TW, Zhang L, et al. Ciliary targeting of olfactory CNG channels requires the CNGB1b subunit and the kinesin-2 motor protein, KIF17. *Curr Biol.* 2006;16(12):1211–1216.
26. Lodowski KH, Lee R, Ropelewski P, Nemet I, Tian G, Imanishi Y. Signals governing the trafficking and mistrafficking of a ciliary GPCR, rhodopsin. *J Neurosci.* 2013;33(34):13621–13638.
27. Wang J, Morita Y, Mazelova J, Deretic D. The Arf GAP ASAP1 provides a platform to regulate Arf4- and Rab11-Rab8-mediated ciliary receptor targeting. *EMBO J.* 2012;31(20):4057–4071.
28. Jin H, White SR, Shida T, et al. The conserved Bardet-Biedl syndrome proteins assemble a coat that traffics membrane proteins to cilia. *Cell.* 2010;141(7):1208–1219.
29. Nickell S, Park PS, Baumeister W, Palczewski K. Three-dimensional architecture of murine rod outer segments determined by cryoelectron tomography. *J Cell Biol.* 2007;177(5):917–925.
30. Fotiadis D, Liang Y, Filipek S, Saperstein DA, Engel A, Palczewski K. Atomic-force microscopy: rhodopsin dimers in native disc membranes. *Nature.* 2003;421(6919): 127–128.
31. Haeri M, Knox BE. Rhodopsin mutant P23H destabilizes rod photoreceptor disk membranes. *PLoS One.* 2012;7(1):e30101.
32. Chuang JZ, Sung CH. The cytoplasmic tail of rhodopsin acts as a novel apical sorting signal in polarized MDCK cells. *J Cell Biol.* 1998;142(5):1245–1256.
33. Fliesler SJ, Basinger SF. Tunicamycin blocks the incorporation of opsin into retinal rod outer segment membranes. *Proc Natl Acad Sci USA.* 1985;82(4):1116–1120.
34. Saliba RS, Munro PM, Luthert PJ, Cheetham ME. The cellular fate of mutant rhodopsin: quality control, degradation and aggresome formation. *J Cell Sci.* 2002;115(Pt 14):2907–2918.
35. Kosmaoglou M, Cheetham ME. Calnexin is not essential for mammalian rod opsin biogenesis. *Mol Vis.* 2008;14:2466–2474.
36. Anukanth A, Khorana HG. Structure and function in rhodopsin. Requirements of a specific structure for the intradiscal domain. *J Biol Chem.* 1994;269(31):19738–19744.
37. Gorbatyuk MS, Knox T, LaVail MM, et al. Restoration of visual function in P23H rhodopsin transgenic rats by gene delivery of BiP/Grp78. *Proc Natl Acad Sci USA.* 2010;107(13):5961–5966.
38. Athanasiou D, Kosmaoglou M, Kanuga N, et al. BiP prevents rod opsin aggregation. *Mol Biol Cell.* 2012;23(18):3522–3531.
39. Athanasiou D, Bevilacqua D, Aguila M, et al. The co-chaperone and reductase ERdj5 facilitates rod opsin biogenesis and quality control. *Hum Mol Genet.* 2014;23: 6594–6606.
40. Tam BM, Noorwez SM, Kaushal S, Kono M, Moritz OL. Photoactivation-induced instability of rhodopsin mutants T4K and T17M in rod outer segments underlies retinal degeneration in X. laevis transgenic models of retinitis pigmentosa. *J Neurosci.* 2014;34(40):13336–13348.
41. Veit M, Sachs K, Heckelmann M, Maretzki D, Hofmann KP, Schmidt MF. Palmitoylation of rhodopsin with S-protein acyltransferase: enzyme catalyzed reaction versus autocatalytic acylation. *Biochim Biophys Acta.* 1998;1394(1):90–98.
42. Wang Z, Wen XH, Ablonczy Z, Crouch RK, Makino CL, Lem J. Enhanced shutoff of phototransduction in transgenic mice expressing palmitoylation-deficient rhodopsin. *J Biol Chem.* 2005;280(26):24293–24300.
43. Maeda A, Okano K, Park PS, et al. Palmitoylation stabilizes unliganded rod opsin. *Proc Natl Acad Sci USA.* 2010;107(18):8428–8433.
44. Schmidt K, Cavodeassi F, Feng Y, Stephens DJ. Early stages of retinal development depend on Sec13 function. *Biol Open.* 2013;2(3):256–266.

45. Kunduri G, Yuan C, Parthibane V, et al. Phosphatidic acid phospholipase A1 mediates ER-Golgi transit of a family of G protein-coupled receptors. *J Cell Biol.* 2014; 206(1):79–95.
46. Deretic D, Williams AH, Ransom N, Morel V, Hargrave PA, Arendt A. Rhodopsin C terminus, the site of mutations causing retinal disease, regulates trafficking by binding to ADP-ribosylation factor 4 (ARF4). *Proc Natl Acad Sci USA.* 2005;102(9): 3301–3306.
47. Reish NJ, Boitet ER, Bales KL, Gross AK. Nucleotide bound to rab11a controls localization in rod cells but not interaction with rhodopsin. *J Neurosci.* 2014;34(45): 14854–14863.
48. Nachury MV, Loktev AV, Zhang Q, et al. A core complex of BBS proteins cooperates with the GTPase Rab8 to promote ciliary membrane biogenesis. *Cell.* 2007;129(6):1201–1213.
49. Mykytyn K, Mullins RF, Andrews M, et al. Bardet-Biedl syndrome type 4 (BBS4)-null mice implicate Bbs4 in flagella formation but not global cilia assembly. *Proc Natl Acad Sci USA.* 2004;101(23):8664–8669.
50. Nishimura DY, Fath M, Mullins RF, et al. Bbs2-null mice have neurosensory deficits, a defect in social dominance, and retinopathy associated with mislocalization of rhodopsin. *Proc Natl Acad Sci USA.* 2004;101(47):16588–16593.
51. Davis RE, Swiderski RE, Rahmouni K, et al. A knockin mouse model of the Bardet-Biedl syndrome 1 M390R mutation has cilia defects, ventriculomegaly, retinopathy, and obesity. *Proc Natl Acad Sci USA.* 2007;104(49):19422–19427.
52. Gilliam JC, Chang JT, Sandoval IM, et al. Three-dimensional architecture of the rod sensory cilium and its disruption in retinal neurodegeneration. *Cell.* 2012;151(5): 1029–1041.
53. Blacque OE, Reardon MJ, Li C, et al. Loss of C. elegans BBS-7 and BBS-8 protein function results in cilia defects and compromised intraflagellar transport. *Genes Dev.* 2004;18(13):1630–1642.
54. Follit JA, San Agustin JT, Xu F, et al. The Golgin GMAP210/TRIP11 anchors IFT20 to the Golgi complex. *PLoS Genet.* 2008;4(12):e1000315.
55. Follit JA, Tuft RA, Fogarty KE, Pazour GJ. The intraflagellar transport protein IFT20 is associated with the Golgi complex and is required for cilia assembly. *Mol Biol Cell.* 2006;17(9):3781–3792.
56. Omori Y, Zhao C, Saras A, et al. Elipsa is an early determinant of ciliogenesis that links the IFT particle to membrane-associated small GTPase Rab8. *Nat Cell Biol.* 2008;10(4):437–444.
57. Keady BT, Le YZ, Pazour GJ. IFT20 is required for opsin trafficking and photoreceptor outer segment development. *Mol Biol Cell.* 2011;22(7):921–930.
58. Li L, Khan N, Hurd T, et al. Ablation of the X-linked retinitis pigmentosa 2 (Rp2) gene in mice results in opsin mislocalization and photoreceptor degeneration. *Invest Ophthalmol Vis Sci.* 2013;54(7):4503–4511.
59. Schrick JJ, Vogel P, Abuin A, Hampton B, Rice DS. ADP-ribosylation factor-like 3 is involved in kidney and photoreceptor development. *Am J Pathol.* 2006;168(4): 1288–1298.
60. Troutt LL, Burnside B. Microtubule polarity and distribution in teleost photoreceptors. *J Neurosci.* 1988;8(7):2371–2380.
61. Tai AW, Chuang JZ, Bode C, Wolfrum U, Sung CH. Rhodopsin's carboxy-terminal cytoplasmic tail acts as a membrane receptor for cytoplasmic dynein by binding to the dynein light chain Tctex-1. *Cell.* 1999;97(7):877–887.
62. Vaughan DK, Fisher SK, Bernstein SA, Hale IL, Linberg KA, Matsumoto B. Evidence that microtubules do not mediate opsin vesicle transport in photoreceptors. *J Cell Biol.* 1989;109(6 Pt 1):3053–3062.

63. Krock BL, Mills-Henry I, Perkins BD. Retrograde intraflagellar transport by cytoplasmic dynein-2 is required for outer segment extension in vertebrate photoreceptors but not arrestin translocation. *Invest Ophthalmol Vis Sci.* 2009;50(11):5463–5471.
64. Insinna C, Baye LM, Amsterdam A, Besharse JC, Link BA. Analysis of a zebrafish dync1h1 mutant reveals multiple functions for cytoplasmic dynein 1 during retinal photoreceptor development. *Neural Dev.* 2010;5:12.
65. Tai AW, Chuang JZ, Sung CH. Cytoplasmic dynein regulation by subunit heterogeneity and its role in apical transport. *J Cell Biol.* 2001;153(7):1499–1509.
66. Inoue H, Ha VL, Prekeris R, Randazzo PA. Arf GTPase-activating protein ASAP1 interacts with Rab11 effector FIP3 and regulates pericentrosomal localization of transferrin receptor-positive recycling endosome. *Mol Biol Cell.* 2008;19(10):4224–4237.
67. Mazelova J, Astuto-Gribble L, Inoue H, et al. Ciliary targeting motif VxPx directs assembly of a trafficking module through Arf4. *EMBO J.* 2009;28(3):183–192.
68. Horgan CP, Hanscom SR, Jolly RS, Futter CE, McCaffrey MW. Rab11-FIP3 binds dynein light intermediate chain 2 and its overexpression fragments the Golgi complex. *Biochem Biophys Res Commun.* 2010;394(2):387–392.
69. Horgan CP, Hanscom SR, Jolly RS, Futter CE, McCaffrey MW. Rab11-FIP3 links the Rab11 GTPase and cytoplasmic dynein to mediate transport to the endosomal-recycling compartment. *J Cell Sci.* 2010;123(Pt 2):181–191.
70. Rosenbaum J. Intraflagellar transport. *Curr Biol.* 2002;12(4):R125.
71. Luby-Phelps K, Fogerty J, Baker SA, Pazour GJ, Besharse JC. Spatial distribution of intraflagellar transport proteins in vertebrate photoreceptors. *Vis Res.* 2008;48(3):413–423.
72. Baker SA, Freeman K, Luby-Phelps K, Pazour GJ, Besharse JC. IFT20 links kinesin II with a mammalian intraflagellar transport complex that is conserved in motile flagella and sensory cilia. *J Biol Chem.* 2003;278(36):34211–34218.
73. Krock BL, Perkins BD. The intraflagellar transport protein IFT57 is required for cilia maintenance and regulates IFT-particle-kinesin-II dissociation in vertebrate photoreceptors. *J Cell Sci.* 2008;121(Pt 11):1907–1915.
74. Pazour GJ, Baker SA, Deane JA, et al. The intraflagellar transport protein, IFT88, is essential for vertebrate photoreceptor assembly and maintenance. *J Cell Biol.* 2002;157(1):103–113.
75. Follit JA, Xu F, Keady BT, Pazour GJ. Characterization of mouse IFT complex B. *Cell Motil Cytoskeleton.* 2009;66(8):457–468.
76. Tsujikawa M, Malicki J. Intraflagellar transport genes are essential for differentiation and survival of vertebrate sensory neurons. *Neuron.* 2004;42(5):703–716.
77. Crouse JA, Lopes VS, Sanagustin JT, Keady BT, Williams DS, Pazour GJ. Distinct functions for IFT140 and IFT20 in opsin transport. *Cytoskeleton (Hoboken).* 2014;71(5):302–310.
78. Boldt K, Mans DA, Won J, et al. Disruption of intraflagellar protein transport in photoreceptor cilia causes Leber congenital amaurosis in humans and mice. *J Clin Invest.* 2011;121(6):2169–2180.
79. Zhao C, Malicki J. Nephrocystins and MKS proteins interact with IFT particle and facilitate transport of selected ciliary cargos. *EMBO J.* 2011;30(13):2532–2544.
80. MacRae DW, Howard RO, Albert DM, Hsia YE. Ocular manifestations of the Meckel syndrome. *Arch Ophthalmol.* 1972;88(1):106–113.
81. Otto EA, Schermer B, Obara T, et al. Mutations in INVS encoding inversin cause nephronophthisis type 2, linking renal cystic disease to the function of primary cilia and left-right axis determination. *Nat Genet.* 2003;34(4):413–420.
82. Scholey JM. Kinesin-2: a family of heterotrimeric and homodimeric motors with diverse intracellular transport functions. *Annu Rev Cell Dev Biol.* 2013;29:443–469.

83. Khanna H, Hurd TW, Lillo C, et al. RPGR-ORF15, which is mutated in retinitis pigmentosa, associates with SMC1, SMC3, and microtubule transport proteins. *J Biol Chem.* 2005;280(39):33580–33587.
84. Dryja TP, Adams SM, Grimsby JL, et al. Null RPGRIP1 alleles in patients with Leber congenital amaurosis. *Am J Hum Genet.* 2001;68(5):1295–1298.
85. Meindl A, Dry K, Herrmann K, et al. A gene (RPGR) with homology to the RCC1 guanine nucleotide exchange factor is mutated in X-linked retinitis pigmentosa (RP3). *Nat Genet.* 1996;13(1):35–42.
86. Hong DH, Yue G, Adamian M, Li T. Retinitis pigmentosa GTPase regulator (RPGRr)-interacting protein is stably associated with the photoreceptor ciliary axoneme and anchors RPGR to the connecting cilium. *J Biol Chem.* 2001;276(15):12091–12099.
87. Zhao Y, Hong DH, Pawlyk B, et al. The retinitis pigmentosa GTPase regulator (RPGR)- interacting protein: subserving RPGR function and participating in disk morphogenesis. *Proc Natl Acad Sci USA.* 2003;100(7):3965–3970.
88. Marszalek JR, Liu X, Roberts EA, et al. Genetic evidence for selective transport of opsin and arrestin by kinesin-II in mammalian photoreceptors. *Cell.* 2000;102(2):175–187.
89. Jimeno D, Feiner L, Lillo C, et al. Analysis of kinesin-2 function in photoreceptor cells using synchronous Cre-loxP knockout of Kif3a with RHO-Cre. *Invest Ophthalmol Vis Sci.* 2006;47(11):5039–5046.
90. Avasthi P, Watt CB, Williams DS, et al. Trafficking of membrane proteins to cone but not rod outer segments is dependent on heterotrimeric kinesin-II. *J Neurosci.* 2009;29(45):14287–14298.
91. Lopes VS, Jimeno D, Khanobdee K, et al. Dysfunction of heterotrimeric kinesin-2 in rod photoreceptor cells and the role of opsin mislocalization in rapid cell death. *Mol Biol Cell.* 2010;21(23):4076–4088.
92. Lobanova ES, Finkelstein S, Skiba NP, Arshavsky VY. Proteasome overload is a common stress factor in multiple forms of inherited retinal degeneration. *Proc Natl Acad Sci USA.* 2013;110(24):9986–9991.
93. Trivedi D, Colin E, Louie CM, Williams DS. Live-cell imaging evidence for the ciliary transport of rod photoreceptor opsin by heterotrimeric kinesin-2. *J Neurosci.* 2012;32(31):10587–10593.
94. Zhao C, Omori Y, Brodowska K, Kovach P, Malicki J. Kinesin-2 family in vertebrate ciliogenesis. *Proc Natl Acad Sci USA.* 2012;109(7):2388–2393.
95. Jimeno D, Lillo C, Roberts EA, Goldstein LS, Williams DS. Kinesin-2 and photoreceptor cell death: requirement of motor subunits. *Exp Eye Res.* 2006;82(2):351–353.
96. Insinna C, Pathak N, Perkins B, Drummond I, Besharse JC. The homodimeric kinesin, Kif17, is essential for vertebrate photoreceptor sensory outer segment development. *Dev Biol.* 2008;316(1):160–170.
97. Weil D, Levy G, Sahly I, et al. Human myosin VIIA responsible for the Usher 1B syndrome: a predicted membrane-associated motor protein expressed in developing sensory epithelia. *Proc Natl Acad Sci USA.* 1996;93(8):3232–3237.
98. Liu X, Vansant G, Udovichenko IP, Wolfrum U, Williams DS. Myosin VIIa, the product of the Usher 1B syndrome gene, is concentrated in the connecting cilia of photoreceptor cells. *Cell Motil Cytoskeleton.* 1997;37(3):240–252.
99. Liu X, Udovichenko IP, Brown SD, Steel KP, Williams DS. Myosin VIIa participates in opsin transport through the photoreceptor cilium. *J Neurosci.* 1999;19(15):6267–6274.
100. Papal S, Cortese M, Legendre K, et al. The giant spectrin betaV couples the molecular motors to phototransduction and Usher syndrome type I proteins along their trafficking route. *Hum Mol Genet.* 2013;22(18):3773–3788.

101. Williams DS, Linberg KA, Vaughan DK, Fariss RN, Fisher SK. Disruption of microfilament organization and deregulation of disk membrane morphogenesis by cytochalasin D in rod and cone photoreceptors. *J Comp Neurol.* 1988;272(2):161–176.
102. Wei Q, Zhang Y, Li Y, Zhang Q, Ling K, Hu J. The BBSome controls IFT assembly and turnaround in cilia. *Nat Cell Biol.* 2012;14(9):950–957.
103. Mikami A, Tynan SH, Hama T, et al. Molecular structure of cytoplasmic dynein 2 and its distribution in neuronal and ciliated cells. *J Cell Sci.* 2002;115(Pt 24):4801–4808.
104. Tian G, Lodowski KH, Lee R, Imanishi Y. Retrograde intraciliary trafficking of opsin during the maintenance of cone shaped photoreceptor outer segments of Xenopus laevis. *J Comp Neurol.* 2014;522:3577–3589.
105. Nemet I, Tian G, Imanishi Y. Organization of cGMP sensing structures on the rod photoreceptor outer segment plasma membrane. *Channels.* 2014;8:528–535.
106. Nemet I, Tian G, Imanishi Y. Submembrane assembly and renewal of rod photoreceptor cGMP-gated channel: insight into the actin-dependent process of outer segment morphogenesis. *J Neurosci.* 2014;34(24):8164–8174.
107. Steinberg RH, Fisher SK, Anderson DH. Disc morphogenesis in vertebrate photoreceptors. *J Comp Neurol.* 1980;190(3):501–508.
108. Miyaguchi K, Hashimoto PH. Evidence for the transport of opsin in the connecting cilium and basal rod outer segment in rat retina: rapid-freeze, deep-etch and horseradish peroxidase labelling studies. *J Neurocytol.* 1992;21(6):449–457.
109. Obata S, Usukura J. Morphogenesis of the photoreceptor outer segment during postnatal development in the mouse (BALB/c) retina. *Cell Tissue Res.* 1992;269(1):39–48.
110. Chuang JZ, Zhao Y, Sung CH. SARA-regulated vesicular targeting underlies formation of the light-sensing organelle in mammalian rods. *Cell.* 2007;130(3):535–547.
111. Carter-Dawson LD, LaVail MM. Rods and cones in the mouse retina. I. Structural analysis using light and electron microscopy. *J Comp Neurol.* 1979;188(2):245–262.
112. Kinney MS, Fisher SK. The photoreceptors and pigment epithelium of the larval Xenopus retina: morphogenesis and outer segment renewal. *Proc R Soc Lond B Biol Sci.* 1978;201(1143):149–167.
113. Arikawa K, Williams DS. Organization of actin filaments and immunocolocalization of alpha-actinin in the connecting cilium of rat photoreceptors. *J Comp Neurol.* 1989;288(4):640–646.
114. Rattner A, Smallwood PM, Williams J, et al. A photoreceptor-specific cadherin is essential for the structural integrity of the outer segment and for photoreceptor survival. *Neuron.* 2001;32(5):775–786.
115. Maw MA, Corbeil D, Koch J, et al. A frameshift mutation in prominin (mouse)-like 1 causes human retinal degeneration. *Hum Mol Genet.* 2000;9(1):27–34.
116. Yang Z, Chen Y, Lillo C, et al. Mutant prominin 1 found in patients with macular degeneration disrupts photoreceptor disk morphogenesis in mice. *J Clin Invest.* 2008;118(8):2908–2916.
117. Han Z, Anderson DW, Papermaster DS. Prominin-1 localizes to the open rims of outer segment lamellae in Xenopus laevis rod and cone photoreceptors. *Invest Ophthalmol Vis Sci.* 2012;53(1):361–373.
118. Cohen AI. Some cytological and initial biochemical observations on photoreceptors in retinas of rds mice. *Invest Ophthalmol Vis Sci.* 1983;24(7):832–843.
119. Nir I, Papermaster DS. Immunocytochemical localization of opsin in the inner segment and ciliary plasma membrane of photoreceptors in retinas of rds mutant mice. *Invest Ophthalmol Vis Sci.* 1986;27(5):836–840.
120. Travis GH, Brennan MB, Danielson PE, Kozak CA, Sutcliffe JG. Identification of a photoreceptor-specific mRNA encoded by the gene responsible for retinal degeneration slow (rds). *Nature.* 1989;338(6210):70–73.

121. Cheng T, Peachey NS, Li S, Goto Y, Cao Y, Naash MI. The effect of peripherin/rds haploinsufficiency on rod and cone photoreceptors. *J Neurosci.* 1997;17(21): 8118–8128.
122. Arikawa K, Molday LL, Molday RS, Williams DS. Localization of peripherin/rds in the disk membranes of cone and rod photoreceptors: relationship to disk membrane morphogenesis and retinal degeneration. *J Cell Biol.* 1992;116(3):659–667.
123. Tian G, Ropelewski P, Nemet I, Lee R, Lodowski KH, Imanishi Y. An unconventional secretory pathway mediates the cilia targeting of peripherin/rds. *J Neurosci.* 2014;34(3):992–1006.
124. Boesze-Battaglia K, Lamba OP, Napoli Jr AA, Sinha S, Guo Y. Fusion between retinal rod outer segment membranes and model membranes: a role for photoreceptor peripherin/rds. *Biochemistry.* 1998;37(26):9477–9487.
125. Rattner A, Chen J, Nathans J. Proteolytic shedding of the extracellular domain of photoreceptor cadherin. Implications for outer segment assembly. *J Biol Chem.* 2004;279(40):42202–42210.
126. Zhang Y, Molday LL, Molday RS, et al. Knockout of GARPs and the beta-subunit of the rod cGMP-gated channel disrupts disk morphogenesis and rod outer segment structural integrity. *J Cell Sci.* 2009;122(Pt 8):1192–1200.
127. Kwok MC, Holopainen JM, Molday LL, Foster LJ, Molday RS. Proteomics of photoreceptor outer segments identifies a subset of SNARE and Rab proteins implicated in membrane vesicle trafficking and fusion. *Mol Cell Proteomics.* 2008;7(6):1053–1066.
128. Dryja TP, McGee TL, Hahn LB, et al. Mutations within the rhodopsin gene in patients with autosomal dominant retinitis pigmentosa. *N Engl J Med.* 1990;323(19): 1302–1307.
129. Tam BM, Xie G, Oprian DD, Moritz OL. Mislocalized rhodopsin does not require activation to cause retinal degeneration and neurite outgrowth in Xenopus laevis. *J Neurosci.* 2006;26(1):203–209.
130. Hollingsworth TJ, Gross AK. The severe autosomal dominant retinitis pigmentosa rhodopsin mutant Ter349Glu mislocalizes and induces rapid rod cell death. *J Biol Chem.* 2013;288(40):29047–29055.
131. Sahly I, Dufour E, Schietroma C, et al. Localization of Usher 1 proteins to the photoreceptor calyceal processes, which are absent from mice. *J Cell Biol.* 2012;199(2):381–399.
132. Alfinito PD, Townes-Anderson E. Activation of mislocalized opsin kills rod cells: a novel mechanism for rod cell death in retinal disease. *Proc Natl Acad Sci USA.* 2002;99(8):5655–5660.
133. Wang J, Zhang N, Beuve A, Townes-Anderson E. Mislocalized opsin and cAMP signaling: a mechanism for sprouting by rod cells in retinal degeneration. *Invest Ophthalmol Vis Sci.* 2012;53(10):6355–6369.
134. Nakao T, Tsujikawa M, Notomi S, Ikeda Y, Nishida K. The role of mislocalized phototransduction in photoreceptor cell death of retinitis pigmentosa. *PLoS One.* 2012;7(4):e32472.
135. Cohen GB, Yang T, Robinson PR, Oprian DD. Constitutive activation of opsin: influence of charge at position 134 and size at position 296. *Biochemistry.* 1993;32(23):6111–6115.
136. Kaushal S, Khorana HG. Structure and function in rhodopsin. 7. Point mutations associated with autosomal dominant retinitis pigmentosa. *Biochemistry.* 1994;33(20): 6121–6128.
137. Illing ME, Rajan RS, Bence NF, Kopito RR. A rhodopsin mutant linked to autosomal dominant retinitis pigmentosa is prone to aggregate and interacts with the ubiquitin proteasome system. *J Biol Chem.* 2002;277(37):34150–34160.

138. Roof DJ, Adamian M, Hayes A. Rhodopsin accumulation at abnormal sites in retinas of mice with a human P23H rhodopsin transgene. *Invest Ophthalmol Vis Sci*. 1994;35(12):4049–4062.
139. Tam BM, Moritz OL. Characterization of rhodopsin P23H-induced retinal degeneration in a Xenopus laevis model of retinitis pigmentosa. *Invest Ophthalmol Vis Sci*. 2006;47(8):3234–3241.
140. Lin JH, Li H, Yasumura D, et al. IRE1 signaling affects cell fate during the unfolded protein response. *Science*. 2007;318(5852):944–949.
141. Chiang WC, Kroeger H, Sakami S, et al. Robust endoplasmic reticulum-associated degradation of rhodopsin precedes retinal degeneration. *Molecular Neurobiology*. 2014; http://dx.doi.org/10.1007/s12035-014-8881-8.
142. Adekeye A, Haeri M, Solessio E, Knox BE. Ablation of the proapoptotic genes CHOP or Ask1 does not prevent or delay loss of visual function in a P23H transgenic mouse model of retinitis pigmentosa. *PLoS One*. 2014;9(2):e83871.
143. Sakami S, Kolesnikov AV, Kefalov VJ, Palczewski K. P23H opsin knock-in mice reveal a novel step in retinal rod disc morphogenesis. *Hum Mol Genet*. 2014;23(7):1723–1741.
144. Liu X, Wu TH, Stowe S, et al. Defective phototransductive disk membrane morphogenesis in transgenic mice expressing opsin with a mutated N-terminal domain. *J Cell Sci*. 1997;110(Pt 20):2589–2597.
145. Li ZY, Jacobson SG, Milam AH. Autosomal dominant retinitis pigmentosa caused by the threonine-17-methionine rhodopsin mutation: retinal histopathology and immunocytochemistry. *Exp Eye Res*. 1994;58(4):397–408.
146. van den Born LI, van Schooneveld MJ, de Jong LA, et al. Thr4Lys rhodopsin mutation is associated with autosomal dominant retinitis pigmentosa of the cone-rod type in a small Dutch family. *Ophthalmic Genet*. 1994;15(2):51–60.
147. Fishman GA, Stone EM, Sheffield VC, Gilbert LD, Kimura AE. Ocular findings associated with rhodopsin gene codon 17 and codon 182 transition mutations in dominant retinitis pigmentosa. *Arch Ophthalmol*. 1992;110(1):54–62.
148. Kijas JW, Cideciyan AV, Aleman TS, et al. Naturally occurring rhodopsin mutation in the dog causes retinal dysfunction and degeneration mimicking human dominant retinitis pigmentosa. *Proc Natl Acad Sci USA*. 2002;99(9):6328–6333.
149. Cideciyan AV, Jacobson SG, Aleman TS, et al. *In vivo* dynamics of retinal injury and repair in the rhodopsin mutant dog model of human retinitis pigmentosa. *Proc Natl Acad Sci USA*. 2005;102(14):5233–5238.
150. White DA, Fritz JJ, Hauswirth WW, Kaushal S, Lewin AS. Increased sensitivity to light-induced damage in a mouse model of autosomal dominant retinal disease. *Invest Ophthalmol Vis Sci*. 2007;48(5):1942–1951.
151. Kunte MM, Choudhury S, Manheim JF, et al. ER stress is involved in T17M rhodopsin-induced retinal degeneration. *Invest Ophthalmol Vis Sci*. 2012;53(7): 3792–3800.
152. Tam BM, Moritz OL. The role of rhodopsin glycosylation in protein folding, trafficking, and light-sensitive retinal degeneration. *J Neurosci*. 2009;29(48):15145–15154.
153. Kang DS, Tian X, Benovic JL. Role of beta-arrestins and arrestin domain-containing proteins in G protein-coupled receptor trafficking. *Curr Opin Cell Biol*. 2014;27:63–71.
154. Laporte SA, Oakley RH, Holt JA, Barak LS, Caron MG. The interaction of beta-arrestin with the AP-2 adaptor is required for the clustering of beta 2-adrenergic receptor into clathrin-coated pits. *J Biol Chem*. 2000;275(30):23120–23126.
155. Goodman Jr OB, Krupnick JG, Santini F, et al. Beta-arrestin acts as a clathrin adaptor in endocytosis of the beta2-adrenergic receptor. *Nature*. 1996;383(6599):447–450.
156. Goodman Jr OB, Krupnick JG, Gurevich VV, Benovic JL, Keen JH. Arrestin/clathrin interaction. Localization of the arrestin binding locus to the clathrin terminal domain. *J Biol Chem*. 1997;272(23):15017–15022.

157. Krupnick JG, Goodman Jr OB, Keen JH, Benovic JL. Arrestin/clathrin interaction. Localization of the clathrin binding domain of nonvisual arrestins to the carboxy terminus. *J Biol Chem.* 1997;272(23):15011–15016.
158. Moaven H, Koike Y, Jao CC, Gurevich VV, Langen R, Chen J. Visual arrestin interaction with clathrin adaptor AP-2 regulates photoreceptor survival in the vertebrate retina. *Proc Natl Acad Sci USA.* 2013;110(23):9463–9468.
159. Park PS. Constitutively active rhodopsin and retinal disease. *Adv Pharmacol.* 2014;70:1–36.
160. Robinson PR, Cohen GB, Zhukovsky EA, Oprian DD. Constitutively active mutants of rhodopsin. *Neuron.* 1992;9(4):719–725.
161. Li T, Franson WK, Gordon JW, Berson EL, Dryja TP. Constitutive activation of phototransduction by K296E opsin is not a cause of photoreceptor degeneration. *Proc Natl Acad Sci USA.* 1995;92(8):3551–3555.
162. Chen J, Shi G, Concepcion FA, Xie G, Oprian D. Stable rhodopsin/arrestin complex leads to retinal degeneration in a transgenic mouse model of autosomal dominant retinitis pigmentosa. *J Neurosci.* 2006;26(46):11929–11937.
163. Shi W, Sports CD, Raman D, Shirakawa S, Osawa S, Weiss ER. Rhodopsin arginine-135 mutants are phosphorylated by rhodopsin kinase and bind arrestin in the absence of 11-cis-retinal. *Biochemistry.* 1998;37(14):4869–4874.
164. Chuang JZ, Vega C, Jun W, Sung CH. Structural and functional impairment of endocytic pathways by retinitis pigmentosa mutant rhodopsin-arrestin complexes. *J Clin Invest.* 2004;114(1):131–140.
165. Ritter LM, Khattree N, Tam B, Moritz OL, Schmitz F, Goldberg AF. *In situ* visualization of protein interactions in sensory neurons: glutamic acid-rich proteins (GARPs) play differential roles for photoreceptor outer segment scaffolding. *J Neurosci.* 2011;31(31):11231–11243.
166. Betzig E, Patterson GH, Sougrat R, et al. Imaging intracellular fluorescent proteins at nanometer resolution. *Science.* 2006;313(5793):1642–1645.
167. Egner A, Hell SW. Fluorescence microscopy with super-resolved optical sections. *Trends Cell Biol.* 2005;15(4):207–215.
168. Huang B, Wang W, Bates M, Zhuang X. Three-dimensional super-resolution imaging by stochastic optical reconstruction microscopy. *Science.* 2008;319(5864):810–813.
169. Huang B, Babcock H, Zhuang X. Breaking the diffraction barrier: super-resolution imaging of cells. *Cell.* 2010;143(7):1047–1058.
170. Gaj T, Gersbach CA, Barbas 3rd CF. ZFN, TALEN, and CRISPR/Cas-based methods for genome engineering. *Trends Biotechnol.* 2013;31(7):397–405.
171. Moritz OL, Tam BM, Papermaster DS, Nakayama T. A functional rhodopsin-green fluorescent protein fusion protein localizes correctly in transgenic Xenopus laevis retinal rods and is expressed in a time-dependent pattern. *J Biol Chem.* 2001;276(30):28242–28251.
172. Jin S, McKee TD, Oprian DD. An improved rhodopsin/EGFP fusion protein for use in the generation of transgenic Xenopus laevis. *FEBS Lett.* 2003;542(1–3):142–146.
173. Zhuo X, Haeri M, Solessio E, Knox BE. An inducible expression system to measure rhodopsin transport in transgenic Xenopus rod outer segments. *PLoS One.* 2013;8(12):e82629.
174. Nakano A. Spinning-disk confocal microscopy—a cutting-edge tool for imaging of membrane traffic. *Cell Struct Funct.* 2002;27(5):349–355.
175. Imanishi Y, Lodowski KH, Koutalos Y. Two-photon microscopy: shedding light on the chemistry of vision. *Biochemistry.* 2007;46(34):9674–9684.

CHAPTER FOUR

Intracellular Trafficking of Neuropeptide Y Receptors

Karin Mörl[1], Annette G. Beck-Sickinger
Faculty of Biosciences, Pharmacy and Psychology, Institute of Biochemistry, Universität Leipzig, Leipzig, Germany
[1]Corresponding author: e-mail address: kmoerl@uni-leipzig.de

Contents

Abstract

The multireceptor multiligand system of neuropeptide Y receptors and their ligands is involved in the regulation of a multitude of physiological and pathophysiological processes. Specific expression patterns, ligand-binding modes, and signaling properties contribute to the complex network regulating distinct cellular responses. Intracellular trafficking processes are important key steps that are regulated in context with accessory proteins. These proteins exert their influence by interacting directly or indirectly with the receptors, causing modification of the receptors, or operating as scaffolds for the assembly of larger signaling complexes. On the intracellular receptor faces, sequence-specific motifs have been identified that play an important role in this process. Interestingly, it is also possible to influence the receptor internalization by modification of the peptide ligand.

1. INTRODUCTION: THE NEUROPEPTIDE Y RECEPTOR FAMILY

In primates, the neuropeptide Y (NPY) receptor family of rhodopsin-like G protein-coupled receptors (GPCRs) includes the Y_2 receptor and

Progress in Molecular Biology and Translational Science, Volume 132
ISSN 1877-1173
http://dx.doi.org/10.1016/bs.pmbts.2015.02.011

three other receptor subtypes (Y_1, Y_4, and Y_5), which are known to signal via pertussis toxin-sensitive, heterotrimeric $G_{i/o}$ proteins.[1,2] All four receptors are activated by three closely related endogenous peptide ligands, NPY, peptide YY (PYY), and pancreatic polypeptide (PP), with distinct, overlapping ligand-binding profiles. Whereas NPY and PYY activate Y_1, Y_2, and Y_5 receptors with high affinities, PP predominantly binds to Y_4 receptors.[3] This multiligand/multireceptor system has been intensively characterized in the past years on the basis of its influences on a wide range of essential physiological key functions, e.g., food intake, memory retention, gastrointestinal transit, and regulation of blood pressure. Furthermore, it is associated with a variety of major human diseases such as obesity, cancer, epilepsy, mood disorders, gastrointestinal, and cardiovascular complications.[4–10] Expression of the receptors in the same tissues has been shown to evoke synergistic or antagonizing effects with respect to these processes.[4,11] Therefore, it is of great interest to unravel specific differences with respect to receptor–ligand interactions and intracellular effects following receptor activation. Irrespective of their common evolutionary origin, it has been shown that Y receptor subtypes are characterized by differences in receptor structure- and ligand-binding modes. The use of amino acid scans, development of receptor subtype-selective agonists as well as the analysis of modified ligands and receptors in structure–affinity and structure–activity studies contributed to the understanding of distinct receptor subtype attributes, binding modes, and partial characterization of distinct receptor binding pockets.[3,12] Furthermore, several agonists have been developed in order to intervene in pathophysiological processes.[13] However, rapid clearance and degradation of the ligands as well as desensitization of the receptors, evolving drug resistance, and adverse effects *in vivo* turned out to be a major drawback.[14–17] With respect to these problems, temporal and spatial aspects of GPCR signaling, receptor desensitization, and alternative signaling profiles attracted attention. It is obvious that receptor density at the cell surface influences signal intensity upon ligand binding. Agonist binding triggers conformational changes in the receptor and consequently stimulates G protein activation. Uncoupling from G proteins occurs within seconds when activated GPCRs undergo rapid phosphorylation by G protein-coupled receptor kinases (GRK) and recruit arrestin.[18,19] This prevents further interaction of G proteins with the receptor and terminates the primary signal transduction. Subsequently, arrestin mediates receptor internalization through clathrin- and dynamin-dependent pathways[20,21] and receptors are either degraded or recycle back to the surface.[22–24] However, alternative signal transduction pathways and trafficking mechanisms have been described.[25–29] These findings highlight that intracellular trafficking contributes to the tight regulation of GPCR responsiveness and signaling. Furthermore, pathological receptor subtype expression on the surface of tumor cells was recognized as a tool to shuttle medicinal therapeutics, like, e.g., cytotoxic compounds, inside the cancer cell by taking advantage of receptor-mediated cointernalization upon ligand binding.[30–32]

Here, we report on recent findings describing intracellular trafficking, focusing on Y_2 receptors, and its distinctive features within the multireceptor family of NPY receptors.

2. EVOLUTION OF THE NPY RECEPTOR FAMILY

Comparative analysis of protein sequences is a valuable tool to identify structurally and functionally important regions, domains, and even single amino acid residues. Reciprocal mutation was, for example, used to identify amino acids in the Y_2 receptor which are important for agonist binding.[33] Therefore, it is interesting to consider the evolutionary relationship of NPY receptor subtypes.

Sequence information on representative NPY receptor genes from amphibians, bony fishes, sharks, lamprey, and mammals, their phylogenetic and taxonomic analysis as well as the consideration of the chromosomal location of Y receptor genes in different species allowed the elucidation of evolutionary relationships. Genome and gene duplication events as well as gene loss apparently contributed to the existence of several subtypes within this multireceptor family. Despite the fact that all receptors are able to bind NPY, PYY, and PP, although with different affinities, surprisingly high structural diversity, and low sequence homology was found. According to their degree of amino acid sequence identity, the existing vertebrate NPY receptors can be sorted to three distinct subfamilies corresponding to three proposed Y receptor ancestor genes. The Y_1 subfamily includes the mammalian Y_1,Y_4, and y_6 receptors and three zebrafish receptors $Y_{a/b/c}$. Notably, the y_6 receptor is not functional in humans and other primates, guinea pig, and pig and is absent in the rat genome and therefore denoted with a lower case "y." The Y_4 receptor and its ligand PP are rapidly evolving and show surprisingly high sequence divergence between species. The Y_2 receptor is the only Y_2 receptor subfamily member in mammals, but it also includes Y_7 receptors identified in frogs and zebra fish. The Y_5

receptor has so far no known relatives. A pharmacologically described Y_3 receptor could not be identified on DNA level so far.[1,34–38]

In mammals, despite the existence of five receptor subtypes, no major evolutionary changes and differences could be attributed to the receptor subtype signal transduction pathways so far. Signaling via $G_{i/o}$ pathways to inhibit cAMP formation, K^+ and Ca^{2+} signaling has been shown for all receptors.[39] In addition, for Y_2 and Y_4 receptors G_q protein coupling and as a consequence activation of phospholipase C_β and increasing production of inositol-1,4,5-phosphate has been described after receptor activation in rabbit smooth muscle cells.[40]

3. INTRACELLULAR TRAFFICKING OF Y RECEPTORS

In the recent years, it became evident that intracellular trafficking pathways following receptor activation and endocytosis are involved in the regulation of signal intensities, signaling properties, and resensitization of GPCRs. It has been shown that GPCR signaling is not only restricted to G protein activation. Some receptors continue to signal from intracellular compartments. Intracellular interacting proteins like, e.g., GRK and arrestin have been shown to modulate GPCR internalization and signaling by recruiting further enzymes and regulatory proteins.[26,27,41] Alternative signaling platforms and pathways as well as the existence of biased agonists emerge as important key findings in this field.[19,42] Depending on the cell type and the receptor, postendocytotic trafficking may lead to receptor degradation or recycling to the cell membrane, thereby influencing resensitization processes.[24,25,43–45] Such differences together with changes in receptor structures, ligand preferences, and expression patterns might account for functional diversity and opposing physiologic effects of Y receptor stimulation, as has been shown, for example, in the regulation of feeding behavior.[4] Other interesting aspects are, for example, the following: Y_5 and Y_1 genes are transcribed in opposite directions from a common promoter region and in the brain the Y_5 receptor is consistently colocalized with the Y_1 receptor, which itself shows broader distribution.[46,47] Since, both receptors are activated by the same ligands it is of great interest to reveal the correlation of different signaling or trafficking processes to the modulation of distinct physiological effects. Furthermore, the investigation of Y_2 effects on cancer progression highlights the variability and sometimes even opposing effects of cell-specific signaling processes: whereas Y_2-mediated signaling has been shown to play a role in promoting neuroblastoma growth and vascularization,[48] inhibition of pancreatic tumor cell growth by Y_2 signaling is discussed.[4,49,50] Therefore, understanding and specific modulation of Y_2 receptor localization, signaling and intracellular trafficking is of great interest for the development of highly effective and long-acting pharmaceuticals.

With regard to study intracellular receptor trafficking processes using noninvasive methods in living cells, the use of fluorescent tags and autofluorescent proteins that allow labeling of proteins *in vivo* without the requirement of any cofactors was established very successfully and found widespread application. It has been shown, for example, that C-terminal fusion of NPY receptors to variants of the green fluorescent protein do not interfere with subcellular receptor trafficking and signaling processes.[51,52]

3.1 Anterograde Transport of Y Receptors

Even if overexpressed in heterologous cell systems, Y receptors are predominantly expressed at the cell surface. Especially, the Y_2 receptor is always almost exclusively located to the cell membrane[51] (Fig. 1). Biosynthesis

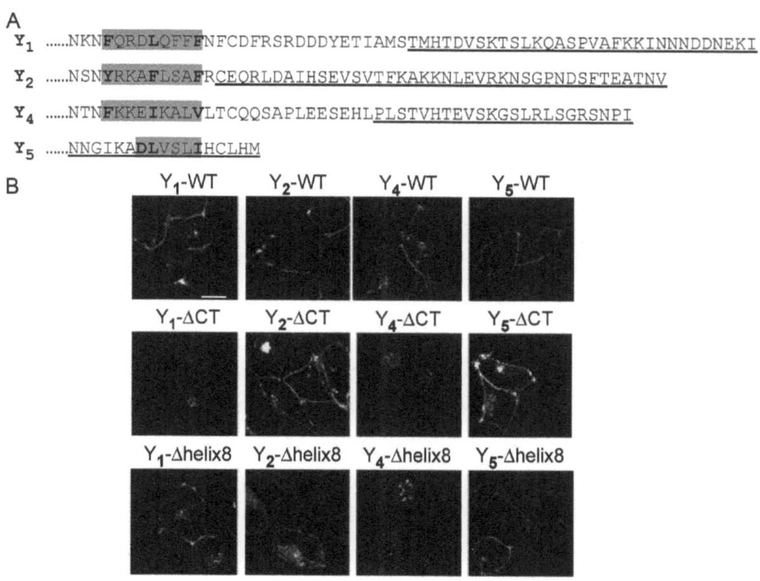

Figure 1 C-terminal sequences regulate cell surface expression of Y receptors: (A) C-terminal sequences of human Y receptors. C-terminal deletion (ΔCT) is depicted by underlined sequences. Helix eight motifs are shaded in gray. (B) Representative fluorescence microscopy images of Y receptors and mutants, C-terminally fused to EYFP (yellow), transiently expressed in HEK293 cells as described in Walther *et al.*[53] Cell nuclei were visualized with Hoechst33342 (blue). ΔCT, deletion of C-terminal sequences underlined in (A); Δhelix8, deletion of helix eight motifs shaded in (A); scale bar: 10 μM. (See the color plate.)

of GPCRs starts with translation to the endoplasmic reticulum (ER) where posttranslational modification like glycosylation, folding, dimerization, and association with accessory proteins takes place. Further, trafficking through the ER–Golgi intermediate complex and Golgi apparatus is important for final maturation and cell surface expression of GPCRs. Distinct N- and C-terminal receptor sequence motifs have been shown to be involved in the regulation of protein export to the cell membrane, either by stabilizing the receptor structure itself or by mediating interactions with specific accessory proteins. It has been shown that such proteins have the potential to regulate receptor trafficking to the cell surface, assist receptor maturation and might even modulate their function. Besides proteins that ensure the anterograde transport, also negative regulation that leads to retention of GPCRs and reduces plasma membrane expression has been observed. Some accessory proteins can promote either of these effects, depending on the GPCRs they associate with.[44,54–56] Quality control systems in the intracellular compartments ensure removal and degradation of misfolded proteins, which are retained and subjected to degradation in proteasomes via ER-associated mechanisms or at later stages transported to endosomes followed by degradation in lysosomes.[57,58]

The export of all Y receptors at ER exit sites depends on Sar1 activity for sorting to budding COPII vesicles.[59,60] As it has been described for other GPCRs,[61] coexpression with a constitutive active Sar1[H79G] mutant causes trapping of Y receptors in the ER.[53] Further transport of GPCRs via the Golgi apparatus has been shown to be mediated by ras-related in brain (rab) proteins 1, 2, 6, and 8.[62] However, in contrast to other GPCRs,[63,64] expression of neither dominant negative rab1[S25N] nor rab8[T22N] were able to retain Y_2 receptors intracellularly (own unpublished results).

By sequential deletion of N-terminal sequences, it could be shown that with respect to cell surface expression the Y_2 receptor tolerates complete deletion of its 49 N-terminal amino acids. However, complete N-terminal deletion caused significantly reduced ligand affinity and activity. Interestingly, replacement of the Y_2 N-terminus by a nine amino acid hemagglutinin (HA) tag at least partially restored specific ligand binding and activation properties. Furthermore, mutation of the potential N-glycosylation site, by replacing asparagine 11 to glutamine did not affect expression at the cell surface and did not change binding characteristics of the receptor. Neither N-terminal glycosylation nor structural stabilization by the N-terminus seems to be a prerequisite for efficient transport of the Y_2 receptor to the membrane. This could also be shown for the Y_5 receptor. In contrast Y_2 and Y_4 receptor export, although it is comparably independent of glycosylation motifs, needs N-terminal stabilization of the overall receptor structure by at least eight amino acid residues to pass the Golgi apparatus. Yet, these short sequences are not sequence restricted since replacement of the N-terminus by arbitrary amino acid tags rescues membrane targeting. Consequently, anterograde transport of Y receptors does not depend on glycosylation or sequence-specific elements in the N-terminus. Concordantly, neither fusion of Y_5 nor Y_1 N-terminal sequences to the Y_2 backbone had a meaningful influence on receptor anterograde transport, affinity, activity, or subtype selectivity for specific agonists.[65]

With respect to C-terminal sequences, intriguing subtype-specific differences have been shown. Whereas the Y_5 receptor to a certain extent tolerates complete deletion of its short 17 amino acid cytoplasmic tail, the cell surface expression of all other Y receptor subtypes depends on distinct C-terminal sequences (Fig. 1). The 53 amino acid Y_2 receptor C-terminus can be shortened by up to 40 amino acids with 13 amino acids remaining, without severely impairing export to the cell membrane. Only total deletion impairs cell surface expression.[53] In contrast, trimming the Y_1 or Y_4 receptor C-terminus to 29 or 27 remaining amino acids already impairs cell surface expression at the plasma membrane to a substantial extend (own unpublished results). However, at least for the Y_1 receptor, this effect is attributed to constitutive internalization rather than impaired export as Y_1 receptors trimmed by further 10 amino acid residues are expressed at the cell surface, suggesting that a sequence motif in this region is responsible for constitutive internalization.[66,67] Further studies will be necessary to clarify if distal C-terminal sequences in Y_1 and Y_4 receptors regulate anterograde transport. Surprisingly, deletion of helix eight, a highly conserved structural motif in the very distal C-terminus of many rhodopsin-like GPCRs, had a differential effect on the anterograde transport of the Y_1, Y_2, and Y_4 receptor subtypes, despite their evolutionary relationship (Fig. 1)[53]: detailed studies using fluorescence microscopy, ELISA, and functional studies have been performed for the Y_2 receptor. Sequential truncation of C-terminal sequences revealed a proximal region within the proposed helix eight to be important for cell surface delivery. If mutations or deletions are introduced in the sequence motif $Y(X)_3F(X)_3F$ the Y_2 receptor accumulates in the ER. Even single, individual mutation of tyrosine or phenylalanine residues within this motif is sufficient to block the receptor exit from the ER. Furthermore, it has been shown that the motif is dependent on its position specific context, since transfer to more distal parts of the C-terminus abrogates its function. Y_4 receptors with a mutated corresponding $F(X)_3L(X)_3F$ motif were at least able to move on to the Golgi apparatus, whereas trafficking of the Y_1 receptor to the cell membrane was completely unaffected by deletion of the corresponding $F(X)_3I(X)_3V$ sequence.[53] Next to helix eight a palmitoylation site is located, which has been shown to be relevant for receptor maturation, structural stabilization, and functionality in several GPCRs.[68,69] Surprisingly, mutation of the putative palmitoylation site in the C-terminus had no influence on Y_2 as well as Y_1 receptor export. However, absent palmitoylation was shown to impair Y_1 signaling, whereas, Y_2 signaling is not affected.[53,70] These data show that conserved characteristic structural features may have a different effect and functional significance even in evolutionary closely related receptor subtypes.

3.2 Internalization of Y Receptors

Internalization of GPCRs is an important regulatory process that can prevent receptors from undergoing excessive stimulation and prolonged activity. After ligand binding and activation for many GPCRs arrestin-dependent internalization via clathrin-coated pits can be observed. GRK, a protein kinase family with seven members in mammals, specifically recognize, and phosphorylate agonist-activated GPCRs.[71] Phosphorylation of serine and threonine residues by in the third intracellular loop (ICL) and in C-terminal tails of the receptors has been shown to play a central role in arrestin recruitment to the receptor.[72,73] Additionally, phosphorylation independent, sequence specific, or structurally conserved amino acid sequences that are only exposed in activated receptors have been proposed to be important for arrestin interaction. For example, the sequence following the DRY motif in the ICL2 has been shown to regulate not only G protein but also arrestin interaction.[74] This sequence is highly conserved with respect to amino acid type, charge, or hydrophobic nature in several hundred rhodopsin-like GPCRs.[74] On the other hand it has been shown that G protein uncoupling can also occur independent of phosphorylation and arrestin-independent internalization occurs for several GPCRs, at least in certain cell types.[75–78]

Despite controversial discussion on Y_2 receptor internalization properties in the beginning, it has been shown that agonist stimulation induces internalization of Y_1, Y_2, and Y_4 receptor subtypes within few minutes, whereas, the Y_5 receptor internalizes significantly slower. Y_1 and Y_2 receptors internalize through clathrin coated pits.[79,80] Notably, species-specific differences might influence internalization properties of the Y_2 receptor as opposed to the Y_1 receptor. For example, it has been shown that, while human and guinea pig Y_1 receptors do not significantly differ in internalization properties, human Y_2 receptors show higher internalization rates as compared to their guinea pig orthologs. Sequence differences in the Y_2 receptor C-termini of different species may account for these variations (Fig. 2A). Interestingly, it has been shown that replacement of threonine 376 (marked with * in Fig. 2A) in the human Y_2 C-terminus by alanine, the corresponding amino acid in guinea pig, receptor internalization is impaired (Fig. 2B).[80] Such species-sequence differences are not found in Y_1 receptor C-termini (Fig. 2A). On the other hand cell surface masking is discussed as a parameter influencing receptor internalization, as described in more detail later on (in Section 3.2.2).[81,82] Furthermore, it has been shown, for example, in radioligand binding, fluorescence correlation spectroscopy and bimolecular fluorescence complementation assays that compared to the Y_1 receptor, Y_2 internalization, and arrestin 3 recruitment displays reduced agonist potency and strongly depends on the agonist concentration used in an experiment.[83–85] Walther *et al.*[80] demonstrated that in the presence of 1 μM NPY Y_2 receptors internalize at comparable rates

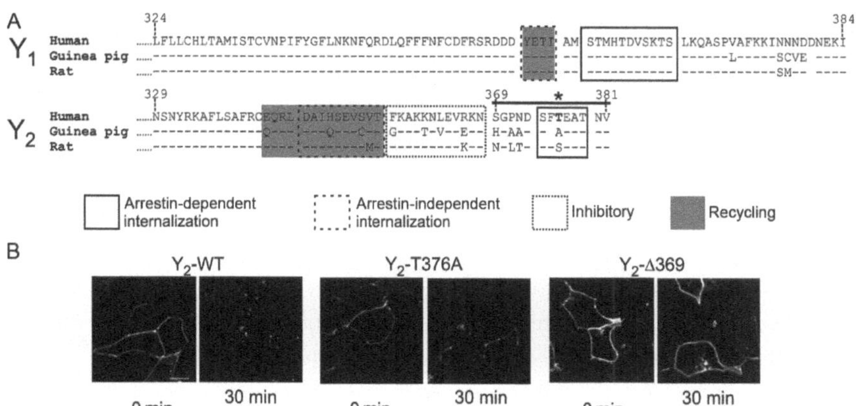

Figure 2 Localization of identified trafficking motifs in C-terminal Y_1 and Y_2 receptor sequences: (A) alignment of human, guinea pig, and rat Y_1 and Y_2 sequences. Sequence motifs identified in human Y_2 receptors are labeled as depicted in the legend. The tyrosine mutated in Y_2-T376A is marked by an asterisk. The sequence truncated in Y_2-Δ369 is labeled by a bar above it. (B) Representative fluorescence microscopy images of human Y_2 receptor wild-type (Y_2-WT), the mutant Y_2-T376A, and the truncated receptor Y_2-Δ369, C-terminally fused to EYFP (yellow) and transiently expressed in HEK293 cells as described in Walther *et al.*,[80] before (0 min, left) and after 30 min (right) stimulation with 1 μM NPY. Cell nuclei were visualized with Hoechst33342 (blue). Scale bar: 10 μM. (See the color plate.)

when exogenously expressed in HEK293 cells as well as after overexpression in SMS-KAN and MHH-NB11 cells, which also endogenously express the receptor. Noteworthy, differences in ligand concentrations like low serum concentrations in the periphery and high concentrations of NPY in the synaptic gap in the brain could influence Y_2 receptor internalization behavior in different tissues, thereby modulating intracellular signaling duration and properties. Furthermore, it has been shown that in all three cell lines Y_2 receptor endocytosis is preferentially mediated by arrestin 3-dependent mechanisms. Arrestin 3 is clearly redistributed to the cell membrane after receptor stimulation and overexpression of arrestin 3 was shown to enhance Y_2 internalization. Following internalization arrestin rapidly dissociates from the receptor and does not travel to late endocytic vesicles with the receptor. A recruitment of arrestin 2 could not be shown.[80]

Activation of human, rat, and guinea pig Y_1 receptors results in rapid internalization. Y_1 receptors recruit arrestin 2 and 3 to a similar degree presumably in a symmetric mode of association and stimulate the formation of AP2 complexes in both cases. Arrestins are cointernalized with Y_1 receptors and transported to late endocytic vesicles in complex with them.[84,86]

Like for the Y_2 receptor, initial studies on Y_4 receptor internalization revealed contradicting results.[87] However, it is well accepted in the meanwhile that the Y_4 receptor recruits arrestin 3 and is rapidly internalized following agonist exposure, however, at slightly reduced rates as compared to Y_1 receptors.[51,83,88,89] Endocytosis of the Y_4 receptor has been shown to be highly sensitive to agonist affinity and efficacy. Thus, full agonists induce higher internalization rates as compared to partial agonist. Interestingly, species differences in peptide efficacy are reflected by a decrease in the internalization rate if receptors and ligands from different species are combined.

In contrast to the other receptor subtypes, only extremely slow receptor internalization via clathrin-coated pit formation has been observed for the Y_5 receptor.[51,90] Using BRET studies, arrestin 3 recruitment could be detected at low levels, which are hardly detected in microscopic studies.[83,91] Nonetheless considerable receptor desensitization could be observed in HEK293 cells.[51]

Radioligand internalization, internalization of fluorescently labeled ligands, cell surface radioligand binding, antibody techniques like cell surface staining or ELISA, and microscopic studies have been applied to study receptor internalization. Neither N-terminal HA tags nor C-terminally fused autofluorescent proteins or tags, used to visualize and track Y receptor trafficking in microscopic studies, have been shown to influence

G protein activation and internalization properties.[51] However, we could show that N-terminal fusion of YFP, although it has no influence on receptor expression and activity, prevents endocytosis of human Y_2 receptors (own unpublished data), whereas human Y_1 receptor internalization apparently is not influenced.[86] Chimeric receptors and deletion of larger sequence stretches have been used to study the functional relevance of ICLs and C-termini. A more detailed analysis of functional relations can be performed by the introduction of directed point mutations in putative, generally short sequence motifs, like, for example, serine/threonine motifs serving as phosphorylation sensors.

3.2.1 Chimeric Receptors

With respect to chimeric receptors, unique subtype-specific features are addressed first. For example, the Y_5 receptor, which in contrast to the other family members only shows very slow internalization upon ligand binding, is characterized by an exceptionally long third ICL and a short C-terminal tail. If either of these is replaced by the corresponding sequence of the Y_2 receptor, the resulting chimeric receptors gain internalization competence.[51] Similarly, deletion of almost the complete ICL3 in the Y_5 receptor also allows internalization after ligand binding (own unpublished data). In addition, it has been shown in microscopic studies as well as in a quantitative radioligand binding assay that Y_2 C-terminal sequences are capable to trigger and even enhance Y_1 receptor internalization.[85,86] These data point to the existence of both negative and positive impact of intracellularly located, subtype-specific sequence elements on receptor trafficking processes.

3.2.2 N-Terminal Sequences

It has been shown that the N-terminus of the Y_2 receptor is not directly involved in ligand binding. As mentioned above deletion of large parts, replacement by an HA tag, mutation of the potential glycosylation side and substitution of aspartate 35 by alanine did not influence cell surface expression, agonist activity, and G protein interaction relative to the wild-type receptor.[65] Acidic and proline rich motifs in the N-terminus of Y_2 receptors have been suggested to interact with extracellular matrix components. Hence, mechanical dispersion as well as detachment of cells by metal ion chelation enhances receptor endocytosis. Cysteine bridging between cell surface proteins and extracellular matrix components apparently plays an important role in the regulation of receptor masking. Additionally, complexation of cholesterol and cholesterol-aggregating macrolides as well as the application of cholesterol-neutral detergents were shown to have an influence on Y_2 masking in CHO cells.[81] Additionally, replacement of Asp35 by alanine in the second acidic zipper PDPEPE motif enhances receptor internalization.[82] This effects could lead to a restriction of the receptors accessibility for the peptide ligands and thereby modulation of Y_2 internalization kinetics. Species differences have been detected within these motifs and could account for the differences observed in Y_2 endocytosis. Also cell-specific properties could influence the interaction with differentially expressed extracellular matrix components and therefore influence Y_2 receptor internalization. For Y_1, Y_4, and Y_5 receptors, no masking could be detected.[81,92] In contrast to these studies, it has been shown that replacement of the Y_2 N-terminus by Y_1 sequences slows down receptor internalization and replacement by Y_5 sequences does not change internalization rates of the Y_2 receptor in COS-7 cells.[65] Interestingly, replacement of the Y_2 N-terminal sequences by an HA tag reduces receptor endocytosis after activation.[65] Obviously, N-terminal sequences in Y_2 are important for stabilizing the receptor conformation with respect to ligand binding, activation, and internalization independent of specific sequence elements.

3.2.3 C-Terminal Sequences

To study the impact of C-terminal amino acid sequences on human Y_2 receptor internalization, a series of truncation mutants was generated by shortening the Y_2 receptor C-terminal tail. All truncated receptors are normally expressed at the plasma membrane and show ligand affinities and activities comparable to wild-type receptors. Three short sequence elements in the C-terminus that regulate human receptor internalization could be ascertained (Fig. 2A). A distal SxTxxT motif has been shown to mediate arrestin 3 dependent internalization. Its deletion via C-terminal truncation by 13 amino acids (Y_2-Δ369) abolishes endocytosis of Y_2 receptors. Phosphorylation of serine and threonine residues in this motif obviously is required for high affinity recruitment of arrestin 3 and therefore internalization. Introduction of a single point mutation within this motif in the mutant Y_2-T376A is sufficient to retain the receptor at the cell surface (Fig. 2B). A proximal DxxxSExSxT motif has the capability to induce arrestin-independent internalization at least in truncated receptor mutants. Again serine and threonine residues are functionally important in this motif. Mutation of this sequence in full-length receptors does not impair internalization and presumably arrestin 3 recruitment to the distal SxTxxT motif described above. There might be specific conditions which allow arrestin-independent endocytosis of full-length Y_2 receptors guided by the DxxxSExSxT motif; however, they have not been identified so far. Interestingly, the basic sequence FKAKKNLEVRKN located between the two internalization motifs is called to account for masking the proximal internalization motif, thereby blocking arrestin-independent internalization.[80]

For the Y_1 receptor, it has been shown that C-terminal truncation of the receptor by 32 amino acids, although it reduces the total amount of cell surface expression, does not interfere with G protein signaling. On the other hand, deletion of C-terminal sequences impaired endocytosis of activated receptors. Serine and threonine residues within a C-terminal (S/T)-(S/T)-ϕ-H-(S/T)-(E/D)-V-(S/T)-x-T motif need to be phosphorylated by GRK to promote arrestin 2 and arrestin 3 binding, internalization, and desensitization (Fig. 2A). It has been shown that multiple phosphorylation events are necessary.[66,67,84,86,93] Upstream of this cluster a second tyrosine-based sequence, the (YETI) motif, hypothesized to directly interact with AP2,[94,95] was associated with constitutive endocytosis to transferrin positive recycling compartments in the absence of distal C-terminal sequences (Fig. 2A). This clearly indicates that the conformation of both the receptor and accessory proteins, as well as the accessibility of consensus motifs plays an important role in the complex regulation of receptor endocytosis and intracellular trafficking.

Overexpression of a dominant negative Rab5a[S34N] or depletion of all three Rab5 isoforms in siRNA experiments could enhance cell surface expression of C-terminally truncated Y_1 receptors, possibly by interfering with Rab5-dependent internalization pathways.[66,67]

These data suggest that arrestin-mediated internalization is the main pathway followed by Y_1 and Y_2 receptors after agonist activation. However, arrestin-independent mechanisms may also exist to regulate and modify receptor endocytosis under certain circumstances.

Like for the other receptors, putative internalization motifs in the Y_4 receptor C-terminal sequences, like the serine/threonine cluster STVHTEVSKG, are hypothesized to play a role in receptor endocytosis.[91] However, none of these motifs has been experimentally confirmed so far.

3.2.4 Sequences in ICLs

Additional sequences in the receptors ICL2 have been shown to be involved in arrestin 3 binding to Y_2 receptors. The sequence around the ERY motif in the Y_2 receptor is well conserved in orthologous sequences and has been analyzed with respect to arrestin binding. Interestingly, whereas predominantly a proline occurs six amino acids downstream of the DRY motif in many GPCRs, including Y_1, Y_4, and Y_5, in the Y_2 receptor, a histidine is found in this position. Experiments with other GPCRs underline that a proline in this position favors arrestin recruitment without changes in the overall phosphorylation of the receptor. Replacement of this histidine by proline is described to enhance Y_2 receptor arrestin recruitment and endocytosis. On the other hand, in the Y_1 receptor, single substitution of proline to histidine did not affect internalization.[74,84,86,93] Further motifs like Yxxϕ, dileucine or triple basic motifs are present in ICL2 and ICL3, and the C-terminus and might contribute to a complex control mechanism also in intracellular trafficking processes of Y_1 receptors.[91,96]

Further studies are necessary to clarify the role of these ICL sequences in more detail.

3.2.5 Arrestin Binding

Arrestin 2 and 3 are ubiquitously expressed and have been identified as important multifunctional adaptor molecules that regulate the signal amplitude and duration, desensitization, internalization, intracellular signaling, and recycling of a large number of GPCRs.[97–100] GPCRs that exist in the activated conformation and are phosphorylated by a GRK efficiently bind arrestin. Multiple phosphorylation within the C-terminus of the receptor is necessary for arrestin binding. In arrestin, a phosphorylation sensor, the polar core, recognizes phosphorylated sequence motifs in the activated receptors. This interaction induces global conformational changes in arrestin that are required for high affinity binding to the receptor. Thereby, the C-terminal tail of arrestin is detached from the polar core[101] and the clathrin- and AP2-binding sites get accessible.[102–104] There is increasing evidence that depending on the phosphorylation pattern in the receptors C-terminus, functionally different complexes with arrestin are formed. GRK subtypes contribute differently to the processes of receptor desensitization, endocytosis, and signaling by the phosphorylation of distinct combinations of phosphorylation sites in the receptor C-terminus and consequently by causing distinct arrestin conformations.[105–107]

Receptor interactions with residues on the concave sides of arrestin, the activation sensor, determine receptor preference. It has been shown that manipulation of 10 nonphosphate binding amino acids within the arrestin sensor generates nonvisual arrestins that specifically bind to particular receptor subtypes. These mutations affect the preassociation as well as the interaction with activated receptors.[108,109]

In the intracellular sequences of NPY receptors, arrestin-binding motifs have been identified and arrestin-dependent internalization has been shown to terminate G protein-mediated signaling. In order to characterize the arrestin interaction with Y_1 and Y_2 receptors in more detail, 15 different arrestin variants with mutations in the receptor-discriminator residues have been tested in BRET assays. If all 10 known receptor-discriminator residues are replaced by alanine, arrestin binding to many receptors is hampered. Interestingly, in COS-7 cells, this mutant shows greatly reduced predocking to inactive Y_1 receptors; however, agonist-induced binding was only slightly reduced and revealed high binding levels as compared to all other receptors tested so far. In contrast, neither basal- nor agonist-induced recruitment could be observed for this mutant when it was combined with Y_2 receptors. Interestingly, several other arrestin mutants were only recruited to activated Y_1 but not Y_2 receptors, with the arrestin mutant Y239T showing the most prominent effect (Fig. 3). Therefore, targeted mutagenesis of arrestin can differentially direct arrestins to interaction with activated Y_1 as opposed to Y_2 receptors, which are more sensitive. Furthermore, it could be shown that predocking mechanisms are different from final recruitment of arrestin to activated receptors and arrestin mutants can be applied to separately study the effects of basal predocking and agonist-induced arrestin binding.[110]

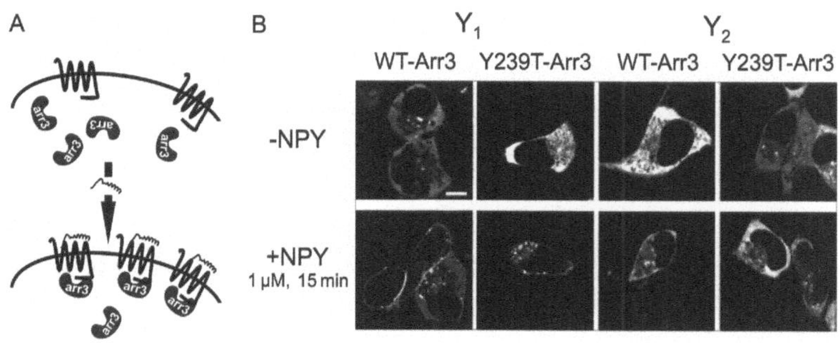

Figure 3 Recruitment of wild-type and mutated (Y239T) arrestin to human Y_1 and Y_2 receptors in transiently transfected COS-7 cells. (A) Schematic illustration of arrestin recruitment to Y receptors. (B) Representative fluorescence microscopy images of Venus-arrestin 3 recruitment to Y_1 (left) and Y_2 (right) receptors. Distribution of wild-type and mutant (Y239T) arrestin is shown prior (upper panel) and after 15 min of stimulation with 1 μM NPY. Scale bar: 10 μM. Experiments were performed as described in Gimenez *et al.*[110] Scale bar: 10 μM.

3.3 Recycling of Y Receptors

After internalization GPCRs are either targeted to endosomes followed by recycling of the receptor back to the membrane or they are translocated to lysosomes for degradation. It has been shown that distinct GRK phosphorylation sites in the receptor C-termini are key determinants for the binding affinity of receptor arrestin complexes. Based on the stability of the complex, two classes of GPCRs have been proposed. Class A GPCRs interact with arrestins in a transient manner and display higher affinity for arrestin 3 as compared to arrestin 2. After ligand induced internalization, the arrestin dissociates from the receptor near the cell membrane and does not traffic to the endocytotic vesicles in complex with the receptor. Thus, the receptor gets accessible for phosphatases and dephosphorylation allows receptor recycling. In contrast, Class B GPCRs tightly bind both arrestins with equivalent affinity, travel to the endosomes as a stable complex. They are more likely targeted for degradation or they are retained in endosomes and recycle relative slowly.[23,100,111] The sorting of GPCRs is a highly regulated process that is not well understood so far. Sorting sequences in the intracellular domains of GPCRs are thought to play an important role in the tight spatiotemporal regulation of these trafficking pathways.

In Y_1 and Y_2 receptors, the first motifs involved in recycling processes have been identified. Interestingly, these sequences are different with respect to amino acid sequence, but in both cases overlap with motifs that have been shown to mediate arrestin receptor-independent internalization in C-terminally truncated receptors (Fig. 2A). After activation of Y_2 receptors, arrestin 3 is recruited to the cell periphery and does not migrate to endocytic vesicles in complex with the receptor. Furthermore, after withdrawal of the ligand, Y_2 receptors reappear at the membrane within 30 min. Within the Y_2 receptor C-terminus, an EQRLDAIHSEVSVT recycling motif has been identified. Interestingly, this sequence overlaps with the proximal internalization motif responsible for arrestin-independent endocytosis. Mutation or deletion of this motif abrogates recycling of the receptor to the plasma membrane in HEK293, SMS-KAN, and MHH-NB-11 cells.[80]

Similarly, Y_1 receptors have been shown to recycle via fast and slow routes to the plasma membrane.[79] Interestingly also here, the tyrosine residue within the YETI motif, which has been associated with constitutive and therefore, presumably arrestin-independent endocytosis, has been shown to be important for receptor recycling to the membrane after internalization. Also, Y_4 receptors have been shown to be targeted to recycling compartments after internalization.[89]

Further elucidation of recycling pathways will reveal new insights in the regulatory mechanisms of intracellular trafficking, resensitization, and reactivation of NPY receptors.

4. MODULATION OF INTERNALIZATION BY LIGAND MODIFICATION

As a consequence of its manifold physiological functions, the Y_2 receptor is regarded as a potential therapeutic target. For example, in the research field of antiobesity therapies selective Y_2 agonists like PYY (3–36) or PYY(13–36), as well as the Y_2/Y_4 selective agonist Obinepitide [Q^{34}]hPP have been shown to reduce food intake and energy expenditure.[16,112–114] However, despite promising results in preclinical and clinical phase I/II trials, all these compounds need further improvement with respect to metabolic stability and side effects to accomplish marketing approval. In addition, Y_2 receptors are known to play a role in tumor growth and vascularization and were recognized as tumor markers overexpressed on the surface of cancer cells.[4,115–117] Selective compounds, like, for example, ^{99m}Tc labeled or *ortho*-carbaborane coupled Y_2 selective analogs have been developed for tumor imaging and targeting.[30,118,119]

Peptide drugs are known to be advantageous with respect to high selectivity, affinity, and rare toxic degradation. However, rapid proteolytic degradation and short circulation times are well-known challenges, counteracted, for example, by conjugation of modified amino acids, lipids, or polyethyleneglycol (PEG).[16,17] Interestingly, it has been shown recently that such modification of peptide ligands may not only influence selectivity but also alter intracellular processes at NPY receptors following activation.[88,120] PEGylation and lipidation of the Y_2/Y_4 specific peptide analogs [K^{22},Q^{34}]hPP or [K^{13},Q^{34}]hPP not only prolong the plasma half-live but also modulate Y_2 and Y_4 receptor signaling and arrestin recruitment in HEK293 and Col-24 cells. Palmitoylation of the ligand, irrespective of unchanged potency and affinity at the receptors, strongly increases arrestin recruitment and receptor-mediated internalization (Fig. 4). Notably, this effect is promoted with increasing fatty acid chain length. In contrast, PEGylation of the peptide analog reduced arrestin recruitment and blocked internalization, thereby biasing Y receptor signaling toward the G protein.[88] These findings illustrate that introduction of specific modifications in the ligand not only modulates receptor signaling due to bioavailability of the peptide, but is also able to modulate intracellular processes like G protein signaling, arrestin recruitment, and internalization.

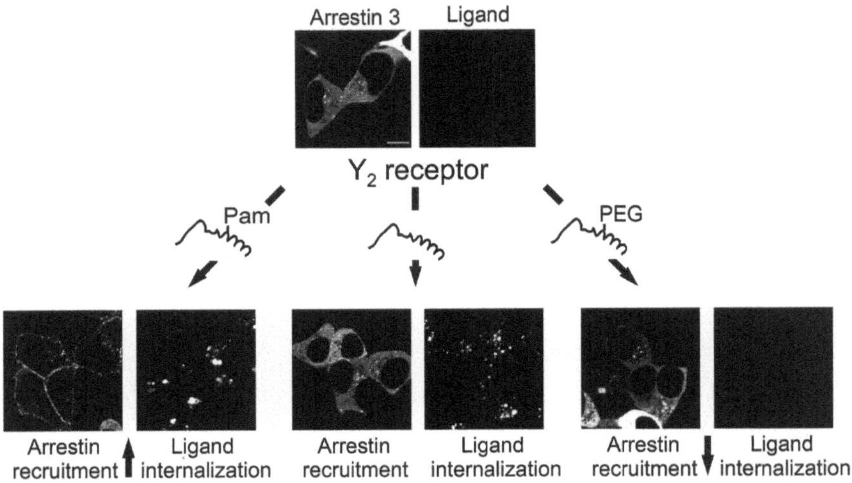

Figure 4 Arrestin recruitment and ligand internalization upon activation of human Y_2 receptors with $[K^{22},Q^{34}]$hPP analogs in HEK293 cells: live-cell images of Y_2 receptor-mediated mCherry-arrestin 3 recruitment before (upper panel) and after (lower panel) 10 min stimulation with 100 nM $[K^{22}$(E-Pam)$,Q^{34}]$hPP (left), 100 nM $[K^{22},Q^{34}]$hPP (middle), and 1 μM $[K^{22}$(PEG)$,Q^{34}]$hPP (right). Ligand internalization after 60 min stimulation with the corresponding TAMRA-labeled peptides. Experiments were performed as described in Mäde *et al.*[88] Pam, palmitoylation; PEG, PEGylation with a 22 kDa moiety; scale bar: 10 μM.

5. CONCLUSIONS

As it became obvious in the recent years, many different protein–protein interactions and protein modifications contribute to a very complex network that regulates intracellular trafficking and signaling pathways of GPCRs. The involved proteins and mechanisms differ between cell types and species, as well as receptors. The studies summarized in this review highlight the existence of receptor subtype, cell type, and species-specific differences in the NPY receptor intracellular trafficking. It is obvious that different motifs in different locations are important for the regulation of anterograde transport, internalization, and recycling of the NPY receptor subtypes. Even structurally conserved motifs, like helix eight or palmitoylation sites can have a distinct impact on the function and intracellular trafficking. Surprisingly, motifs with different consensus sequences in Y_1 and Y_2 receptor have been described to regulate similar aspects, like, for example, arrestin-mediated internalization. Further, research in this field will open new insights with respect to understanding the versatile physiological outcomes modified specifically by the distinct NPY receptor subtypes. Furthermore, it could also be shown that modification of intracellular components like arrestin, but also of the ligands, is able to specifically influence receptor subtype characteristic intracellular processes. Based on these findings important innovative aspects for the generation of peptide drugs with extended therapeutic potential will expand the possible alternatives for molecular intervention.

REFERENCES

1. Larhammar D, Salaneck E. Molecular evolution of NPY receptor subtypes. *Neuropeptides*. 2004;38(4):141–151.
2. Michel MC, Beck-Sickinger A, Cox H, et al. XVI. International Union of Pharmacology recommendations for the nomenclature of neuropeptide Y, peptide YY, and pancreatic polypeptide receptors. *Pharmacol Rev*. 1998;50(1):143–150.
3. Pedragosa-Badia X, Stichel J, Beck-Sickinger AG. Neuropeptide Y receptors: how to get subtype selectivity. *Front Endocrinol*. 2013;4:5.
4. Zhang L, Bijker MS, Herzog H. The neuropeptide Y system: pathophysiological and therapeutic implications in obesity and cancer. *Pharmacol Ther*. 2011;131(1):91–113.
5. Jacques D, Sader S, Perreault C, Abdel-Samad D, Provost C. Roles of nuclear NPY and NPY receptors in the regulation of the endocardial endothelium and heart function. *Can J Physiol Pharmacol*. 2006;84(7):695–705.
6. Cox HM. Peptide YY: a neuroendocrine neighbor of note. *Peptides*. 2007; 28(2):345–351.
7. Pons J, Lee EW, Li L, Kitlinska J. Neuropeptide Y: multiple receptors and multiple roles in cardiovascular diseases. *Curr Opin Investig Drugs*. 2004;5(9):957–962.
8. Kormos V, Gaszner B. Role of neuropeptides in anxiety, stress, and depression: from animals to humans. *Neuropeptides*. 2013;47(6):401–419.
9. Decressac M, Barker RA. Neuropeptide Y and its role in CNS disease and repair. *Exp Neurol*. 2012;238(2):265–272.
10. Borbély E, Scheich B, Helyes Z. Neuropeptides in learning and memory. *Neuropeptides*. 2013;47(6):439–450.
11. Morales-Medina JC, Dumont Y, Quirion R. A possible role of neuropeptide Y in depression and stress. *Brain Res*. 2010;1314:194–205.
12. Cabrele C, Beck-Sickinger AG. Molecular characterization of the ligand-receptor interaction of the neuropeptide Y family. *J Pept Sci*. 2000;6(3):97–122.
13. Walther C, Mörl K, Beck-Sickinger AG. Neuropeptide Y receptors: ligand binding and trafficking suggest novel approaches in drug development. *J Pept Sci*. 2011;17(4):233–246.
14. Vlieghe P, Lisowski V, Martinez J, Khrestchatisky M. Synthetic therapeutic peptides: science and market. *Drug Discov Today*. 2010;15(1–2):40–56.
15. Mäde V, Els-Heindl S, Beck-Sickinger AG. Automated solid-phase peptide synthesis to obtain therapeutic peptides. *Beilstein J Org Chem*. 2014;10:1197–1212.
16. Bellmann-Sickert K, Beck-Sickinger AG. Peptide drugs to target G protein-coupled receptors. *Trends Pharmacol Sci*. 2010;31(9):434–441.
17. Ahrens VM, Bellmann-Sickert K, Beck-Sickinger AG. Peptides and peptide conjugates: therapeutics on the upward path. *Future Med Chem*. 2012;4(12):1567–1586.
18. Moore CAC, Milano SK, Benovic JL. Regulation of receptor trafficking by GRKs and arrestins. *Annu Rev Physiol*. 2007;69:451–482.

19. Lohse MJ, Hoffmann C. Arrestin interactions with G protein-coupled receptors. *Handb Exp Pharmacol.* 2014;219:15–56.
20. Doherty GJ, McMahon HT. Mechanisms of endocytosis. *Annu Rev Biochem.* 2009;78:857–902.
21. Wolfe BL, Trejo J. Clathrin-dependent mechanisms of G protein-coupled receptor endocytosis. *Traffic.* 2007;8(5):462–470.
22. Cottrell GS. Roles of proteolysis in regulation of GPCR function. *Br J Pharmacol.* 2013;168(3):576–590.
23. Hanyaloglu AC, von Zastrow M. Regulation of GPCRs by endocytic membrane trafficking and its potential implications. *Annu Rev Pharmacol Toxicol.* 2008;48:537–568.
24. Jean-Alphonse F, Bowersox S, Chen S, Beard G, Puthenveedu MA, Hanyaloglu AC. Spatially restricted G protein-coupled receptor activity via divergent endocytic compartments. *J Biol Chem.* 2014;289(7):3960–3977.
25. Jalink K, Moolenaar WH. G protein-coupled receptors: the inside story. *Bioessays.* 2010;32(1):13–16.
26. Shukla AK, Xiao K, Lefkowitz RJ. Emerging paradigms of β-arrestin-dependent seven transmembrane receptor signaling. *Trends Biochem Sci.* 2011;36(9):457–469.
27. Penela P, Murga C, Ribas C, Lafarga V, Mayor F. The complex G protein-coupled receptor kinase 2 (GRK2) interactome unveils new physiopathological targets. *Br J Pharmacol.* 2010;160(4):821–832.
28. Ribas C, Penela P, Murga C, et al. The G protein-coupled receptor kinase (GRK) interactome: role of GRKs in GPCR regulation and signaling. *Biochim Biophys Acta.* 2007;1768(4):913–922.
29. DeFea KA. Beta-arrestins as regulators of signal termination and transduction: how do they determine what to scaffold? *Cell Signal.* 2011;23(4):621–629.
30. Ahrens VM, Frank R, Stadlbauer S, Beck-Sickinger AG, Hey-Hawkins E. Incorporation of ortho-carbaboranyl-Nε-modified L-lysine into neuropeptide Y receptor Y1- and Y2-selective analogues. *J Med Chem.* 2011;54(7):2368–2377.
31. Kwekkeboom D, Krenning EP, de Jong M. Peptide receptor imaging and therapy. *J Nucl Med.* 2000;41(10):1704–1713.
32. Reubi JC, Mäcke HR, Krenning EP. Candidates for peptide receptor radiotherapy today and in the future. *J Nucl Med.* 2005;46(suppl 1):67S–75S.
33. Berglund MM, Fredriksson R, Salaneck E, Larhammar D. Reciprocal mutations of neuropeptide Y receptor Y2 in human and chicken identify amino acids important for antagonist binding. *FEBS Lett.* 2002;518(1–3):5–9.
34. Larhammar D. Structural diversity of receptors for neuropeptide Y, peptide YY and pancreatic polypeptide. *Regul Pept.* 1996;65(3):165–174.
35. Larhammar D. Evolution of neuropeptide Y, peptide YY and pancreatic polypeptide. *Regul Pept.* 1996;62(1):1–11.
36. Larhammar D, Wraith A, Berglund MM, Holmberg SK, Lundell I. Origins of the many NPY-family receptors in mammals. *Peptides.* 2001;22(3):295–307.
37. Larhammar D, Blomqvist AG, Söderberg C. Evolution of neuropeptide Y and its related peptides. *Comp Biochem Physiol C.* 1993;106(3):743–752.
38. Larsson TA, Olsson F, Sundstrom G, et al. Early vertebrate chromosome duplications and the evolution of the neuropeptide Y receptor gene regions. *BMC Evol Biol.* 2008;8:184.
39. Holliday ND, Michel MC, Cox HM. NPY receptor subtypes and their signal transduction. In: *Neuropeptide Y and Related Peptides.* Berlin: Springer; 2004:45–73.
40. Misra S, Murthy KS, Zhou H, Grider JR. Coexpression of Y1, Y2, and Y4 receptors in smooth muscle coupled to distinct signaling pathways. *J Pharmacol Exp Ther.* 2004;311(3):1154–1162.

41. Evron T, Daigle TL, Caron MG. GRK2: multiple roles beyond G protein-coupled receptor desensitization. *Trends Pharmacol Sci*. 2012;33(3):154–164.
42. Khoury E, Clément S, Laporte SA. Allosteric and biased g protein-coupled receptor signaling regulation: potentials for new therapeutics. *Front Endocrinol*. 2014;5:68.
43. Calebiro D, Nikolaev VO, Persani L, Lohse MJ. Signaling by internalized G-protein-coupled receptors. *Trends Pharmacol Sci*. 2010;31(5):221–228.
44. Jean-Alphonse F, Hanyaloglu AC. Regulation of GPCR signal networks via membrane trafficking. *Mol Cell Endocrinol*. 2011;331(2):205–214.
45. Lefkowitz RJ, Whalen EJ. beta-arrestins: traffic cops of cell signaling. *Curr Opin Cell Biol*. 2004;16(2):162–168.
46. Naveilhan P, Hassani H, Lucas G, et al. Reduced antinociception and plasma extravasation in mice lacking a neuropeptide Y receptor. *Nature*. 2001;409(6819):513–517.
47. Parker RM, Herzog H. Regional distribution of Y-receptor subtype mRNAs in rat brain. *Eur J Neurosci*. 1999;11(4):1431–1448.
48. Kitlinska J. Neuropeptide Y, (NPY) in neuroblastoma: effect on growth and vascularization. *Peptides*. 2007;28(2):405–412.
49. Heisler T, Towfigh S, Simon N, Liu C, McFadden DW. Peptide YY augments gross inhibition by vitamin E succinate of human pancreatic cancer cell growth. *J Surg Res*. 2000;88(1):23–25.
50. Liu CD, Balasubramaniam A, Saxton RE, Paiva M, McFadden DW. Human pancreatic cancer growth is inhibited by peptide YY and BIM-43004-1. *J Surg Res*. 1995;58(6):707–712.
51. Böhme I, Stichel J, Walther C, Mörl K, Beck-Sickinger AG. Agonist induced receptor internalization of neuropeptide Y receptor subtypes depends on third intracellular loop and C-terminus. *Cell Signal*. 2008;20(10):1740–1749.
52. Böhme I, Mörl K, Bamming D, Meyer C, Beck-Sickinger AG. Tracking of human Y receptors in living cells—a fluorescence approach. *Peptides*. 2007;28(2):226–234.
53. Walther C, Lotze J, Beck-Sickinger AG, Mörl K. The anterograde transport of the human neuropeptide Y2 receptor is regulated by a subtype specific mechanism mediated by the C-terminus. *Neuropeptides*. 2012;46(6):335–343.
54. Cooray SN, Chan L, Webb TR, Metherell L, Clark AJL. Accessory proteins are vital for the functional expression of certain G protein-coupled receptors. *Mol Cell Endocrinol*. 2009;300(1–2):17–24.
55. Dong C, Filipeanu CM, Duvernay MT, Wu G. Regulation of G protein-coupled receptor export trafficking. *Biochim Biophys Acta*. 2007;1768(4):853–870.
56. Roux BT, Cottrell GS. G protein-coupled receptors: what a difference a "partner" makes. *Int J Mol Sci*. 2014;15(1):1112–1142.
57. Achour L, Labbé-Jullié C, Scott MGH, Marullo S. An escort for GPCRs: implications for regulation of receptor density at the cell surface. *Trends Pharmacol Sci*. 2008;29(10):528–535.
58. Sitia R, Braakman I. Quality control in the endoplasmic reticulum protein factory. *Nature*. 2003;426(6968):891–894.
59. Miller EA, Schekman R. COPII—a flexible vesicle formation system. *Curr Opin Cell Biol*. 2013;25(4):420–427.
60. Sato K, Nakano A. Mechanisms of COPII vesicle formation and protein sorting. *FEBS Lett*. 2007;581(11):2076–2082.
61. Dong C, Zhou F, Fugetta EK, Filipeanu CM, Wu G. Endoplasmic reticulum export of adrenergic and angiotensin II receptors is differentially regulated by Sar1 GTPase. *Cell Signal*. 2008;20(6):1035–1043.
62. Wang G, Wu G. Small GTPase regulation of GPCR anterograde trafficking. *Trends Pharmacol Sci*. 2012;33(1):28–34.

63. Dong C, Yang L, Zhang X, et al. Rab8 interacts with distinct motifs in alpha2B- and beta2-adrenergic receptors and differentially modulates their transport. *J Biol Chem.* 2010;285(26):20369–20380.
64. Filipeanu CM, Zhou F, Claycomb WC, Wu G. Regulation of the cell surface expression and function of angiotensin II type 1 receptor by Rab1-mediated endoplasmic reticulum-to-Golgi transport in cardiac myocytes. *J Biol Chem.* 2004;279(39):41077–41084.
65. Lindner D, Walther C, Tennemann A, Beck-Sickinger AG. Functional role of the extracellular N-terminal domain of neuropeptide Y subfamily receptors in membrane integration and agonist-stimulated internalization. *Cell Signal.* 2009;21(1):61–68.
66. Holliday ND, Lam C-W, Tough IR, Cox HM. Role of the C terminus in neuropeptide Y Y1 receptor desensitization and internalization. *Mol Pharmacol.* 2005;67(3):655–664.
67. Lecat S, Ouédraogo M, Cherrier T, et al. Contribution of a tyrosine-based motif to cellular trafficking of wild-type and truncated NPY Y(1) receptors. *Cell Signal.* 2011;23(1):228–238.
68. Chini B, Parenti M. G-protein-coupled receptors, cholesterol and palmitoylation: facts about fats. *J Mol Endocrinol.* 2009;42(5):371–379.
69. Qanbar R, Bouvier M. Role of palmitoylation/depalmitoylation reactions in G-protein-coupled receptor function. *Pharmacol Ther.* 2003;97(1):1–33.
70. Holliday ND, Cox HM. Control of signalling efficacy by palmitoylation of the rat Y1 receptor. *Br J Pharmacol.* 2003;139(3):501–512.
71. Premont RT, Gainetdinov RR. Physiological roles of G protein-coupled receptor kinases and arrestins. *Annu Rev Physiol.* 2007;69:511–534.
72. Ferguson SS. Evolving concepts in G protein-coupled receptor endocytosis: the role in receptor desensitization and signaling. *Pharmacol Rev.* 2001;53(1):1–24.
73. Oakley RH, Laporte SA, Holt JA, Barak LS, Caron MG. Molecular determinants underlying the formation of stable intracellular G protein-coupled receptor-beta-arrestin complexes after receptor endocytosis. *J Biol Chem.* 2001;276(22):19452–19460.
74. Marion S, Oakley RH, Kim K-M, Caron MG, Barak LS. A beta-arrestin binding determinant common to the second intracellular loops of rhodopsin family G protein-coupled receptors. *J Biol Chem.* 2006;281(5):2932–2938.
75. Chen Z, Gaudreau R, Le Gouill C, Rola-Pleszczynski M, Staňková J. Agonist-induced internalization of leukotriene B(4) receptor 1 requires G-protein-coupled receptor kinase 2 but not arrestins. *Mol Pharmacol.* 2004;66(3):377–386.
76. Paing MM, Stutts AB, Kohout TA, Lefkowitz RJ, Trejo J. beta-Arrestins regulate protease-activated receptor-1 desensitization but not internalization or Down-regulation. *J Biol Chem.* 2002;277(2):1292–1300.
77. Van Koppen CJ, Kaiser B. Regulation of muscarinic acetylcholine receptor signaling. *Pharmacol Ther.* 2003;98(2):197–220.
78. Vines CM, Revankar CM, Maestas DC, et al. N-formyl peptide receptors internalize but do not recycle in the absence of arrestins. *J Biol Chem.* 2003;278(43):41581–41584.
79. Gicquiaux H, Lecat S, Gaire M, et al. Rapid internalization and recycling of the human neuropeptide Y Y(1) receptor. *J Biol Chem.* 2002;277(8):6645–6655.
80. Walther C, Nagel S, Gimenez LE, Morl K, Gurevich VV, Beck-Sickinger AG. Ligand-induced internalization and recycling of the human neuropeptide Y2 receptor is regulated by its carboxyl-terminal tail. *J Biol Chem.* 2010;285(53):41578–41590.
81. Parker MS, Sah R, Parker SL. Surface masking shapes the traffic of the neuropeptide Y Y2 receptor. *Peptides.* 2012;37(1):40–48.
82. Parker SL, Parker MS, Wong YY, Sah R, Balasubramaniam A, Sallee F. Importance of a N-terminal aspartate in the internalization of the neuropeptide Y Y2 receptor. *Eur J Pharmacol.* 2008;594(1–3):26–31.

83. Berglund MM, Schober DA, Statnick MA, McDonald PH, Gehlert DR. The use of bioluminescence resonance energy transfer 2 to study neuropeptide Y receptor agonist-induced beta-arrestin 2 interaction. *J Pharmacol Exp Ther.* 2003; 306(1):147–156.
84. Kilpatrick LE, Briddon SJ, Holliday ND. Fluorescence correlation spectroscopy, combined with bimolecular fluorescence complementation, reveals the effects of β-arrestin complexes and endocytic targeting on the membrane mobility of neuropeptide Y receptors. *Biochim Biophys Acta.* 2012;1823(6):1068–1081.
85. Lundell I, Rabe Bernhardt N, Johnsson A-K, Larhammar D. Internalization studies of chimeric neuropeptide Y receptors Y1 and Y2 suggest complex interactions between cytoplasmic domains. *Regul Pept.* 2011;168(1–3):50–58.
86. Ouedraogo M, Lecat S, Rochdi MD, et al. Distinct motifs of neuropeptide Y receptors differentially regulate trafficking and desensitization. *Traffic.* 2008;9(3):305–324.
87. Voisin T, Goumain M, Lorinet AM, Maoret JJ, Laburthe M. Functional and molecular properties of the human recombinant Y4 receptor: resistance to agonist-promoted desensitization. *J Pharmacol Exp Ther.* 2000;292(2):638–646.
88. Mäde V, Babilon S, Jolly N, et al. Peptide modifications differentially alter G protein-coupled receptor internalization and signaling bias. *Angew Chem Int Ed Engl.* 2014;53(38):10067–10071.
89. Tough IR, Holliday ND, Cox HM. Y(4) receptors mediate the inhibitory responses of pancreatic polypeptide in human and mouse colon mucosa. *J Pharmacol Exp Ther.* 2006;319(1):20–30.
90. Parker SL, Parker MS, Buschauer A, Balasubramaniam A. Ligand internalization by cloned neuropeptide Y Y5 receptors excludes Y2 and Y4 receptor-selective peptides. *Eur J Pharmacol.* 2003;474(1):31–42.
91. Babilon S, Mörl K, Beck-Sickinger AG. Towards improved receptor targeting: anterograde transport, internalization and postendocytic trafficking of neuropeptide Y receptors. *Biol Chem.* 2013;394(8):921–936.
92. Parker SL, Kane JK, Parker MS, Berglund MM, Lundell IA, Li MD. Cloned neuropeptide Y (NPY) Y1 and pancreatic polypeptide Y4 receptors expressed in Chinese hamster ovary cells show considerable agonist-driven internalization, in contrast to the NPY Y2 receptor. *Eur J Biochem.* 2001;268(4):877–886.
93. Kilpatrick LE, Briddon SJ, Hill SJ, Holliday ND. Quantitative analysis of neuropeptide Y receptor association with beta-arrestin2 measured by bimolecular fluorescence complementation. *Br J Pharmacol.* 2010;160(4):892–906.
94. Ohno H, Aguilar RC, Fournier MC, Hennecke S, Cosson P, Bonifacino JS. Interaction of endocytic signals from the HIV-1 envelope glycoprotein complex with members of the adaptor medium chain family. *Virology.* 1997;238(2):305–315.
95. Owen DJ, Setiadi H, Evans PR, McEver RP, Green SA. A third specificity-determining site in mu 2 adaptin for sequences upstream of Yxx phi sorting motifs. *Traffic.* 2001;2(2):105–110.
96. Pandey KN. Functional roles of short sequence motifs in the endocytosis of membrane receptors. *Front Biosci (Landmark Ed).* 2009;14:5339–5360.
97. Gurevich EV, Gurevich VV. Arrestins: ubiquitous regulators of cellular signaling pathways. *Genome Biol.* 2006;7(9):236.
98. Kang DS, Tian X, Benovic JL. Role of β-arrestins and arrestin domain-containing proteins in G protein-coupled receptor trafficking. *Curr Opin Cell Biol.* 2014; 27:63–71.
99. Lefkowitz RJ, Shenoy SK. Transduction of receptor signals by beta-arrestins. *Science.* 2005;308(5721):512–517.
100. Shenoy SK, Lefkowitz RJ. β-Arrestin-mediated receptor trafficking and signal transduction. *Trends Pharmacol Sci.* 2011;32(9):521–533.

101. Vishnivetskiy SA, Schubert C, Climaco GC, Gurevich YV, Velez MG, Gurevich VV. An additional phosphate-binding element in arrestin molecule. Implications for the mechanism of arrestin activation. *J Biol Chem.* 2000;275(52):41049–41057.
102. Han M, Gurevich VV, Vishnivetskiy SA, Sigler PB, Schubert C. Crystal structure of beta-arrestin at 1.9 A: possible mechanism of receptor binding and membrane translocation. *Structure.* 2001;9(9):869–880.
103. Krupnick JG, Goodman OB, Keen JH, Benovic JL. Arrestin/clathrin interaction. Localization of the clathrin binding domain of nonvisual arrestins to the carboxy terminus. *J Biol Chem.* 1997;272(23):15011–15016.
104. Laporte SA, Oakley RH, Holt JA, Barak LS, Caron MG. The interaction of beta-arrestin with the AP-2 adaptor is required for the clustering of beta 2-adrenergic receptor into clathrin-coated pits. *J Biol Chem.* 2000;275(30):23120–23126.
105. Butcher AJ, Tobin AB, Kong KC. Examining site-specific GPCR phosphorylation. *Methods Mol Biol.* 2011;746:237–249.
106. Nobles KN, Xiao K, Ahn S, et al. Distinct phosphorylation sites on the β(2)-adrenergic receptor establish a barcode that encodes differential functions of β-arrestin. *Sci Signal.* 2011;4(185):ra51.
107. Tobin AB, Butcher AJ, Kong KC. Location, location, location. . .site-specific GPCR phosphorylation offers a mechanism for cell-type-specific signalling. *Trends Pharmacol Sci.* 2008;29(8):413–420.
108. Gimenez LE, Kook S, Vishnivetskiy SA, Ahmed MR, Gurevich EV, Gurevich VV. Role of receptor-attached phosphates in binding of visual and non-visual arrestins to G protein-coupled receptors. *J Biol Chem.* 2012;287(12):9028–9040.
109. Vishnivetskiy SA, Gimenez LE, Francis DJ, et al. Few residues within an extensive binding interface drive receptor interaction and determine the specificity of arrestin proteins. *J Biol Chem.* 2011;286(27):24288–24299.
110. Gimenez LE, Babilon S, Wanka L, Beck-Sickinger AG, Gurevich VV. Mutations in arrestin-3 differentially affect binding to neuropeptide Y receptor subtypes. *Cell Signal.* 2014;26(7):1523–1531.
111. Marchese A, Paing MM, Temple BRS, Trejo J. G protein-coupled receptor sorting to endosomes and lysosomes. *Annu Rev Pharmacol Toxicol.* 2008;48:601–629.
112. Degen L, Oesch S, Casanova M, et al. Effect of peptide YY3-36 on food intake in humans. *Gastroenterology.* 2005;129(5):1430–1436.
113. Pedersen SL, Sasikumar PG, Chelur S, et al. Peptide hormone isoforms: N-terminally branched PYY3-36 isoforms give improved lipid and fat-cell metabolism in diet-induced obese mice. *J Pept Sci.* 2010;16(11):664–673.
114. Sato N, Ogino Y, Mashiko S, Ando M. Modulation of neuropeptide Y receptors for the treatment of obesity. *Expert Opin Ther Pat.* 2009;19(10):1401–1415.
115. Körner M, Reubi JC. Neuropeptide Y receptors in primary human brain tumors: overexpression in high-grade tumors. *J Neuropathol Exp Neurol.* 2008;67(8):741–749.
116. Körner M, Waser B, Reubi JC. High expression of neuropeptide Y1 receptors in Ewing sarcoma tumors. *Clin Cancer Res.* 2008;14(16):5043–5049.
117. Lu C, Everhart L, Tilan J, et al. Neuropeptide Y and its Y2 receptor: potential targets in neuroblastoma therapy. *Oncogene.* 2010;29(41):5630–5642.
118. Langer M, Kratz F, Rothen-Rutishauser B, Wunderli-Allenspach H, Beck-Sickinger AG. Novel peptide conjugates for tumor-specific chemotherapy. *J Med Chem.* 2001;44(9):1341–1348.
119. Langer M, La Bella R, Garcia-Garayoa E, Beck-Sickinger AG. 99mTc-labeled neuropeptide Y analogues as potential tumor imaging agents. *Bioconjug Chem.* 2001; 12(6):1028–1034.
120. Mäde V, Bellmann-Sickert K, Kaiser A, Meiler J, Beck-Sickinger AG. Position and length of fatty acids strongly affect receptor selectivity pattern of human pancreatic polypeptide analogues. *ChemMedChem.* 2014;9:2463–2474.

CHAPTER FIVE

Insights into Serotonin Receptor Trafficking: Cell Membrane Targeting and Internalization

Michèle Darmon[*,†,1], **Sana Al Awabdh**[*,†], **Michel-Boris Emerit**[*,†], **Justine Masson**[*,†]
[*]INSERM U894, Centre de Psychiatrie et Neurosciences, Paris, France
[†]Centre de Psychiatrie et Neurosciences, Université Paris Descartes, Sorbonne Paris Cité, Paris, France
[1]Corresponding author: e-mail address: michele.darmon@inserm.fr

Contents

Abstract

Serotonin receptors (5-HTRs) mediate both central and peripheral control on numerous physiological functions such as sleep/wake cycle, thermoregulation, food intake, nociception, locomotion, sexual behavior, gastrointestinal motility, blood coagulation, and cardiovascular homeostasis. Six families of the G-protein-coupled receptors

Progress in Molecular Biology and Translational Science, Volume 132
ISSN 1877-1173
http://dx.doi.org/10.1016/bs.pmbts.2015.02.009

comprise most of serotonin receptors besides the conserved 5-HT_3R Cys-loop type which belongs to the family of Cys-loop ligand-gated cation channel receptors. Many of these receptors are targets of pharmaceutical drugs, justifying the importance for elucidating their coupling, signaling and functioning. Recently, special interest has been focused on their trafficking inside cell lines or neurons in conjunction with their interaction with partner proteins. In this review, we describe the trafficking of 5-HTRs including their internalization, desensitization, or addressing to the plasma membrane depending on specific mechanisms which are peculiar for each class of serotonin receptor.

1. INTRODUCTION

Serotonin or 5-hydroxytryptamine (5-HT) besides its vasoconstrictor properties[1] has mainly a neurotransmitter function in brain.[2] In mammals, 5-HT is synthesized in serotoninergic neurons from the raphe nuclei in central nervous system (CNS) or from the intestine in periphery, by hydroxylation of L-tryptophan to 5-hydroxytryptophan (5-HTP) by tryptophan hydroxylase, TPH-2,[3] that was shown to be neuronal and constitute the rate-limiting step of 5-HT synthesis.[4] After decarboxylation of 5-HTP by the nonspecific aromatic amino-acid decarboxylase,[5] 5-HT is then concentrated in vesicles by the electrogenic antiporter (exchanger) of protons and monoamines VMAT2[6,7] and then released by exocytosis from terminals. The action of serotonin on pre- and post-serotoninergic receptors is limited in time and intensity by two mechanisms of inactivation of serotonin and another resulting from the desensitization of some receptors. The first mechanism relies on the recapture of 5-HT by the plasma membrane serotonin transporter,[8] whose activity depends on ionic membrane gradients of Na^+, K^+, and Cl^-, and may be blocked by selective serotonin reuptake inhibitors (SSRI),[9] and some tricyclic antidepressants.[10] The second mechanism is the degradation of 5-HT by monoamine oxydase.[11] Blockade of uptake or degradation has been widely used to increase the extracellular concentration of serotonin in several treatments of depression. The third mechanism including the desensitization of receptors could be targeted by drugs and therefore could constitute a new way of increasing serotonin effects.

In the CNS, 5-HT plays a role in numerous physiological functions, resulting both from its large innervations in the CNS and from the existence of seven different classes of serotonin receptors. 5-HT is implicated in some vital functions such as the control of sleep/wake cycle,[12] thermoregulation,[13] eating,[14] pain,[15] migraine,[16] and neurovegetative function like arterial pressure and heart rate regulation,[17] and sexual behavior of male[18] and female[19] rats, or locomotion.[20] 5-HT also plays a role in cognition.[21] But 5-HT is deeply involved in the control of mood and emotions. Indeed, some pathologies such as depression or general anxiety are linked, at least in part, to a dysfunction of the serotoninergic system.[22] In therapeutics, numerous antidepressants are inhibitors of serotonin reuptake,[23] and anxiolytic properties of serotonin agonists are recognized.[24] Finally, some atypical antipsychotics, among them clozapine, efficient in the treatment of schizophrenia, are potent antagonists of some serotonin receptors 5-HT_{2A},[25] 5-HT_6, or 5-HT_7[26] receptors, suggesting a noxious role of serotonin in this disease.

The classification of serotonin receptors is a long story with multiple steps leading to the identification of 14 receptors divided into 7 families in rodents, in which 6 families are G-protein-coupled receptors (GPCRs) and 1 is a channel: (1) 5-HT_1 receptors (5-HT_1Rs: 1A, 1B, 1C, 1D, and 1E) negatively coupled to adenylyl cyclase, intronless, and with the highest affinity for serotonin; (2) 5-HT_2 receptors (2A, 2B, and 2C) activating the phospholipase C cascade; (3) 5-HT_3 receptors, the only ligand-gated channel receptor (3A and 3B, with a recently increased diversity in human: 3C, 3D, and 3E); (4) 5-HT_4 receptors positively coupled to adenylyl cyclase (two principal splice variants 4_S and 4_L); (5) 5-HT_5 receptors (5A and 5B); (6) 5-HT_6 receptors; and (7) 5-HT_7 receptors. The exact number of serotonin receptors varies from species to species because some genes are identified in human but not in rodents (5-$HT_{3C,3D,3E}$ and 5-HT_{1E}). In addition, splicing and editing increase the number of receptors: 5-HT_4 N-splice variants and 5-HT_{2C} N-edited forms.

Although the specific pharmacology and coupling of these receptors have been described extensively, much less is known about their trafficking, internalization, desensitization, or addressing to the plasma membrane. This review will describe the current knowledge on the trafficking of serotonin receptors.

2. TRAFFICKING OF THE 5-HT_1R

Following the proposition of at least two different classes of high-affinity 5-HT_1Rs, based on differences in regional pharmacological properties of [^{3}H]5-HT binding,[27] this family has been growing with the cloning of other related receptors. In fact, five receptors are members of the 5-HT_1R family: 5-HT_{1A}, 5-HT_{1B}, 5-HT_{1D}, 5-HT_{1E}, and 5-HT_{1F}. The 5-HT_1Rs couple primarily through Gi/o-proteins to inhibit adenylyl cyclase activity. These receptors are closely related: their genes exhibit 40–63% identity and are all intronless.

2.1 The 5-HT$_{1A}$ Receptor

Serotonin 5-HT$_{1A}$ receptors are exclusively localized in the CNS and are major targets for psychotropic drugs. They are found as autoreceptors on serotoninergic neurons from raphe nuclei and as heteroreceptors in limbic areas: the lateral septum, the CA1 area of Ammon's horn, the dentate gyrus in the hippocampus, and frontal and entorhinal cortices.[28] In many central diseases, implication of the 5-HT$_{1A}$R has been proposed, in particular in mood disorders such as depression and anxiety, but also schizophrenia.[29]

2.1.1 5-HT$_{1A}$R Desensitization upon SSRI Treatment In Vivo

Stimulation of the 5-HT$_{1A}$R by an agonist reduces the firing activity of serotoninergic neurons in nucleus raphe dorsalis (NRD), as well as the release of 5-HT, and the firing of postsynaptic neurons such as pyramidal neurons in hippocampus (CA1). When located on 5-HT neurons, 5-HT$_{1A}$Rs (autoreceptors) are known to be subject to desensitization after a prolonged treatment with antidepressant, as it can be monitored using extracellular electrophysiological recordings. Conversely, 5-HT$_{1A}$Rs do not desensitize after a prolonged antidepressant treatment when they are located on non-5-HT neurons such as hippocampal neurons.[30] Chronic treatments using SSRI achieve efficacy only after a 2- to 3-week lapse in depressed patients, and the 5-HT$_{1A}$ autoreceptor slowly developing desensitization is thought to be responsible for this delay. The favored hypothesis relies on a desensitization related to receptor internalization. Indeed, the 5-HT$_{1A}$ autoreceptor undergoes an internalization process that can be visualized by electron microscopy in NRD, after *in vivo* administration of the 5-HT$_{1A}$-specific agonist 8-OH-DPAT[31] or after acute administration of fluoxetine, the prototype of SSRI antidepressant.[32] No internalization can be visualized on heteroreceptors, especially those localized in the hippocampus. The same kind of data have been obtained, using PET imaging in NRD, with [^{18}F]MPPF in rats[32] and in human.[33]

2.1.2 5-HT$_{1A}$R Internalization in Cell Lines

In vitro models have been generated in order to get more insights into the mechanisms responsible for such an internalization/desensitization process.

Using stably transfected CHO or HEK 293 cell lines, it has been shown that the Gi coupling of the 5-HT_{1A}R leads to rapid activation of the MAP kinase pathway via a mechanism dependent upon both Ras activation and clathrin-mediated endocytosis.[34] This activation was attenuated by chelation of intracellular Ca^{2+} and Ca^{2+}/calmodulin (CaM) inhibitors. The 5-HT_{1A}R has been shown to reside in membrane rafts with palmitoylation serving as a targeting signal responsible for its retention in membrane rafts. More importantly, the raft localization of the 5-HT_{1A}R seems to be involved in receptor-mediated signaling and relocation of $G\alpha_{i3}$ into the cytosol.[35]

Constitutive endocytosis and plasma membrane recycling of tagged-5-HT_{1A}R were also observed in stably transfected LLC-PK1 cell line.[36] The 5-HT_{1A}R followed a clathrin-dependent internalization pathway via a monensin-sensitive mechanism that involved fast-recycling vesicles expressing the Rab4 protein. In this cell line, the constitutive endocytosis was not related to constitutive activity because inverse agonists could not block it.

2.1.3 5-HT_{1A}R Internalization in Neuronal Cultures

Constitutive internalization was also monitored in neurons. After internalization, vesicles endowed with 5-HT_{1A}R were visualized trafficking into the soma to be targeted back to the cell membrane or transported inside the neurons from the soma into dendrites. These 5-HT_{1A}R constitutive endocytosis/recycling events could be involved in "retargeting" of receptors to specific postsynaptic sites so as to modulate receptor function as it has been shown for the Delta/Notch-like EGF-related receptor (DNER).[37] Indeed, this receptor exhibits a somatodendritic localization in hippocampal neurons, with only a small proportion found on the cell surface in a punctate pattern, while the majority resides in intracellular organelles. Disruption of endocytosis induces an increase in cell surface expression of DNER, resulting in a rather uniformly distribution of the receptor in the axon and dendrites, then broadly and evenly localized on the cell surface. The somatodendritic targeting of DNER requires a clathrin-independent endocytosis, not implicated in the somatodendritic targeting of other polarized receptors.[37]

Agonist-dependent 5-HT_{1A}R internalization does exist in neuronal cultures but depends on agonist efficacy and neuronal phenotype. Acute stimulation by full agonists only triggers internalization of 5-HT_{1A}R in

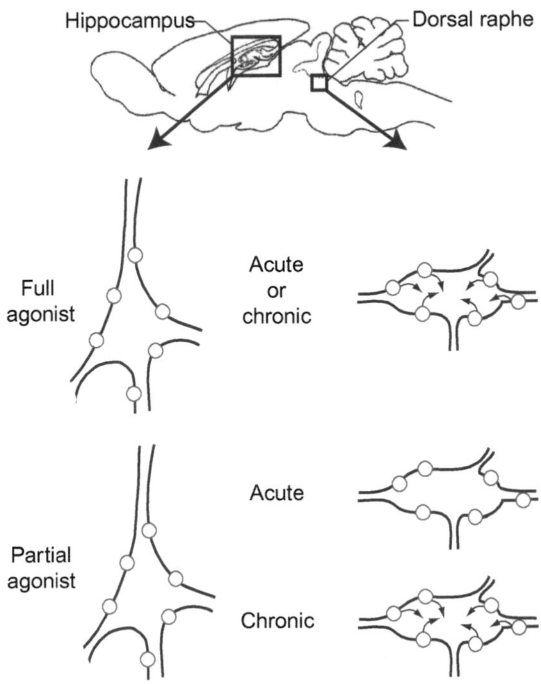

Figure 1 Peculiar internalization of the 5-$HT_{1A}R$ depending on neuronal type and full or partial agonist. Using primary neuronal cultures of hippocampal and dorsal raphe, internalization of the receptor was monitored using a specific antibody directed against the Flag tag added at the N-terminal part of the 5-$HT_{1A}R$.[36] The receptor undergoes agonist-dependent internalization only in raphe neurons with either acute or chronic stimulation with a full agonist, and not in hippocampal neurons. Using stimulation with a partial agonist such as 8-OH-DPAT, internalization is effective only after chronic treatment.

serotonergic raphe neurons but not in hippocampal neurons. After sustained exposure, even partial agonists can induce 5-$HT_{1A}R$ endocytosis in raphe serotonergic neurons, whereas only a small proportion of hippocampal neurons exhibit 5-$HT_{1A}R$ endocytosis (Fig. 1). The differential occurrence of 5-$HT_{1A}R$ internalization in serotonergic versus non-serotoninergic neurons supports the idea that endocytosis might underlie the differential 5-$HT_{1A}R$ desensitization in serotonergic versus non-serotoninergic neurons under SSRI antidepressant therapy.[36]

2.2 5-$HT_{1A}R$ Addressing

Because the 5-$HT_{1A}R$ exhibits a clear somatodendritic localization, the mechanism of its addressing was studied first in cell lines and then in neurons and compared to the addressing of the 5-HT_{1B}R that exhibits a clear axonal localization. Indeed, despite their relatively high degree of homology in their amino-acid sequences and their common coupling features, the 5-HT_{1A} and 5-HT_{1B} receptors exhibit a different localization in neurons. Both are auto- and heteroreceptors that modulate the activity of several neuronal systems. As autoreceptors on serotoninergic neurons, both receptors mediate a negative control on serotonin release using different mechanisms resulting from their different localization. Upon stimulation, the 5-HT_{1A}R inhibits neuronal firing, whereas the 5-HT_{1B}R drives an inhibitory feedback regulation of 5-HT release from nerve terminals. The matching of the respective distributions of 5-HT_{1A} mRNA and protein throughout the CNS suggests that this receptor is expressed at the somatodendritic level.[38] In contrast, in case of the 5-HT_{1B}R, the regional distributions of the mRNA and the protein are markedly different, leading to the conclusion that this receptor is transported from cell bodies, where it is synthesized, to axon terminals.[39]

2.2.1 5-HT_{1A}R and 5-HT_{1B}R Addressing in Polarized Cell Lines

This different localization was reproduced in polarized epithelial cell lines like LLC-PK1[40] or MDCK cells[41] which, after confluence, reproduce a polarized epithelium with basolateral and apical compartments. The 5-HT_{1A}R was predominantly located in basolateral membranes, while the 5-HT_{1B}R was observed mainly in intracellular vesicles.[40] The subcellular localization of chimeras of both receptors transfected in LLC-PK1 cells showed that a targeting signal within the third intracellular loop of 5-HT_{1A}R was essential for the plasma membrane targeting, whereas that of the 5-HT_{1B}R was essential for its intracellular vesicular localization.[42] Moreover, the C-terminal segment of the 5-HT_{1A}R was essential for its basolateral confinement, whereas that of the 5-HT_{1B}R enabled an apical localization.

2.2.2 5-HT_{1A}R and 5-HT_{1B}R Addressing in Neuronal Cultures

When transfected in hippocampal neurons, the membrane-bound 5-HT_{1A}R was mainly somatodendritic, whereas the 5-HT_{1B}R was addressed to the membrane of axons, while a high proportion remained in intracellular somatodendritic vesicles.[43,44]

Using chimeras, we showed that the third intracellular loop of the 5-HT_{1B}R plays a crucial role for its axonal localization,[43,44] whereas the C-terminus of the 5-HT_{1A}R is essential for its somatodendritic localization.

2.2.3 Trafficking of the 5-HT$_{1A}$R with Yif1B, Its Partner Protein

In order to describe the mechanisms responsible for the somatodendritic localization of the 5-HT$_{1A}$R, a two-hybrid screen was performed using the small 17 aa C-terminal region as a bait. Yif1B was identified for the first time, as an intracellular protein interacting with the 5-HT$_{1A}$R.[45] This protein is homologous to the yeast Yif1p, previously shown to be implicated in vesicular trafficking between the endoplasmic reticulum (ER) and the Golgi apparatus.[46] Yif1B is a membrane-bound ubiquitous protein, highly expressed in the brain and specifically in raphe 5-HT$_{1A}$R-expressing neurons. Colocalization of Yif1B and 5-HT$_{1A}$R was observed in small vesicles involved in transient intracellular trafficking. More importantly, inhibition of endogenous expression of Yif1B in primary neuron cultures prevented the addressing of 5-HT$_{1A}$R to distal portions of the dendrites, without affecting other receptors, such as the GPCR somatostatin sst2AR receptor, the purinergic P2X$_2$R, or the channel 5-HT$_{3A}$R. Yif1B scaffolds a trafficking complex mediating the intracellular traffic of the rat 5-HT$_{1A}$R toward dendrites. 5-HT$_{1A}$R–Yif1B interaction was shown to be direct, involving a tribasic motif in the C-tail of the 5-HT$_{1A}$R on which Yif1B binds directly with high affinity[47] ($K_D \sim 37$ nM) through a triacidic motif. The complex involves Yip1A, Rab6, and Kif5B as new partners of the 5-HT$_{1A}$R/Yif1B complex, whose expression in neurons is crucial for the dendritic targeting of the 5-HT$_{1A}$R. Live videomicroscopy revealed that 5-HT$_{1A}$R, Yif1B, Yip1A, and Rab6 traffic in vesicles exiting the soma toward the dendritic tree, sustaining their role in 5-HT$_{1A}$R dendritic targeting but also exhibit bidirectional motions.[47] As previously described, the 5-HT$_{1A}$R/Yif1B complex contained the dynein and p150 which is a key component of the dynactin complex.[47] Like Yif1B, Yip1A, Rab6, and Kif5B, the dynein is also crucial for the 5-HT$_{1A}$R dendritic targeting (Fig. 2). These data suggest that Kif5B and the dynein enable the 5-HT$_{1A}$R bidirectional movements with dynactin coordinating the interaction and the switch of the two opposite molecular motors for the traffic of vesicles along dendritic microtubules (Fig. 2, unpublished data, Darmon et al.). We propose a new dendritic trafficking pathway model in which Yif1B is the scaffold protein recruiting the 5-HT$_{1A}$R in a complex including Yip1A and Rab6, with dynein/dynactin complex and Kif5B as crucial molecular motors for the 5-HT$_{1A}$R dendritic addressing. This targeting pathway opens new insights for GPCRs trafficking in neurons.

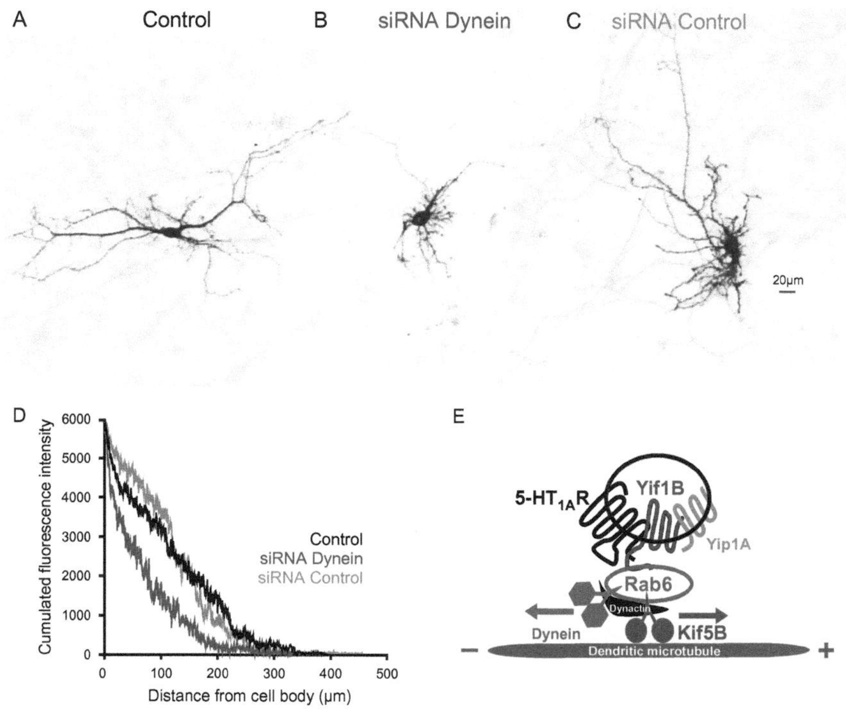

Figure 2 Dynein depletion disturbs 5-HT_{1A}R targeting toward the distal part of dendrites. Immunofluorescence of neurons transfected with the 5-HT_{1A}R–eGFP alone (A), cotransfected with siRNA control (B), or siRNA against endogenous dynein (C). Immunolabeling was performed with anti-GFP antibody to enhance the GFP signal on rat hippocampal neurons transfected at DIV7 and visualized 24 h after transfection. In this condition, the dendritic tree and axon were not perturbed by dynein depletion (as visualized by tubulin labeling, not shown). (D) The graph represents the cumulated fluorescence intensities of the 5-HT_{1A}R (cumulated fluorescence intensity, arbitrary unit) along the longest dendrite of monitored neurons (μm), in control condition (black) compared to siRNA control (blue) or to siRNA dynein (red). (E) Schematic representation of the Yif1B-scaffolding complex involved in 5-HT1AR trafficking toward the dendrites. Yif1B-dependent transport would involve Yif1B as the scaffold protein assembling the 5-HT1AR, Yip1A, and Rab6 in the same vesicles trafficking along the dendritic microtubules using two opposite molecular motors, the conventional kinesin Kif5B and the dynein for their bidirectional movements. The dynactin subunit p150 would enable the switch between the molecular motors and also the bidirectional transport observed in the dendrites. Rab6 might play the role of an intermediate protein for the interaction of these molecular motors with the Yif1B–scaffolding complex. Scale bar: 20 μm. (See the color plate.)

2.3 The 5-HT_{1B}R

The 5-HT_{1B}R is the primary target of antimigraine drugs like triptans when localized in the trigeminal ganglia.[48] It is also involved in cognitive processes such as learning and memory[49] and is implicated in the pathophysiology of neurological disorders such as obsessive compulsive disorder, drug addiction, depression, anxiety, aggression, and sleep. The 5-HT_{1B}R regulates serotonin transmission via presynaptic receptors but can also affect transmitter release at heterosynaptic sites.

2.3.1 Constitutive Activity and Constitutive Internalization in 5-HT_{1B}R Targeting

We showed that the vesicular somatodendritic localization of the 5-HT_{1B}R results from an activation-dependent constitutive endocytosis, necessary for its axonal targeting.[44] Indeed, inverse-agonist treatment, which prevents constitutive activation, leads to atypical accumulation of newly synthesized 5-HT_{1B}R on the somatodendritic plasma membrane. Using receptor chimeras composed of different domains from the third intracellular loop of the 5-HT_{1A}R and 5-HT_{1B}R, we showed that the complete third intracellular loop of 5-HT_{1B}R is necessary and sufficient for constitutive activation and efficient axonal targeting, both sensitive to inverse-agonist treatment.

These results suggest that activation and targeting of 5-HT_{1B}R are intimately interconnected in neurons as it has been described for another GPCR, the cannabinoid CB1 receptor.[50] The constitutive somatodendritic endocytosis preceding the axonal targeting of a GPCR can be compared to the transcytotic delivery described in intestinal epithelial cells.[51]

2.3.2 Trafficking of the 5-HT_{1B}R with p11, Its Partner Protein

In a yeast two-hybrid screen using the third intracellular loop of the 5-HT_{1B}R, p11 was identified as a partner protein of the 5-HT_{1B}R.[52] p11, also called S100A10, 42C, calpactin I light chain, and annexin II light chain, is a member of the S100 EF-hand (helix-loop-helix) protein family.[53] Svenningsson and colleagues showed that p11 increases localization of 5-HT_{1B}R at the cell surface. Antidepressant treatments or electroconvulsive therapy increases p11 expression in rodent brains, but p11 expression is decreased in an animal model of depression and in brain tissues from depressed patients. Overexpression of p11 increases 5-HT_{1B}R function in cells, in relation with the increase of 5-HT_{1B}R at the cell surface, and recapitulates certain behaviors seen after antidepressant treatment in mice. Accordingly, p11 knockout mice exhibit a depression-like phenotype and have reduced responsiveness to 5-HT$_{1B}$R agonists and reduced behavioral reactions to an antidepressant.[52] Thus, the dynamic modulation of 5-HT$_{1B}$R function by p11 may be involved in molecular adaptations occurring in neuronal networks that are dysfunctional in depression-like states.

Moreover, p11 has been shown to play a role in the influence of the serotonin system in the symptomatology of Parkinson disease. Indeed, striatum receives a strong serotonin innervation and chronic administration of L-DOPA increases 5-HT$_{1B}$R and p11 expression in dopamine-denervated striatonigral neurons, in a rodent model using unilaterally 6-OHDA-lesioned rats.[54] Administration of the selective 5-HT$_{1B}$ agonist CP94253 counteracts L-DOPA-induced abnormal involuntary movements and rotational behavior in WT mice in a p11-dependent mechanism since these effects were not found in p11 KO mice.[54]

In addition, it has been shown that the level of hippocampal p11 determines the bidirectionality of 5-HT$_{1B}$R action on memory processing and modulates hippocampal functionality. 5-HT$_{1B}$R agonist stimulation induces reverse effects on emotional memory when comparing WT (impairment) and p11 KO mice (enhancement).[55] All these data underline the importance of partner protein such as p11 in the regulation of the serotonin 5-HT$_{1B}$R function.

2.4 The 5-HT$_{1D}$R

The 5-HT$_{1D}$R is closely related to the 5-HT$_{1B}$R which was originally thought to be primarily or exclusively expressed in rodent (hamster, mouse, rat) tissues, while the 5-HT$_{1D}$R would be expressed in other species (human, cow, dog, guinea pig). Indeed, both receptors are expressed in all species with a similar, although not identical, pharmacology. The 5-HT$_{1B}$R expression levels are higher than the 5-HT$_{1D}$R in substantia nigra and globus pallidus, whereas both 5-HT$_{1B}$R and 5-HT$_{1D}$R types have been found to be expressed in raphe serotoninergic neurons and in the trigeminal ganglion, sustaining their implication in headache. The major pharmacological distinction is that most β-adrenergic receptor antagonists bind with high affinity to the 5-HT$_{1B}$R, but not to the 5-HT$_{1D}$R. In contrast, the currently available antimigraine drugs of the triptan family do not distinguish between 5-HT$_{1B}$R and 5-HT$_{1D}$R. Actually, both 5-HT$_{1B}$R and 5-HT$_{1D}$R immunoreactivities are found in human trigeminal ganglia, where the receptors colocalize with calcitonin gene-related peptide, substance P, and NOS which are involved in migraine-related mechanisms. However, recent data with selective 5-HT_{1D}R agonists confirmed that it is the 5-HT_{1B}R that mediates the vasoconstriction produced by triptans.[48]

Little is known on the trafficking of this receptor. The 5-HT_{1D}R is able to form hetero-oligomers with the 5-HT_{1B}R rather than homodimers,[56] but also with the 5-HT_{1A}R in transfected HEK cells[57] based on coimmunoprecipitation studies. These heteromers may require a cotranslational or specific cellular mechanism, since they do not appear when mixing membranes expressing only one type of each receptor. However, the amount of oligomers formed with 5-HT_{1B} and 5-HT_{1D} is increased when solubilized receptors are incubated with the agonists serotonin and 5-carboxamidotryptamine.[56]

To date, no data have been published on the trafficking of 5-HT_{1E} or 5-HT_{1F} receptors.

3. TRAFFICKING OF THE 5-HT_2R

The three 5-HT_2R types, 5-HT_{2A}R, 5-HT_{2B}R, 5-HT_{2C}R, share a high homology (from 46% to 50% sequence identity) and similar signaling pathways. They are generally coupled to $G\alpha_{q/11}$-proteins and activate PLC that hydrolyzes phosphatidylinositol biphosphate into inositol 1,4,5-triphosphate (IP3) and diacylglycerol (DAG).[58] Intracellular IP3 serves as a second messenger in the cytosol to stimulate the release of Ca^{2+} from the endoplasmic reticulum, whereas DAG remains membrane-bound to activate protein kinase C. The 5-HT_2R family can also activate phospholipase A2 to evoke the release of arachidonic acid or stimulate MAPK, specifically ERK1 and ERK2.

Although the three 5-HT_2R types are similar in structure and pharmacological profiles, their trafficking relying on different interacting proteins is quite different.

3.1 The 5-HT_{2A}R

The 5-HT_{2A}Rs mediates contractile responses to 5-HT in many vascular smooth muscle preparations, e.g., tracheal, bronchial, uterine, urinary smooth muscle, and ileum. 5-HT_{2A}R exerts also a mitogen activity with the ability of 5-HT_{2A}R agonists to induce smooth muscle cell growth and to potentiate the mitogenic activity of other growth factors.

In the CNS, 5-HT_{2A}Rs are abundant in the cerebral cortex, in insula, and in some nuclei of the brainstem and the limbic system and are involved in a number of psychiatric disorders, including schizophrenia, depression, and anxiety. 5-HT_{2A}Rs mediate the effects of hallucinogens and are the target of a number of commonly prescribed medications including atypical antipsychotics, antidepressants and anxiolytics. Indeed, 5-HT_{2A}R blockade has been claimed to contribute to the efficacy of atypical antipsychotics to reduce both positive and negative symptoms of schizophrenia.[59]

3.1.1 5-HT_{2A}R Internalization

On activation by exposure to their agonist, 5-HT_{2A}R increases IP3 levels and undergoes desensitization and internalization in stably transfected cell lines such as NIH-3T3, HEK293, or C6 glioma.[60–63] Both agonist- and antagonist-induced desensitization of the 5-HT_{2A}R involves receptor internalization through a clathrin- and dynamin-dependent process with a dual mechanism of early and late desensitization by the antagonist ketanserin.[60] Internalized receptors also recycle to the surface, thus participating to a resensitization.[63] Several atypical antipsychotic drugs with high 5-HT_{2A}R affinities induce a redistribution of 5-HT_{2A}R both *in vitro and in vivo,* causing a decrease in labeling of apical dendrites in the medial prefrontal cortex. It is conceivable that the loss of 5-HT_{2A}R from the apical dendrites of pyramidal neurons is important for the beneficial effects of atypical antipsychotic drugs and other 5-HT_{2A}R antagonists in schizophrenia.[64] This internalization was shown to be β-arrestin dependent in rat C6 glioma cells[65] or in rat cortical neurons,[66] whereas it was arrestin independent in HEK cells.[67] Differences in trafficking between rat and human 5-HT_{2A}R led to identification of a primate-specific tripeptide ASK motif in the C-terminus that confers GRK-2 and β-arrestin interactions and slower recycling kinetics to the human 5-HT_{2A}R in comparison to rat 5-HT_{2A}R.[68]

3.1.2 Scaffolding Proteins and 5-HT_{2A}R

At the end of the C-terminus, the 5-HT_{2A}R possesses a canonical Type I PDZ-binding domain (CSV) that was shown to bind directly to postsynaptic density-95 (PSD-95), which promotes 5-HT_{2A}R clustering on the plasma membrane and signal transduction. The augmentation of 5-HT_{2A}R signaling by PSD-95 was not accompanied by alterations in the kinetics of 5-HT_{2A}R desensitization but was associated with the inhibition of agonist-induced 5-HT_{2A}R internalization in HEK-293 cells.[69]

This PDZ-binding domain of the 5-HT_{2A}R represents a necessary but not sufficient signal for the selective targeting of 5-HT_{2A}R to dendrites in pyramidal neurons from dissociated cortical cultures.[70]

In vivo disruption of 5-HT_{2A}R/PSD95 interaction, using a cell-penetrating peptidyl mimetic 5-HT_{2A}R C-terminus, induces an anti-hyperalgesic effect in diabetic neuropathic rats.[71] Indeed in persistent neuropathic pain, the analgesic effects of SSRIs are strongly dependent of spinal 5-HT_{2A}R activation that act by suppressing allodynia and mediate antinociceptive actions. Hence, disruption of 5-HT_{2A}R/PSD95 interaction might be a valuable strategy to design novel treatments for neuropathic pain and to increase the effectiveness of SSRIs.[71] The same peptide was also shown to relieve mechanical hyperalgesia in rats suffering a subchronic inflammatory pain induced by Carrageenan.[72]

3.1.3 *Caveolin and 5-HT_{2A}R: Signaling and Trafficking in Lipid Microdomains*

Endogenous 5-HT_{2A}R coimmunoprecipitate with caveolin-1 (Cav-1), a scaffolding protein enriched in caveolae, in preparations of C6 glioma cells or rat brain synaptic membranes. Moreover, it has been shown that 5-HT_{2A}R/Cav-1 association promotes the binding of the 5-HT_{2A}R with G_q at the plasma membrane.[73] Interaction with Cav-1 in C6 glioma cells was required for 5-HT_{2A}R-mediated signal transduction as measured by calcium flux assays through G_q coupling. A growing body of evidence indicates that caveolae regulate many GPCR signaling cascades by partitioning GPCRs, heterotrimeric G proteins, and their various effectors in membrane microdomains. Indeed, phospholipase Cβ was shown to be enriched in caveolae that also complex the voltage-gated potassium channels (KV1.5). Therefore, caveolin interactions with 5-HT_{2A}R may scaffold the receptor with G_q and PLC in lipid rafts or caveolae to facilitate 5-HT-mediated signaling through lipid microdomain organization.[74]

Association of 5-HT_{2A}R with the ubiquitin ligase (c-Cbl) plays a role in the receptor recycling. Downregulation of c-Cbl by RNA interference blocked efficient recycling of 5-HT_{2A}R to the plasma membrane by trapping 5-HT_{2A}R in early sorting endosomes positive for antigen1 and Rab11.[75]

3.2 The 5-HT_{2B}R

5-HT_{2B}R expression, initially thought to be restricted to the stomach fundus,[76,77] is also present in the CNS, more specifically in the cortex, hippocampus amygdala and the cerebellum. At the periphery, the 5-HT_{2B}R is expressed in gastrointestinal and cardiovascular tissues, including stomach, gut, pulmonary smooth muscle, pulmonary and vascular endothelial cells, and cardiomyocytes. In CHO cells stably expressing human 5-HT_{2B}R, it exhibits the most dramatic degree of desensitization in comparison with

5-$HT_{2A}R$ and 5-$HT_{2C}R$. Prior exposure to 5-HT reduces subsequent response to 5-HT by 80%, with an extremely rapid time course (5 min).[78]

Serotonin 5-$HT_{2B}Rs$ are often coexpressed with 5-$HT_{1B}Rs$ in meningeal tissues and endothelial and smooth muscle cells, with an inhibitory effect of 5-$HT_{2B}R$ on 5-$HT_{1B}R$ signaling.[79] Upon coexpression, serotonin-induced internalization of 5-$HT_{2B}R$ is accelerated fivefold and becomes insensitive to a 5-$HT_{2B}R$ antagonist. In this context, 5-$HT_{2B}Rs$ do internalize in response to a 5-$HT_{1B}R$ agonist. In contrast, coexpression does not render 5-$HT_{1B}R$ internalization sensitive to a 5-$HT_{2B}R$ agonist.

Internalization of 5-$HT_{1B}R$ (expressed alone) is entirely clathrin independent and Cav dependent, whereas that of 5-$HT_{2B}R$ (expressed alone) is Cav1 independent and clathrin dependent. Upon coexpression, serotonin-induced 5-$HT_{2B}R$ internalization becomes partially Cav1 dependent, and serotonin-induced 5-$HT_{1B}R$ internalization entirely Cav1 independent, in a protein kinase Cε-dependent fashion. This asymmetric, agonist-dependent, cross-regulation of 5-$HT_{1B}R$ and 5-$HT_{2B}R$ internalizations reveals a probable cross talk between the two receptors, imposing an alternate internalization pathway for the 5-$HT_{2B}R$.[79] It has been shown that 5-$HT_{2B}R$ internalization depends on its C-terminus. Indeed, the truncated mutant R393X of the 5-$HT_{2B}R$, identified in a patient diagnosed with pulmonary hypertension, exhibits a lack of internalization upon agonist stimulation, as observed by confocal microscopy. The truncation of most of the C-terminus of the 5-$HT_{2B}R$ removes putative phosphorylation sites and thus PDZ-dependent intracellular trafficking motifs, likely explaining the absence of internalization.[80]

3.3 The 5-$HT_{2C}R$

The 5-$HT_{2C}Rs$ are mainly expressed in the CNS and are especially abundant in epithelial cells of the choroid plexus where they control cerebrospinal fluid production.[81] Lower levels of expression are observed in limbic areas, hippocampus amygdala, basal ganglia, and the mesocortical/mesolimbic pathways. 5-$HT_{2C}Rs$ are also present in hypothalamic proopiomelanocortin/cocaine amphetamine-regulated transcript neurons within the arcuate nucleus which regulates feeding behavior. Pharmacological investigations and knockout mouse models showed that 5-$HT_{2C}Rs$ are implicated in body weight regulation and obesity.[82] In addition, 5-$HT_{2C}Rs$ have been implicated in psychiatric and neurological diseases (depression, schizophrenia, autism and Parkinson's disease).[83] Selective 5-$HT_{2C}R$

antagonists are endowed with anxiolytic-like properties, and addiction-related behaviors might be alleviated by 5-HT_{2C}R stimulation, through the resulting inhibition of the firing of dopaminergic neurons in the ventral tegmental area.[84] The 5-HT_{2C}R is the only GPCR whose mRNA undergoes adenosine-to-inosine editing, leading to amino-acid substitutions within the second intracellular loop, and the generation of a great number of isoforms ranging from the unedited (INI) to the fully edited (VGV) one and exhibiting different regional distributions.

3.3.1 5-HT_{2C}R Internalization and Constitutive Activity

The 5-HT_{2C}R exhibits agonist-induced internalization. With partial agonists such as DOI or mCPP, only 50% internalization of the level reached by full agonist is observed. The inverse agonists (methoxygramine) and neutral antagonists (mianserin) fail to induce internalization[85] of the 5-HT_{2C}R, a difference with the 5-HT_{2A}R which was shown to internalize upon antagonist stimulation.[60]

The corticotrophin-releasing factor, CRF, acted through CRFR1 to enhance 5-HT_2R-mediated signaling and anxiety behaviors, thereby linking CRF-mediated stress responses to anxiety and depression. Activation of CRFR1 enhanced both 5-HT_{2A}R- and 5-HT_{2C}R-mediated inositol phosphate formation in HEK cells. In mice, preadministration of CRF into the prefrontal cortex enhanced 5-HT_2R-mediated anxiety behaviors in response to 2,5-dimethoxy-4-iodoamphetamine. CRFR1-mediated increase in 5-HT_{2C}R signaling was independent of the activity of second messenger-dependent protein kinases activated by either receptor. The molecular mechanism underlying the sensitization of 5-HT_{2C}R signaling by CRFR1 requires agonist-stimulated CRFR1 endocytosis and recycling, which resulted in increased cell surface expression of 5-HT_{2C}R and increased second messenger responses to 5-HT treatment. It was shown to depend on 5-HT_{2C}R recycling via rapid recycling endosomes by allowing the recruitment of internalized 5-HT_2R to the plasma membrane. This increase relied also on intact PDZ-binding motifs at the carboxyl-terminal tails of both CRFR1 and 5-HT_2Rs.[86]

Phosphatidylinositol 4,5-bisphosphate (PtdIns(4,5)P2) has been shown to be crucial for endocytosis of 5-HT_{2C}R.[87] Trafficking of the receptors to Rab5-positive early endosomes was completely abolished in the absence of PtdIns(4,5)P2, whereas ligand-induced interaction of 5-HT_{2C}R with β-arrestin-2 was unaffected by PtdIns(4,5)P2 depletion. This suggests that in the absence of PtdIns(4,5)P2, 5-HT_{2C}R move into clathrin-coated membrane structures, but these are not cleaved efficiently and hence cannot reach the early endosomal compartment.[87]

3.3.2 Impact of Editing in 5-HT$_{2C}$R Internalization and Constitutive Activity

Constitutive activity at G-protein-dependent signaling of the unedited (INI) isoform seems to correlate with a high level of agonist-independent internalization, which explains why the (INI) isoform (at least when expressed at high levels) is almost entirely intracellular, even before any stimulation by an agonist. The tetracyclic antidepressants mirtazapine and mianserin, but not other clinically established antidepressants, suppress constitutive activity at recombinant and native 5-HT$_{2C}$R.[88]

5-HT$_{2C\text{-}INI}$Rs also activate the extracellular signal-regulated kinase (ERK) 1/2 pathway, independently of receptor coupling to G proteins.[89] In contrast, no increase in ERK1/2 phosphorylation was measured in cells expressing fully edited (5-HT$_{2C\text{-}VGV}$) receptors. Basal activity in HEK 293 cells was unaffected by cellular depletion of G$\alpha_{q/11}$ and Gα_{13} proteins, but strongly reduced in cells depleted of β-arrestin and in cells expressing a dominant-negative CaM or a 5-HT$_{2C\text{-}INI\text{-}R376/377/A}$ receptor mutant not capable of interacting with CaM.[90] The antidepressants mirtazapine and mianserin likewise reduce basal ERK activation and constitutive activity of the 5-HT$_{2C}$R, probably as a result of their inverse agonist effect on β-arrestin recruitment. CaM was identified as an essential protein contributing to both recruitment of β-arrestin 2 by 5-HT$_{2C}$R and receptor-mediated ERK1,2 signaling, independently of activation of heterotrimeric G proteins. CaM binds to a prototypic Ca^{2+}-dependent CaM-binding motif located in the proximal region of the 5-HT$_{2C}$R C-terminus upon receptor activation by 5-HT. Mutation of this motif inhibited both β-arrestin recruitment by 5-HT$_{2C}$R and receptor-operated ERK 1,2 signaling in HEK cells, which was independent of G proteins and dependent on β-arrestins.[89]

In contrast, fully edited receptors, 5-HT$_{2C\text{-}VGV}$, which have little or no constitutive activity, are largely membrane-bound, while partially edited receptors, 5-HT$_{2C\text{-}vsv}$ which display an intermediate level of constitutive activity, are detected both at the cell surface and in intracellular vesicles.[91] 5-HT$_{2C}$R editing results in a decrease in agonist potency and/or efficacy to activate G-protein-mediated signaling, and this effect is roughly proportional to the extent of editing. Thus, the level of membrane localization, which is inversely correlated to the level of constitutive activity, depends on the level of 5-HT$_{2C}$R editing.

Similarly, the more extended editing is, the lower is 5-HT$_{2C}$R agonist-induced PLC signaling and downstream intracellular Ca^{2+} release.

The kinetics of Ca^{2+} release is also altered, with the peak Ca^{2+} concentration reached more slowly following stimulation of the VGV isoform versus the unedited INI receptor.[92] Another consequence of mRNA editing concerns the recruitment of β-arrestin1 and 2 by 5-HT$_{2C}$R variants. Indeed, the ability of variants to spontaneously associate with β-arrestin is strongly correlated with their degree of constitutive activity, which is directly dependent on the extent of editing. The nonedited 5-HT$_{2C\text{-}INI}$ receptor binds to β-arrestin in an agonist-independent manner, a process that results in constitutive receptor internalization and its predominant localization in intracellular compartments. Moreover, constitutive interaction with β-arrestin can be reversed by inverse agonists, which promote receptor redistribution to the plasma membrane. In contrast, the fully edited 5-HT$_{2C\text{-}VGV}$ isoform, which displays the lowest degree of constitutive activity, does not spontaneously associate with β-arrestin, is mainly localized at the cell surface under basal conditions, and only undergoes agonist-dependent endocytosis.[90]

3.3.3 Impact of Editing on Mice Behavior

Although INI mice grow normally, VGV mice have a severely reduced fat mass, despite compensatory hyperphagia, as a result of constitutive activation of the sympathetic nervous system and increased energy expenditure.[93] Mutant mice expressing only the fully edited VGV 5-HT$_{2C}$R consistently display enhanced motor responses to 5-HT$_{2C}$R ligands associated with an increased 5-HT$_{2C}$R density.[93] VGV mice also display increased anxiety in the elevated plus-maze and present increased energy expenditure and loss of fat mass associated with a decrease in cholesterol. These mice display aggressive impulsive behaviors not present in WT mice in the social interaction test, increased anxiogenic effect in response to a 5-HT$_{2C}$R agonist and a stronger freezing reaction to an innate fear stimulus.[94]

3.3.4 5-HT$_{2C}$R Dimerization with Ghrelin Receptor

A novel heterodimer between the ghrelin receptor, GHS-R1a, and the 5-HT$_{2C}$R was identified. Interestingly, dimerization of the GHS-R1a with the unedited 5-HT$_{2C\text{-}INI}$ receptor significantly reduced GHS-R1a agonist-mediated calcium influx, while dimerization with the partially edited 5-HT$_{2C\text{-}VSV}$ isoform did not change calcium influx. Accordingly, GHS-R1a-mediated calcium influx was completely restored following pharmacological blockade of the $5\text{-HT}_{2C}R$ and only in cells coexpressing the GHS-R1a receptor and the unedited $5\text{-HT}_{2C\text{-INI}}$ receptor.[95]

3.3.5 5-HT$_{2C}$R Interaction with Scaffolding Proteins

In addition to editing, interaction with scaffolding proteins controls also the $5\text{-HT}_{2C}R$ desensitization. The prototypic scaffolding protein PSD-95 and another membrane-associated guanylate kinase, MAGUK p55 subfamily member 3 (MPP3), oppositely regulate desensitization of the receptor response in both heterologous cells and mice cortical neurons in primary culture. PSD-95 increased desensitization of the $5\text{-HT}_{2C}R$-mediated Ca^{2+} response, whereas MPP3 prevented desensitization of the Ca^{2+} response.[96]

In conclusion, the $5\text{-HT}_{2C}R$ trafficking exhibits a high diversity in its internalization, desensitization, or signaling. This diversity relies on numerous interactions with proteins of the signaling cascade or with other receptors and also depend on the extent of edition of its mRNA.

4. TRAFFICKING OF THE 5-HT$_4$R

The 5-HT_4R was first identified by its positive coupling to cAMP production in colliculi and hippocampal neurons in culture.[97] 5-HT_4R is mainly present in limbic structures (hypothalamus, nucleus accumbens, amygdala, hippocampus, and septum), islands of Calleja, olfactory tubercle, and basal ganglia (striatum, globus pallidus, and substantia nigra). Rat, murine, and human 5-HT_4R are expressed as splice variants from a unique gene. For most of them, splicing gives rise to alternate C-terminal domains named a, b, c, after the same amino-acid residue (L358). All splice variants are able to activate adenylyl cyclase in transfected cell lines, albeit with different efficacy and potency. These differences could result from the coupling to different G proteins: G_s, G_{13}, G_q, and G_i.

4.1 5-HT$_4$R Internalization and Desensitization

The human 5-HT_{4b} isoform ($h5\text{-HT}_{4b}R$) couples to $G_{i/o}$-proteins in addition to its well-documented G_s coupling, whereas the 5-HT_{4a} receptor couples only to G_s-proteins.[98] These two splice variants exhibit differential internalizations: the agonist stimulation results in a time-dependent internalization of the $h5\text{-HT}_{4b}R$, but not of the $h5\text{-HT}_{4a}R$. The $h5\text{-HT}_{4b}R$ internalization is PTX insensitive, revealing that coupling to G is not necessary.[99]

For the murine 5-HT_4R (m5-HT_4R) also, differences in desensitization kinetics were observed between splice variants. Serotonin induced a rapid desensitization of the adenylyl cyclase response mediated by the m5-HT_{4d} receptor that desensitized with a faster rate than the one induced by the m5-HT_{4e} receptor in CHO cells.[100]

In addition to differences in internalization between spliced forms, differences in uncoupling were observed between endogenous 5-HT_4R expressed in colliculi and those expressed in COS-7 or HEK cells. 5-HT_4R undergoes rapid and profound homologous uncoupling in neurons. However, no significant uncoupling was observed in COS-7 or HEK293 cells, which expressed no the G-protein coupled receptor kinase GRK2 or a weak amount of it. GRK2 requirements for uncoupling and endocytosis are also very different. Indeed, β-arrestin/dynamin-dependent endocytosis was observed in HEK293 cells without any need of GRK2 overexpression. A cluster of serines and threonines, common to all variants, was an absolute requirement for β-arrestin/dynamin-dependent receptor endocytosis, but not for receptor uncoupling. Thus, uncoupling and endocytosis of 5-HT_4R require different GRK2 concentrations and involve distinct molecular events.[101]

4.2 5-HT_4R Interaction with p11 and Antidepressant Treatment

The protein p11 was identified as a 5-HT_4R-interacting protein by two-hybrid and coimmunoprecipitation studies. As it was shown for the 5-$HT_{1B}R$, p11 increases also 5-HT_4R surface expression in COS-7 cells. In addition, less 5-HT_4R expression is detected at the cell surface in the p11 KO mice. p11 facilitates 5-HT_4R signaling, since cotransfection of p11 with 5-HT_4R potentiates cAMP production more than in cells overexpressing 5-HT_4R alone. The 5-HT_4R partial agonist RS67333 has been shown to produce antidepressant-like effects in rodent models of depression. The behavioral antidepressant actions of this agonist were not efficient in p11 KO mice in three tests: the tail suspension test, the forced swim test, and the open-field test.[102] Since most p11-containing cells in cerebral cortex, hippocampus, cerebellum, and caudate–putamen contain also 5-$HT_{1B}R$ and/or 5-HT_4R,[103] it indicates a crucial role for p11 in modulating central actions of serotonin via these receptor subtypes.

5. TRAFFICKING OF THE 5-HT_6R

The 5-HT_6R was identified by cloning from a rat library, as a novel 5-HT receptor with high affinity for typical and atypical antipsychotics, including clozapine. This receptor is most prominently expressed in the caudate nucleus, the olfactory tubercle, the striatum, the hippocampus, and the nucleus accumbens.[104,105] 5-HT_6R activation stimulates AC activity by interacting with G_s-protein, thereby activating the downstream PKA pathway in various transfected cell lines, notably mouse neuroblastoma cell lines, as well as in mouse striatal neurons and pig caudate membranes.[106] Numerous antipsychotic and antidepressant drugs inhibit 5-HT_6R-stimulated adenylyl cyclase. A growing body of evidence supports the use of serotonin 5-HT_6R antagonists as a promising strategy for treating cognitive dysfunction.

5.1 5-HT_6R Localization at the Primary Cilium and Dendrites Outgrowth

In addition to a dendritic membrane localization in neurons of the striatum, nucleus accumbens, olfactory tubercle, and islands of Calleja of the rat brain, 5-HT_6R-like immunoreactivity has also been found in association with the primary cilium of neurons in the same brain regions.[107] The primary cilium is a solitary organelle projecting from the surface of neurons. This cilium lacks the central pair of microtubules needed to generate a motile force.[108] Growing evidence suggests that the primary cilium functions as a signaling center in neurons.[109] Primary cilia play critical roles in early embryonic development and organogenesis in vertebrates, by providing a unique cellular domain that facilitates signal transduction in response to morphogens and growth factors. In the primary cilia of neurons, 5-HT_6R labeling is associated with the membrane and not with microtubules.[107] The function of 5-HT_6R localized within the primary cilia seems associated with dendrite outgrowth.[110] Neuronal overexpression of 5-HT_6R in cortical neurons induces the formation of long and often forked cilia *in vivo* after electroporation in mouse embryo cortex *in utero*. The GPCR SSTR3 and Type III adenylyl cyclase (ACIII), proteins normally enriched in neuronal cilia, were rarely detected in 5HT_6-elongated cilia. The changes in cilia structure were accompanied by changes in neuronal morphology. Specifically, disruption of normal ciliogenesis in developing neocortical neurons, by overexpressing cilia 5-HT_6R, significantly impaired dendrite outgrowth.[110]

In a model of superciliated cells resulting from overexpression of the kinase Plk4,[111] 5-HT_6R localization was diluted into the two supernumerary primary cilia, which also exhibited reduced ciliary concentration of Smoothened in response to Sonic hedgehog stimulation and reduced Shh pathway transcriptional activation. This dilution of 5-HT_6R and reduction of Shh signaling results from partitioning of proteins over two cilia.[111]

5.2 5-HT_6R Interaction with MAP1B Protein

The human 5-HT_6R interacts directly with the C-terminus of the human light chain 1 subunit of MAP1B protein (MAP1B-LC1), a ubiquitous microtubule-associated protein highly expressed in the brain. MAP1B-LC1 is involved in the desensitization and trafficking of 5-HT_6R.[112] Coexpression of MAP1B-LC1 with 5-HT_6R regulates serotonin signaling by specifically controlling the coupling activities of 5-HT_6R, but not those of 5-HT_4R or 5-HT_7R. In HEK cells, cotransfection of $G\alpha_{15}$ with 5-HT_6R increases the association of $G\alpha_S$-coupled receptors with phospholipase C and subsequently increases intracellular Ca^{2+} release. Ectopic expression of MAP1B-LC1 significantly increases 5-HT_6R-induced intracellular Ca^{2+} release as well as 5-HT-induced cAMP level. MAP1B-LC1 overexpression in HEK cells enhances the surface expression of 5-HT_6R and decreases its endocytosis.[112]

6. TRAFFICKING OF THE 5-HT_7R

The 5-HT_7R was identified by cloning from kidney and brain tissues. Several splice variants have been described that differ in their C-terminal intracellular tail (variants are designated a, b and c or d in rat or human, respectively). These variants differ to some extent regarding pharmacological profiles but use similar signal transduction mechanisms and functional coupling.[113] In the CNS, the widespread distribution of 5-HT_7R is suggestive of multiple central roles. The 5-HT_7R is involved in phase-shifting of the circadian rhythm and age-dependent changes in circadian timing. It also plays a key role in the induction of sleep and thermoregulation.

The 5-HT_7R is involved in both central and peripheral endocrine-related controls. The 5-HT_7R mediates the stimulating effect of 5-HT on the release of both vasopressin and oxytocin, and 5-HT_7R agonists have been reported to inhibit the release of luteinizing hormone. In addition, in the adrenal gland, the 5-HT_7R has been shown to mediate 5-HT-induced aldosterone release.[114] Peripherally, 5-HT_7R are present on granulosa–lutein cells, where they stimulate progesterone production. 5-HT_7R also mediate smooth muscle relaxation in peripheral blood vessels (veins and arteries) pig oviduct,[115] human colonic circular smooth muscle, and cerebral arteries.[114]

6.1 Differential Internalization of 5-HT$_7$R Variants

Among human isoforms, the 5-HT$_{7d}$ isoform exhibits a pattern of receptor trafficking that differs from 5-HT$_{7a}$ or 5-HT$_{7b}$ isoforms in HEK cells, in response to agonists. Surface 5-HT$_{7d}$R are constitutively internalized in the absence of agonist. Moreover, the 5-HT$_{7d}$R displays this internalization in the presence of a 5-HT$_7$-specific antagonist. In addition, the human 5-HT$_{7d}$R shows a diminished efficacy in stimulation of cAMP-responsive reporter gene activity in transfected cells, compared to 5-HT$_{7a}$ or 5-HT$_{7b}$ receptors expressed at similar levels.[116]

6.2 Heterodimerization of 5-HT$_{1A}$R and 5-HT$_7$R in Signaling and Trafficking

5-HT$_{1A}$R and 5-HT$_7$R form heterodimers both *in vitro* and *in vivo.*[117] Functionally, heterodimerization decreases 5-HT$_{1A}$R-mediated activation of G$_i$ protein without affecting 5-HT$_7$R-mediated signaling. Moreover, heterodimerization markedly decreases the ability of the 5-HT$_{1A}$R to activate G-protein-gated inwardly rectifying potassium channels in oocytes. In neuronal cultures also, heterodimerization reduces the ability of endogenous 5-HT$_{1A}$R to activate potassium channels. 5-HT$_{1A}$R/5-HT$_7$R heterodimerization specifically attenuates the ability of 5-HT$_{1A}$R to activate G$_i$ protein, whereas 5-HT$_7$R-mediated activation of G$_s$ protein is not affected by the coexpression with 5-HT$_{1A}$R. Erk phosphorylation in cells coexpressing 5-HT$_{1A}$R and 5-HT$_7$R was continuously enhanced in comparison to cells expressing 5-HT$_{1A}$R alone, via activation of the 5-HT$_{1A}$ in the heterodimer.

Heterodimerization is crucially involved in initiation of the serotonin-mediated 5-HT$_{1A}$R internalization. No internalization of 5-HT$_{1A}$R, even with prolonged treatment with serotonin, could be visualized in NIE-115-transfected cells, in contrast to 5-HT$_7$R-expressing cells treated with serotonin. Cotransfection of 5-HT$_{1A}$R and 5-HT$_7$R led to pronounced agonist-mediated cointernalization of 5-HT$_{1A}$R, which was not blocked by the 5-HT$_{1A}$R antagonist WAY100635. By contrast, pharmacological blockade of 5-HT$_7$R with SB269970 completely abolished agonist-induced 5-HT$_{1A}$R. These results suggest that 5-HT$_7$R-mediated signaling is necessary for initiation of the cointernalization of 5-HT$_{1A}$R.[117]

7. CONCLUSION

In conclusion, besides the well-known implications of 5-HTRs in the CNS disorders via their stimulation and activation of multiple signaling pathways, their trafficking contributes to the diversity of drug action on serotonin GPCRs. Their trafficking implies agonist-dependent or -independent endocytosis then recycling to the plasma membrane or degradation. Moreover, endocytosis and recycling are important for regulating their desensitization and resensitization and for modulating their signaling via G-protein-independent signal transduction pathways. In addition, some 5-HTRs have been shown to internalize and/or change their neuronal localization[44] upon agonist stimulation but moreover upon antagonist stimulation and also after prolonged treatments.[36] These new data suggest that previously undescribed mechanisms may modify drug effects and increase the complexity of analyzing drug action during long-term treatments.

In addition, the targeting of 5-HTR to the plasma membrane or to specific subcompartments is essential for the specificity of their function and coupling. In this context, new roles have been assigned to recently identified partner proteins that regulate their addressing to the plasma membrane such as Yif1B for the $5\text{-HT}_{1A}R$[45] or p11 for the $5\text{-HT}_{1B}R$[52] and 5-HT_4R.[102] The anchoring of 5-HTRs into the plasma membrane via various PDZ proteins may influence their function as it has been shown for the $5\text{-HT}_{2C}R$ in which desensitization is increased by interaction with PSD-95 and decreased by interaction with MPP3. Additional investigations are required in order to integrate the role of these partner proteins into the function of these receptors and use them as targets for new drug treatments.

REFERENCES

1. Rapport MM, Green AA, Page IH. Crystalline serotonin. *Science*. 1948;108(2804): 329–330.
2. Amin AH, Crawford TBB, Gaddum JH. The distribution of substance P and serotonin in the central nervous system of the dog. *J Physiol Lond*. 1954;126:596–618.
3. Côté F, Thévenot E, Fligny C, et al. Disruption of the nonneuronal tph1 gene demonstrates the importance of peripheral serotonin in cardiac function. *Proc Natl Acad Sci USA*. 2003;100(23):13525–13530.
4. Hamon M, Bourgoin S, Artaud F, Glowinski J. The role of intraneuronal 5-HT and of tryptophan hydroxylase activation in the control of 5-HT synthesis in rat brain slices incubated in K^+-enriched medium. *J Neurochem*. 1979;33(5):1031–1042.
5. Kühn DM, Wolf WA, Youdim MBH. Serotonin neurochemistry revisited: a new look at some old axioms. *Neurochem Int*. 1986;8:141–154.

6. Erickson JD, Eiden LE, Hoffman BJ. Expression cloning of a reserpine-sensitive vesicular monoamine transporter. *Proc Natl Acad Sci USA*. 1992;89(22):10993–10997.
7. Liu Y, Peter D, Roghani A, et al. A cDNA that suppresses MPP+ toxicity encodes a vesicular amine transporter. *Cell*. 1992;70(4):539–551.
8. Schanberg SM. A study of the transport of 5-hydroxytryptophan and 5-hydroxytryptamine (serotonin) into brain. *J Pharmacol Exp Ther*. 1963;139:191–200.
9. Wong DT, Bymaster FP, Horng JS, Molloy BB. A new selective inhibitor for uptake of serotonin into synaptosomes of rat brain: 3-(p-trifluoromethylphenoxy). N-Methyl-3-phenylpropylamine. *J Pharmacol Exp Ther*. 1975;193(3):804–811.
10. Shaskan EG, Snyder SH. Kinetics of serotonin accumulation into slices from rat brain: relationship to catecholamine uptake. *J Pharmacol Exp Ther*. 1970;175(2):404–418.
11. Shore PA, Mead JA, Kuntzman RG, Spector S, Brodie BB. On the physiologic significance of monoamine oxidase in brain. *Science*. 1957;126(3282):1063–1064.
12. Jouvet M, Bobillier P, Pujol JF, Renault J. Suppression of sleep and decrease of cerebral serotonin caused by lesion of the raphe system in the cat. *C R Acad Sci Hebd Seances Acad Sci D*. 1967;264(2):360–362.
13. Brittain RT, Handley SL. Temperature changes produced by the injection of catecholamines and 5-hydroxytryptamine into the cerebral ventricles of the conscious mouse. *J Physiol*. 1967;192(3):805–813.
14. Fuxe K, Farnebo LO, Hamberger B, Ogren SO. On the in vivo and in vitro actions of fenfluramine and its derivatives on central monoamine neurons, especially 5-hydroxytryptamine neurons, and their relation to the anorectic activity of fenfluramine. *Postgrad Med J*. 1975;51(suppl 1):35–45.
15. Sparkes CG, Spencer PS. Antinociceptive activity of morphine after injection of biogenic amines in the cerebral ventricles of the conscious rat. *Br J Pharmacol*. 1971;42(2):230–241.
16. Graham JR. Methysergide for prevention of headache: experience in five hundred patients over three years. *N Engl J Med*. 1964;270:67–72.
17. Laguzzi R, Reis DJ, Talman WT. Modulation of cardiovascular and electrocortical activity through serotonergic mechanisms in the nucleus tractus solitarius of the rat. *Brain Res*. 1984;304(2):321–328.
18. Ahlenius S, Larsson K, Svensson L. Further evidence for an inhibitory role of central 5-HT in male rat sexual behavior. *Psychopharmacology (Berl)*. 1980;68(3):217–220.
19. Meyerson BJ, Lewander T. Serotonin synthesis inhibition and estrous behavior in female rats. *Life Sci I*. 1970;9(12):661–671.
20. Mabry PD, Campbell BA. Serotonergic inhibition of catecholamine-induced behavioral arousal. *Brain Res*. 1973;49(2):381–391.
21. Domeney AM, Costall B, Gerrard PA, Jones DN, Naylor RJ, Tyers MB. The effect of ondansetron on cognitive performance in the marmoset. *Pharmacol Biochem Behav*. 1991;38(1):169–175.
22. Thiebot MH. Are serotonergic neurons involved in the control of anxiety and in the anxiolytic activity of benzodiazepines? *Pharmacol Biochem Behav*. 1986;24(5):1471–1477.
23. Kasper S, Fuger J, Moller HJ. Comparative efficacy of antidepressants. *Drugs*. 1992;43-(suppl 2):11–22, discussion 22–13.
24. Peroutka SJ. Selective interaction of novel anxiolytics with 5-hydroxytryptamine1A receptors. *Biol Psychiatry*. 1985;20(9):971–979.
25. Meltzer HY. Role of serotonin in the action of atypical antipsychotic drugs. *Clin Neurosci*. 1995;3(2):64–75.
26. Roth BL, Craigo SC, Choudhary MS, et al. Binding of typical and atypical antipsychotic agents to 5-hydroxytryptamine-6 and 5-hydroxytryptamine-7 receptors. *J Pharmacol Exp Ther*. 1994;268(3):1403–1410.

27. Nelson DL, Herbet A, Bourgoin S, Glowinski J, Hamon M. Characteristics of central 5-HT receptors and their adaptive changes following intracerebral 5,7-dihydroxytryptamine administration in the rat. *Mol Pharmacol.* 1978;14(6):983–995.
28. Lanfumey L, Hamon M. Central 5-HT(1A) receptors: regional distribution and functional characteristics. *Nucl Med Biol.* 2000;27(5):429–435.
29. Gray L, Scarr E, Dean B. Serotonin 1a receptor and associated G-protein activation in schizophrenia and bipolar disorder. *Psychiatry Res.* 2006;143(2–3):111–120.
30. Le Poul E, Laaris N, Doucet E, Laporte AM, Hamon M, Lanfumey L. Early desensitization of somato-dendritic 5-HT1A autoreceptors in rats treated with fluoxetine or paroxetine. *Naunyn-Schmiedebergs Arch Pharmacol.* 1995;352(2):141–148.
31. Riad M, Watkins KC, Doucet E, Hamon M, Descarries L. Agonist-induced internalization of serotonin-1a receptors in the dorsal raphe nucleus (autoreceptors) but not hippocampus (heteroreceptors). *J Neurosci.* 2001;21(21):8378–8386.
32. Riad M, Zimmer L, Rbah L, Watkins KC, Hamon M, Descarries L. Acute treatment with the antidepressant fluoxetine internalizes 5-HT1A autoreceptors and reduces the in vivo binding of the PET radioligand [18F]MPPF in the nucleus raphe dorsalis of rat. *J Neurosci.* 2004;24(23):5420–5426.
33. Sibon I, Benkelfat C, Gravel P, et al. Decreased [18F]MPPF binding potential in the dorsal raphe nucleus after a single oral dose of fluoxetine: a positron-emission tomography study in healthy volunteers. *Biol Psychiatry.* 2008;63(12):1135–1140.
34. Della Rocca GJ, Mukhin YV, Garnovskaya MN, et al. Serotonin 5-HT1A receptor-mediated Erk activation requires calcium/calmodulin-dependent receptor endocytosis. *J Biol Chem.* 1999;274(8):4749–4753.
35. Renner U, Glebov K, Lang T, et al. Localization of the mouse 5-hydroxytryptamine(1A) receptor in lipid microdomains depends on its palmitoylation and is involved in receptor-mediated signaling. *Mol Pharmacol.* 2007;72(3):502–513.
36. Bouaziz E, Emerit MB, Vodjdani G, et al. Neuronal phenotype dependency of agonist-induced internalization of the 5-HT1A serotonin receptor. *J Neurosci.* 2014;34(1):282–294.
37. Kurisu J, Fukuda T, Yokoyama S, Hirano T, Kengaku M. Polarized targeting of DNER into dendritic plasma membrane in hippocampal neurons depends on endocytosis. *J Neurochem.* 2010;113(6):1598–1610.
38. Miquel MC, Doucet E, Boni C, et al. Central serotonin$_{1A}$ receptors: respective distribution of encoding mRNA, receptor protein and binding sites by *in situ* hybridization histochemistry, radioimmunohistochemistry and autoradiographic mapping in the rat brain. *Neurochem Int.* 1991;19:453–465.
39. Boschert U, Amara DA, Segu L, Hen R. The mouse 5-hydroxytryptamine1B receptor is localized predominantly on axon terminals. *Neuroscience.* 1994;58(1):167–182.
40. Langlois X, ElMestikawy S, Arpin M, Triller A, Hamon M, Darmon M. Differential addressing of 5-HT1A and 5-HT1B receptors in transfected LLC-PK1 epithelial cells: a model of receptor targeting in neurons. *Neuroscience.* 1996;74(2):297–302.
41. Ghavami A, Stark KL, Jareb M, Ramboz S, Segu L, Hen R. Differential addressing of 5-HT1A and 5-HT1B receptors in epithelial cells and neurons. *J Cell Sci.* 1999;112(Pt 6):967–976.
42. Darmon M, Langlois X, Suffisseau L, Fattaccini CM, Hamon M. Differential membrane targeting and pharmacological characterization of chimeras of rat serotonin 5-HT1A and 5-HT1B receptors expressed in epithelial LLC-PK1 cells. *J Neurochem.* 1998;71(6):2294–2303.
43. Jolimay N, Franck L, Langlois X, Hamon M, Darmon M. Dominant role of the cytosolic C-terminal domain of the rat 5-HT$_{1B}$ receptor in axonal-apical targeting. *J Neurosci.* 2000;20(24):9111–9118.

44. Carrel D, Simon A, Emerit MB, et al. Axonal targeting of the 5-HT1B serotonin receptor relies on structure-specific constitutive activation. *Traffic.* 2011; 12(11):1501–1520.
45. Carrel D, Masson J, Al Awabdh S, et al. Targeting of the 5-HT1A serotonin receptor to neuronal dendrites is mediated by Yif1B. *J Neurosci.* 2008;28(32):8063–8073.
46. Matern H, Yang X, Andrulis E, Sternglanz R, Trepte HH, Gallwitz D. A novel Golgi membrane protein is part of a GTPase-binding protein complex involved in vesicle targeting. *EMBO J.* 2000;19(17):4485–4492.
47. Al Awabdh S, Miserey-Lenkei S, Bouceba T, et al. A new vesicular scaffolding complex mediates the G-protein-coupled 5-HT1A receptor targeting to neuronal dendrites. *J Neurosci.* 2012;32(41):14227–14241.
48. Hannon J, Hoyer D. Molecular biology of 5-HT receptors. *Behav Brain Res.* 2008;195(1):198–213.
49. Malleret G, Hen R, Guillou JL, Segu L, Buhot MC. 5-HT1B receptor knock-out mice exhibit increased exploratory activity and enhanced spatial memory performance in the Morris water maze. *J Neurosci.* 1999;19(14):6157–6168.
50. Leterrier C, Laine J, Darmon M, Boudin H, Rossier J, Lenkei Z. Constitutive activation drives compartment-selective endocytosis and axonal targeting of type 1 cannabinoid receptors. *J Neurosci.* 2006;26(12):3141–3153.
51. Hauri HP, Matter K. Protein traffic in intestinal epithelial cells. *Semin Cell Biol.* 1991;2(6):355–364.
52. Svenningsson P, Chergui K, Rachleff I, et al. Alterations in 5-HT_{1B} receptor function by p11 in depression-like states. *Science.* 2006;311(5757):77–80.
53. Donato R. Functional roles of S100 proteins, calcium-binding proteins of the EF-hand type. *Biochim Biophys Acta.* 1999;1450(3):191–231.
54. Zhang X, Andren PE, Greengard P, Svenningsson P. Evidence for a role of the 5-HT1B receptor and its adaptor protein, p11, in L-DOPA treatment of an animal model of Parkinsonism. *Proc Natl Acad Sci USA.* 2008;105(6):2163–2168.
55. Eriksson TM, Alvarsson A, Stan TL, et al. Bidirectional regulation of emotional memory by 5-HT1B receptors involves hippocampal p11. *Mol Psychiatry.* 2013;18(10):1096–1105.
56. Lee SP, Xie Z, Varghese G, Nguyen T, O'Dowd BF, George SR. Oligomerization of dopamine and serotonin receptors. *Neuropsychopharmacology.* 2000;23(4 suppl): S32–S40.
57. Salim K, Fenton T, Bacha J, et al. Oligomerization of G-protein-coupled receptors shown by selective co-immunoprecipitation. *J Biol Chem.* 2002;277(18):15482–15485.
58. Conn PJ, Sanders-Bush E. Selective 5HT-2 antagonists inhibit serotonin stimulated phosphatidylinositol metabolism in cerebral cortex. *Neuropharmacology.* 1984;23(8): 993–996.
59. Masson J, Emerit MB, Hamon M, Darmon M. Serotonergic signaling: multiple effectors and pleiotropic effects. *WIREs Membr Transp Signal.* 2012;1(6):685–713.
60. Hanley NR, Hensler JG. Mechanisms of ligand-induced desensitization of the 5-hydroxytryptamine(2A) receptor. *J Pharmacol Exp Ther.* 2002;300(2):468–477.
61. Willins DL, Alsayegh L, Berry SA, et al. Serotonergic antagonist effects on trafficking of serotonin 5-HT2A receptors in vitro and in vivo. *Ann N Y Acad Sci.* 1998;861:121–127.
62. Berry SA, Shah MC, Khan N, Roth BL. Rapid agonist-induced internalization of the 5-hydroxytryptamine2A receptor occurs via the endosome pathway in vitro. *Mol Pharmacol.* 1996;50(2):306–313.
63. Bhattacharyya S, Puri S, Miledi R, Panicker MM. Internalization and recycling of 5-HT2A receptors activated by serotonin and protein kinase C-mediated mechanisms. *Proc Natl Acad Sci USA.* 2002;99(22):14470–14475.

64. Willins DL, Berry SA, Alsayegh L, et al. Clozapine and other 5-hydroxytryptamine-2A receptor antagonists alter the subcellular distribution of 5-hydroxytryptamine-2A receptors in vitro and in vivo. *Neuroscience*. 1999;91(2):599–606.
65. Gelber EI, Kroeze WK, Willins DL, et al. Structure and function of the third intracellular loop of the 5-hydroxytryptamine2A receptor: the third intracellular loop is alpha-helical and binds purified arrestins. *J Neurochem*. 1999;72(5):2206–2214.
66. Schmid CL, Raehal KM, Bohn LM. Agonist-directed signaling of the serotonin 2A receptor depends on beta-arrestin-2 interactions in vivo. *Proc Natl Acad Sci USA*. 2008;105(3):1079–1084.
67. Bhatnagar A, Willins DL, Gray JA, Woods J, Benovic JL, Roth BL. The dynamin-dependent, arrestin-independent internalization of 5-hydroxytryptamine 2A (5-HT2A) serotonin receptors reveals differential sorting of arrestins and 5-HT2A receptors during endocytosis. *J Biol Chem*. 2001;276(11):8269–8277.
68. Bhattacharya A, Sankar S, Panicker MM. Differences in the C-terminus contribute to variations in trafficking between rat and human 5-HT(2A) receptor isoforms: identification of a primate-specific tripeptide ASK motif that confers GRK-2 and beta arrestin-2 interactions. *J Neurochem*. 2010;112(3):723–732.
69. Xia Z, Gray JA, Compton-Toth BA, Roth BL. A direct interaction of PSD-95 with 5-HT2A serotonin receptors regulates receptor trafficking and signal transduction. *J Biol Chem*. 2003;278(24):21901–21908.
70. Xia Z, Hufeisen SJ, Gray JA, Roth BL. The PDZ-binding domain is essential for the dendritic targeting of 5-HT2A serotonin receptors in cortical pyramidal neurons in vitro. *Neuroscience*. 2003;122(4):907–920.
71. Pichon X, Wattiez AS, Becamel C, et al. Disrupting 5-HT(2A) receptor/PDZ protein interactions reduces hyperalgesia and enhances SSRI efficacy in neuropathic pain. *Mol Ther*. 2010;18(8):1462–1470.
72. Wattiez AS, Pichon X, Dupuis A, et al. Disruption of 5-HT2A receptor-PDZ protein interactions alleviates mechanical hypersensitivity in carrageenan-induced inflammation in rats. *PLoS One*. 2013;8(9):e74661.
73. Bhatnagar A, Sheffler DJ, Kroeze WK, Compton-Toth B, Roth BL. Caveolin-1 interacts with 5-HT2A serotonin receptors and profoundly modulates the signaling of selected Galphaq-coupled protein receptors. *J Biol Chem*. 2004;279(33):34614–34623.
74. Allen JA, Yadav PN, Roth BL. Insights into the regulation of 5-HT2A serotonin receptors by scaffolding proteins and kinases. *Neuropharmacology*. 2008;55(6):961–968.
75. Baldys A, Raymond JR. Role of c-Cbl carboxyl terminus in serotonin 5-HT2A receptor recycling and resensitization. *J Biol Chem*. 2011;286(28):24656–24665.
76. Foguet M, Hartikka JA, Schmuck K, Lübbert H. Long-term regulation of serotonergic activity in the rat via activation of protein kinase A. *EMBO J*. 1993;12:903–910.
77. Kursar JD, Nelson DL, Wainscott DB, Cohen ML, Baez M. Molecular cloning, functional expression, and pharmacological characterization of a novel serotonin receptor (5-hydroxytryptamine2F) from rat stomach fundus. *Mol Pharmacol*. 1992;42(4):549–557.
78. Porter RH, Malcolm CS, Allen NH, Lamb H, Revell DF, Sheardown MJ. Agonist-induced functional desensitization of recombinant human 5-HT2 receptors expressed in CHO-K1 cells. *Biochem Pharmacol*. 2001;62(4):431–438.
79. Janoshazi A, Deraet M, Callebert J, et al. Modified receptor internalization upon coexpression of 5-HT1B receptor and 5-HT2B receptors. *Mol Pharmacol*. 2007;71(6):1463–1474.
80. Deraet M, Manivet P, Janoshazi A, et al. The natural mutation encoding a C terminus-truncated 5-hydroxytryptamine 2B receptor is a gain of proliferative functions. *Mol Pharmacol*. 2005;67(4):983–991.

81. Sanders-Bush E, Breeding M. Putative selective 5-HT-2 antagonists block serotonin 5-HT-1c receptors in the choroid plexus. *J Pharmacol Exp Ther.* 1988;247(1):169–173.
82. Miller KJ. Serotonin 5-ht2c receptor agonists: potential for the treatment of obesity. *Mol Interv.* 2005;5(5):282–291.
83. Rosenzweig-Lipson S. New horizons for selective 5-HT2C receptor ligands in psychiatric/neurological disorders. *Neuropsychopharmacology.* 2011;36(1):363–364.
84. Masson J, Bouaziz E, Emerit MB, Hamon M, Darmon M. Brain region-dependent internalisation of 5-HT1A receptors: molecular mechanisms and relevance for antidepressant therapy. *Eur Neuropsychopharmacol.* 2012;22:S128.
85. Schlag BD, Lou Z, Fennell M, Dunlop J. Ligand dependency of 5-hydroxytryptamine 2C receptor internalization. *J Pharmacol Exp Ther.* 2004;310(3):865–870.
86. Magalhaes AC, Holmes KD, Dale LB, et al. CRF receptor 1 regulates anxiety behavior via sensitization of 5-HT2 receptor signaling. *Nat Neurosci.* 2010;13(5):622–629.
87. Toth DJ, Toth JT, Gulyas G, et al. Acute depletion of plasma membrane phosphatidylinositol 4,5-bisphosphate impairs specific steps in endocytosis of the G-protein-coupled receptor. *J Cell Sci.* 2012;125(pt 9):2185–2197.
88. Chanrion B, Mannoury la Cour C, Gavarini S, et al. Inverse agonist and neutral antagonist actions of antidepressants at recombinant and native 5-hydroxytryptamine2C receptors: differential modulation of cell surface expression and signal transduction. *Mol Pharmacol.* 2008;73(3):748–757.
89. Labasque M, Reiter E, Becamel C, Bockaert J, Marin P. Physical interaction of calmodulin with the 5-HT2C receptor C-terminus is essential for G protein-independent, arrestin-dependent, receptor signaling. *Mol Biol Cell.* 2008;19:4640–4650.
90. Labasque M, Meffre J, Carrat G, Becamel C, Bockaert J, Marin P. Constitutive activity of serotonin 2C receptors at G protein-independent signaling: modulation by RNA editing and antidepressants. *Mol Pharmacol.* 2010;78(5):818–826.
91. Marion S, Weiner DM, Caron MG. RNA editing induces variation in desensitization and trafficking of 5-hydroxytryptamine 2c receptor isoforms. *J Biol Chem.* 2004; 279(4):2945–2954.
92. Werry TD, Loiacono R, Sexton PM, Christopoulos A. RNA editing of the serotonin 5HT2C receptor and its effects on cell signalling, pharmacology and brain function. *Pharmacol Ther.* 2008;119(1):7–23.
93. Kawahara Y, Grimberg A, Teegarden S, et al. Dysregulated editing of serotonin 2C receptor mRNAs results in energy dissipation and loss of fat mass. *J Neurosci.* 2008;28(48):12834–12844.
94. Martin CB, Ramond F, Farrington DT, et al. RNA splicing and editing modulation of 5-HT(2C) receptor function: relevance to anxiety and aggression in VGV mice. *Mol Psychiatry.* 2013;18(6):656–665.
95. Schellekens H, van Oeffelen WE, Dinan TG, Cryan JF. Promiscuous dimerization of the growth hormone secretagogue receptor (GHS-R1a) attenuates ghrelin-mediated signaling. *J Biol Chem.* 2013;288(1):181–191.
96. Gavarini S, Becamel C, Altier C, et al. Opposite effects of PSD-95 and MPP3 PDZ proteins on serotonin 5-hydroxytryptamine2C receptor desensitization and membrane stability. *Mol Biol Cell.* 2006;17(11):4619–4631.
97. Dumuis A, Bouhelal R, Sebben M, Cory R, Bockaert J. A nonclassical 5-hydroxytryptamine receptor positively coupled with adenylate cyclase in the central nervous system. *Mol Pharmacol.* 1988;34(6):880–887.
98. Pindon A, van Hecke G, van Gompel P, Lesage AS, Leysen JE, Jurzak M. Differences in signal transduction of two 5-HT4 receptor splice variants: compound specificity and dual coupling with Galphas- and Galphai/o-proteins. *Mol Pharmacol.* 2002; 61(1):85–96.

99. Pindon A, Van Hecke G, Josson K, et al. Internalization of human 5-HT4a and 5-HT4b receptors is splice variant dependent. *Biosci Rep.* 2004;24(3):215–223.
100. Mialet J, Fischmeister R, Lezoualc'h F. Characterization of human 5-HT4(d) receptor desensitization in CHO cells. *Br J Pharmacol.* 2003;138(3):445–452.
101. Barthet G, Gaven F, Framery B, et al. Uncoupling and endocytosis of 5-hydroxytryptamine 4 receptors. Distinct molecular events with different GRK2 requirements. *J Biol Chem.* 2005;280(30):27924–27934.
102. Warner-Schmidt JL, Flajolet M, Maller A, et al. Role of p11 in cellular and behavioral effects of 5-HT4 receptor stimulation. *J Neurosci.* 2009;29(6):1937–1946.
103. Egeland M, Warner-Schmidt J, Greengard P, Svenningsson P. Co-expression of serotonin 5-HT(1B) and 5-HT(4) receptors in p11 containing cells in cerebral cortex, hippocampus, caudate-putamen and cerebellum. *Neuropharmacology.* 2011;61(3):442–450.
104. Ruat M, Traiffort E, Arrang JM, et al. A novel rat serotonin (5-HT6) receptor: molecular cloning, localization and stimulation of cAMP accumulation. *Biochem Biophys Res Commun.* 1993;193(1):268–276.
105. Gerard C, Martres MP, Lefevre K, et al. Immuno-localization of serotonin 5-HT6 receptor-like material in the rat central nervous system. *Brain Res.* 1997; 746(1–2):207–219.
106. Sleight AJ, Boess FG, Bos M, Bourson A. The putative 5-ht6 receptor: localization and function. *Ann N Y Acad Sci.* 1998;861:91–96.
107. Brailov I, Bancila M, Brisorgueil MJ, Miquel MC, Hamon M, Verge D. Localization of 5-HT(6) receptors at the plasma membrane of neuronal cilia in the rat brain. *Brain Res.* 2000;872(1–2):271–275.
108. Davenport JR, Yoder BK. An incredible decade for the primary cilium: a look at a once-forgotten organelle. *Am J Physiol Renal Physiol.* 2005;289(6):F1159–F1169.
109. Ikeuchi Y, de la Torre-Ubieta L, Matsuda T, Steen H, Okazawa H, Bonni A. The XLID protein PQBP1 and the GTPase Dynamin 2 define a signaling link that orchestrates ciliary morphogenesis in postmitotic neurons. *Cell Rep.* 2013;4(5):879–889.
110. Guadiana SM, Semple-Rowland S, Daroszewski D, et al. Arborization of dendrites by developing neocortical neurons is dependent on primary cilia and type 3 adenylyl cyclase. *J Neurosci.* 2013;33(6):2626–2638.
111. Mahjoub MR, Stearns T. Supernumerary centrosomes nucleate extra cilia and compromise primary cilium signaling. *Curr Biol.* 2012;22(17):1628–1634.
112. Kim SH, Kim DH, Lee KH, et al. Direct interaction and functional coupling between human 5-HT6 receptor and the light chain 1 subunit of the microtubule-associated protein 1B (MAP1B-LC1). *PLoS One.* 2014;9(3):e91402.
113. Vanhoenacker P, Haegeman G, Leysen JE. 5-HT7 receptors: current knowledge and future prospects. *Trends Pharmacol Sci.* 2000;21(2):70–77.
114. Hedlund PB, Sutcliffe JG. Functional, molecular and pharmacological advances in 5-HT7 receptor research. *Trends Pharmacol Sci.* 2004;25(9):481–486.
115. Inoue M, Kitazawa T, Cao J, Taneike T. 5-HT7 receptor-mediated relaxation of the oviduct in nonpregnant proestrus pigs. *Eur J Pharmacol.* 2003;461(2–3):207–218.
116. Guthrie CR, Murray AT, Franklin AA, Hamblin MW. Differential agonist-mediated internalization of the human 5-hydroxytryptamine 7 receptor isoforms. *J Pharmacol Exp Ther.* 2005;313(3):1003–1010.
117. Renner U, Zeug A, Woehler A, et al. Heterodimerization of serotonin receptors 5-HT1A and 5-HT7 differentially regulates receptor signalling and trafficking. *J Cell Sci.* 2012;125(pt 10):2486–2499.

CHAPTER SIX

Calcium-Sensing Receptor: Trafficking, Endocytosis, Recycling, and Importance of Interacting Proteins

Kausik Ray[1]
Scientific Review Branch, NIDCD, National Institutes of Health, Bethesda, MD, USA
[1]Corresponding author: e-mail address: rayk@nidcd.nih.gov

Contents

Abstract

The cloning of the extracellular calcium-sensing receptor (CaSR) provided a new paradigm in G-protein-coupled receptor (GPCR) signaling in which principal physiological ligand is a cation, namely, extracellular calcium (Ca_o^{2+}). A wealth of information has accumulated in the past two decades about the CaSR's structure and function, its contribution to pathology in disorders of calcium in humans, and CaSR-based therapeutics. The CaSR unlike many other GPCRs must function in the presence of its ligand, thus understanding the mechanisms such as anterograde trafficking and endocytic pathways of this receptor are complex and fallen behind other classical GPCRs. Factors controlling CaSR signaling include various proteins affecting the expression of the CaSR as well as modulation of its trafficking to and from the cell surface. The dimeric cell-surface CaSR links to various heterotrimeric G-proteins ($G_{q/11}$, $G_{i/o}$, $G_{12/13}$, and G_s) to regulate intracellular second messengers, lipid kinases, various protein kinases, and transcription factors that are part of the machinery enabling the receptor to modulate the functions of the wide variety of cells in which it is expressed. This chapter describes key features of CaSR structure and function and discusses novel mechanisms by which the level of cell-surface receptor expression

Progress in Molecular Biology and Translational Science, Volume 132 2015 Published by Elsevier Inc.
ISSN 1877-1173
http://dx.doi.org/10.1016/bs.pmbts.2015.02.006

can be regulated including forward trafficking during biosynthesis, desensitization, internalization and recycling from the cell surface, and degradation. These processes are impacted by its interactions with several proteins in addition to signaling molecules *per se* (i.e., G-proteins, protein kinases, inositol phosphates, etc.) and include small molecular weight G-proteins (Sar1, Rabs, ARF, P24A, RAMPs, filamin A, 14-3-3 proteins, calmodulin, and caveolin-1). Moreover, CaSR signaling seems compartmentalized in cell-type-specific manner, and caveolin and filamin A likely act as scaffolds that bind signaling components and other key cellular elements (e.g., the cytoskeleton) to facilitate the interaction of the receptor with its signaling pathways. Regulatory mechanisms are still evolving to understand how defects in trafficking of CaSR contribute to pathology in disorders of calcium homeostasis.

ABBREVIATIONS

ECD extracellular ligand-binding domain
ELISA intact cell enzyme-linked immunoassay
ER endoplasmic reticulum
hCaSR human Ca^{+}-sensing or calcium receptor
HEK293 human embryonic kidney 293
N-linked asparagine-linked

1. INTRODUCTION

G-protein-coupled receptors (GPCRs) are the largest family of integral membrane proteins that couple via heterotrimeric G-proteins to a variety of effectors to modulate cellular function and are among the best-studied, functionally diverse family of cell-surface proteins requiring synthesis at the endoplasmic reticulum (ER), followed by interaction with the quality control system that target misfolded receptors to the proteosomal degradation pathway.[1,2] Successful passage through the quality control checkpoint is followed by chaperones and small GTP-binding protein-mediated anterograde or forward trafficking to the sites of action, generally at the plasma membrane and/or endocytic membranes, where GPCRs signal through arrestin-scaffolded complexes.[3] The extracellular calcium (Ca_o^{2+})-sensing or calcium receptor (CaSR) belongs to the family-C/3 GPCR gene family consisting of eight metabotropic glutamate (mGluR1–8) receptors, two heterodimeric gamma-aminobutyric acid receptors, the calcium-sensing receptor (CaSR), three taste (T1R1–3) receptors, and a promiscuous L-alpha-amino acid receptor.[4] Like other family-C/3 GPCRs, human Ca^{+}-sensing or calcium receptor (hCaSR) contains a uniquely large extracellular amino-terminal ligand-binding domain (ECD) of 600 amino acids connected to a seven transmembrane helical domain (TMD), prototypical for all GPCRs responsible for G-protein activation and a relatively long carboxyl-terminal tail (C-tail). While the trafficking of rhodopsin-like family-A/1 GPCRs has been the subject of extensive investigation, cellular processing mechanisms of family-C/3 GPCRs remain largely unexplored.

This family of GPCRs is characterized by large ECDs, which bind the endogenous orthosteric agonists. Whereas the agonists of most GPCRs are typically polypeptides, amino acid metabolites, or other small biological molecules, agonists of the CaSR include Ca_o^{2+}, other divalent and trivalent cations including Sr^{2+}, Ba^{2+}, Co^{2+}, Fe^{2+}, Gd^{3+}, Al^{3+}, etc., as well as organic polycations and cationic peptides including polyarginine.[5] The CaSR is also positively modulated by L-amino acids and glutathione analogs and negatively modulated by protons and high ionic strength. In addition, the CaSR responds to various organic polycations including the polyamines spermine and spermidine, aminoglycoside antibiotics such as neomycin and gentamicin, and basic polypeptides including polylysine and polyarginine.[6,7,8] Despite this promiscuous pharmacology, the name "calcium-sensing receptor" remains appropriate, given the receptor's fundamental role in extracellular calcium homeostasis in human by controlling the rate of parathyroid hormone (PTH) secretion from the parathyroid gland and the rate of calcium reabsorption by the kidney. This physiological role of the receptor in calcium homeostasis and the relevance of CaSR mutations causing human disorders are revealed by the phenotype of murine CaSR by disruption of exon-5 of null mice.[9] The CaSR is well-suited to this role as it senses extracellular calcium at the millimolar (mM) concentrations found in most extracellular fluid compartments. The receptor is typically inactive at extracellular calcium level below 0.2 mM but active above threshold calcium levels that vary from around 0.5–2 mM, depending on the cell context and level of receptor expression. In human, the CaSR thus plays an essential role in maintaining systemic calcium homeostasis and is continuously exposed to extracellular calcium in serum yet remains highly sensitive to small changes in extracellular calcium necessary for the normal physiological regulation of PTH secretion.[5] Human calcium homeostasis disorders are primarily associated with inactivating mutations of the hCaSR for autosomal dominant loss-of-function disorders like familial hypocalciuric hypercalcemia (FHH) and neonatal severe hyperparathyroidism (NSHPT); in contrast, activating mutations are responsible for the gain-of-function disorder autosomal-dominant hypocalcemia (ADH). The calcium homeostasis system functions by mechanisms as follows—low extracellular calcium directly stimulates PTH synthesis and secretion by reducing the activity of the CaSR which induces $1,25(OH)_2D_3$ synthesis by PTH and decreases calcitonin (CT) secretion. A key component of this homeostatic system is the sensor CaSR, which detects even minute (e.g., 1–2%) deviations of Ca_o^{2+} from its normal level and triggers signaling motions that alter the functions of other elements of the system, such as kidney, intestine, and bone, so as to restore Ca_o^{2+} to normal. Modulation of the receptor thus can play a role in both calcium homeostasis and fluid regulation via the kidney. Interestingly, in a host of other tissues, including skin, gut, brain, intestine, bone, and breast, the receptor also regulates cellular processes such as secretion, differentiation, and gene expression.[10] The recent findings provide evidence of the role of CaSR on gastric function, intestinal fluid, and salt transport and its modulation.[11] The presence of CaSR is confirmed in all vertebrate classes by the high similarity in nucleotide sequences across all the vertebrate classes.[12] Others tetrapods (e.g., birds, amphibians, mammals, and reptiles) have evolved a CaSR-based fully orchestrated homeostatic system intended to maintain a nearly constant level of the extracellular ionized calcium concentration.[13,14] More unexpected, perhaps, is the expression of a gene homologous to CaSR in animals that do not have parathyroid glands, such as teleost, elasmobranch fish, and dogfish shark.[15,16] The CaSR is expressed in several segments of the shark kidney tubule and, importantly, in many osmoregulatory segments, including rectal gland, intestine, stomach, olfactory epithelium, and gill chloride cells. This hormone-like first messenger property of extracellular calcium on CaSR is unique and contrasts with the role of calcium as a key intracellular signaling molecule.

2. CaSR STRUCTURE AND FUNCTION

The CaSR possesses a large amino-terminal ECD which represents a "Venus flytrap-like" (VFT) structure consisting of lobe 1 and lobe 2 of the ECD and contains disulfide-linked dimerization sites for two molecules of CaSR monomers and orthosteric sites for binding ligand. The VFT domain is linked to the canonical GPCR heptahelical seven TMD by a short globular cysteine-rich linker (Fig. 1). The CaSR responds over a narrower range of Ca_o^{2+} than would be anticipated for a protein with a single binding site for calcium. This positive cooperativity likely results from the CaSR

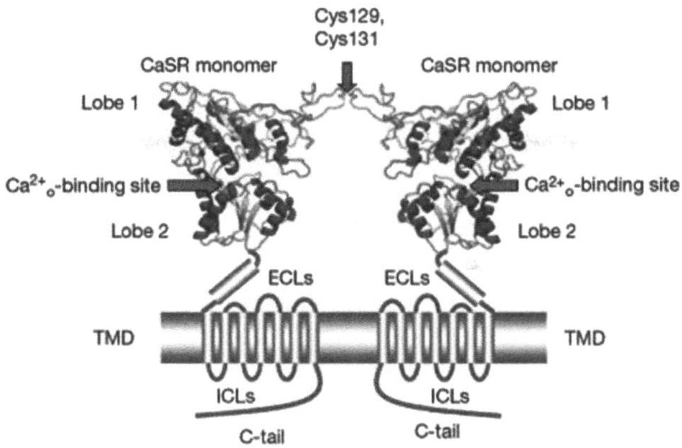

Figure 1 Schematic representation of the dimeric structure of the CaSR based on the known 3D crystal structure of the ECD of mGluR1. The dimer of the ECD is linked by two intermolecular disulfide linkages between cysteine 129-cysteine 129 and cysteine 131-cysteine 131, with each monomer assuming a Venus flytrap-like (VFT) conformation by lobe 1 and lobe 2. Each monomer is assumed to have at least one binding site for calcium in the crevice between the two lobes. Additional binding sites for calcium are likely present elsewhere in the ECD and TMD crevices. Calcimimetics, in contrast, bind to a site within the TMD, with the amino group of the drug in the linker between the two hydrophobic ends anchored to Glu837. The ICL and the long C-tail contain binding sites and determinants for binding of several intracellular proteins. ECL, extracellular loop; TMD, transmembrane domain; ICL, intracellular loop; C-tail, carboxyl-terminal tail. *Reproduced with copyright permission from Encyclopedia of Biological Chemistry; license content authors—W.A. Cramer, E. Yamashita, D. Baniulis, J. Whitelegge, S.S. Hasan.* (See the color plate.)

having at least two binding sites for calcium on each monomer.[18] Due to a lack of three-dimensional crystal structure and low calcium-binding affinities, the actual number of calcium-binding sites on the ECD is unsolved. By modeling CaSR structure using computational algorithms based on the geometric description, surface electrostatic potentials, and other functional studies, the VFT domain has been shown to contain three to five calcium-binding sites.[17–19] Synthetic allosteric modulators that bind in the heptahelical TMD and extracellular loops of the CaSR have also been identified, which include calcimimetics, such as NPS-R568, cinacalcet, Calindol and related phenylalkylamines, and calcilytics, such as NPS 2143, Calhex-231, and ronacaleret.[20–24] Ca_o^{2+} can also activate the CaSR via the heptahelical or extracellular loop domains, as evidenced by its activity at CaSR constructs that lack the amino-terminal VFT domain of the receptor.[25–27]

A majority of the 215 C-tail residues (Lys863–Ser1078) of the hCaSR can be truncated without perturbing G-protein signaling response in heterologous cell expression systems. Disease-causing mutations in the hCaSR

C-tail are relatively rare.[28–31] Nevertheless, the proximal C-tail possesses multiple important determinants that regulate functional response of the receptor. Protein kinase C (PKC)-mediated phosphorylation of Thr888 inhibits extracellular calcium-induced release of intracellular Ca^{2+} stores. High extracellular Ca_o^{2+} also induces stoichiometric binding of calmodulin (CaM) to the C-tail of CaSR (residues 871–898) that may interfere with PKC phosphorylation of Thr888 and thus stabilize cell-surface expression by reducing internalization of the receptor.[29,32] Truncations at the carboxyl-terminus residues can cause either gain of function or loss of function of the CaSR. A large in-frame naturally occurring deletion in the CaSR C-tail, Ser895–Val1075, leads to increased cell-surface expression and gain of function in subjects with ADH, causing a left-shift in the plasma calcium set point and leading to hypocalcemia and hypercalciuria.[33] Mutations of a proximal putative ER retention signal region (Arg890 through Arg898) increased cell-surface expression of the receptor.[34] A truncation of hCaSR at Ala877 likewise exhibited an increased level of cell-surface expression.[28,29] The truncations at residues 874 and 876 of CaSR and point mutations at R886 lead to FHH/NSHPT by modulating cell-surface receptor expression and also inhibiting responsiveness to extracellular calcium.[28,29,35–38] Given these observations, it is apparent that the hCaSR C-tail contains multiple determinants for regulating cell-surface level, desensitization, internalization, and downregulation of the receptor.

During its intracellular biosynthesis, CaSR requires homodimerization and the cell-surface expression is controlled by multiple checkpoints along the secretory pathway to ensure only properly folded and posttranslationally modified receptor access the plasma membrane. The CaSR is dimerized in the ER through two disulfide bonds, between the cysteine 129 and 131 of each monomer[39]; however, noncovalently bound dimers might still be expressed at the cell surface when the disulfide bonds are lacking.[10] The receptor's ECD contains nine potential sites of N-linked glycosylation, which is performed in the Golgi apparatus and is required for the normal expression of the protein at the cell surface. The fully glycosylated monomer has a molecular mass of 150–160 kDa, indicating a CaSR carbohydrate content of 35–40 kDa/monomer. While it is important for trafficking of the receptor to the plasma membrane, N-linked glycosylation is probably not critical for its biological activity.[5,40] CaSR dimerization in the ER also plays an essential role in regulating receptor signaling activity.[39,41] Mutation of several highly conserved cysteine residues found within the ECDs of CaSR, mGluR1 and mGluR5, responsible for intra- and interchain disulfide linkages, either abolish or dramatically reduce CaSR cell-surface expression, likely by disrupting trafficking of the receptors from the ER.[42–44] Thus, during biosynthesis, CaSR navigates a multifaceted quality control process that monitors folding and proper assembly of the receptor within the ER. Several missense mutations causing FHH/NSPHT reduce hCaSR function by impairing cell-surface expression of the receptor and by mostly remaining trapped, intracellularly. Regulation of GPCR signaling, trafficking, and localization are regulated by different GPCR interacting proteins and this is true for CaSR also.[45] Figure 2 illustrates the CaSR life cycle from synthesis to trafficking to the plasma membrane and current knowledge about interactions (chemicals and proteins) that impede or

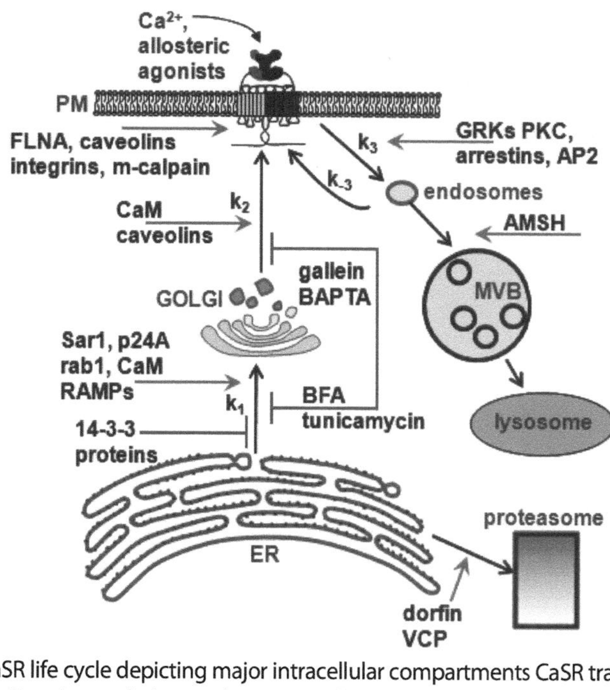

Figure 2 CaSR life cycle depicting major intracellular compartments CaSR traverses during its life cycle. Proteins and chemical compounds which impede trafficking are indicated with red lines, and protein interactions that promote trafficking of CaSR are indicated with green arrows. AP-2, adaptor protein-2; AMSH, associated molecule with the SH3 domain of STAM; BAPTA, 1,2-bis(90-aminophenoxy)ethane-*N*,*N*,*N*′,*N*′-tetraacetic acid; BFA, brefeldin A; CAM, calmodulin; ER, endoplasmic reticulum; FLNA, filamin A; GRKs, G-protein-coupled receptor kinases; MVB, multivesicular body; PM, plasma membrane; RAMPs, receptor-activity-modifying proteins; VCP/p97, valosin-containing protein. *Reproduced with copyright permission from Elsevier Press: Best Practice & Research Clinical Endocrinology & Metabolism 2013; 27:303–313; Author—Gerda Breitweiser; Title—The calcium sensing receptor life cycle; trafficking, cell surface expression, and degradation. Review article.* (See the color plate.)

contribute to CaSR trafficking, desensitization/recycling, and degradation mechanisms, which are the focus of this chapter.

3. ANTEROGRADE TRAFFICKING OF CaSR THROUGH SECRETORY PATHWAY

The large CaSR ECD faces the luminal side of the ER so that proper folding may be achieved by disulfide linkages, posttranslational modifications, and possibly by other quality control mechanisms. ER exit for many cell-surface transmembrane proteins involve selective concentration of cargo in vesicular or tubular complexes destined for Golgi. This complex formation involves interactions with the ER-derived coat protein complex II (COPII), COPII-associated proteins, or cargo receptor in the ER. Assembly and disassembly of COPII-coated vesicles from ER membranes require the recruitment of the Sar1, a small GTP-binding protein to the ER outer membrane. Sar1 cycles between a GTP-bound form stimulated by the GTP–GDP exchange factor Sec12p and a GDP-bound form stimulated by GTPase activation by Sec23p. Binding of Sar1–GTP to the ER outer membranes induces sequential recruitment of coat complexes Sec23p–Sec24p and Sec13p–Sec31p and formation of ER exit sites (reviewed in Ref. 46). While trafficking of many cargo proteins in the secretory pathway utilizes conventional COPII-complex in most cell types, nonconventional trafficking has also been reported for some plasma membrane bound proteins.[47,48] The molecular mechanism(s) underlying GPCR export into ER-derived COPII-coated vesicles are largely unexplored. Recent studies suggest differential and selective ER export of α_2-adrenergic receptor (AR), β_2-AR, and angiotensin-1 receptor is modulated by Sar1. ER export of α-adrenergic and angiotensin II receptors have been found to be differentially regulated by Sar1.[49] Recognition of export signals involved in the ER exit and role of C-tail sequences in regulating GPCR export have been demonstrated for a relatively small number of GPCRs (reviewed in Ref. 50). A number of studies suggest that ER export signals diacidic motif (DXE) and diphenylalanine (FF) motifs play an important role in the COPII-mediated transport of transmembrane proteins.[51] Several other motifs such as $(FX)_3F(X)_3F$, $FN(X)_2LL(X)_3L$, and $F(X)_6LL$ have also been reported to be required for GPCR export trafficking. Unlike the DXE and FF motifs, there is little evidence that these motifs associate with the components of COPII vesicles for GPCR transport from ER to the Golgi. Using both Sar1H79G and siRNA specific for Sar1, disruption of endogenous Sar1 function is found to significantly inhibit cell-surface expression of the CaSR, and transient expression of Sar1H79G leads to retention of CaSR in the ER mostly as immature high-mannose type receptor in heterologously expressed cell lines.[52,53] Confocal microscopic analysis revealed that Sar1H79G arrested CaSR in distinct perinuclear compartments. A significant reduction in cell-surface expression of the CaSR is also confirmed quantitatively by intact cell-surface ELISA and by analysis of CaSR signaling efficacy. In contrast, expression of wild-type Sar1 did not significantly inhibit the cell-surface expression of the receptor or its functional activity. This suggests that Sar1 may be limiting for CaSR transport from the ER to the cell surface in this transient expression paradigm. The ER export of CaSR is likely mediated in ER-derived COPII-coated vesicles; however, no direct evidence is available to suggest that CaSR cargo protein is recruited to COPII-coat vesicles via interactions with any known export signals. In contrast, for a number of mammalian proteins such as the vesicular stomatitis glycoprotein, potassium channels, and cystic fibrosis transmembrane conductance regulator, at least two distinct types of ER export signals, a FF hydrophobic motif and/or a DXE have been identified. Deletion or mutation of these export motifs results in either retention of these transmembrane proteins in the ER or a delayed exit from the ER, resulting in a lowered steady-state level cell-surface expression (reviewed in Ref. 51). Similar motif is present in the CaSR C-tail; however, mutation of the putative clusters of di-acidic sequences or even truncation of the majority of CaSR C-tail does not interfere with cell-surface trafficking of the receptor. As stated earlier, while the C-tail of the CaSR is shown to contain several regulatory elements, a large portion of the CaSR C-tail seems dispensable for forward trafficking of the receptor. Therefore, CaSR requires a diversity of complex signals and may not employ a simple, well-defined export motif like DXE or FF. For many secretory proteins, these trafficking signals include hetero- or homo-oligomerization, glycosylation, and interaction with adaptor proteins in the ER lumen.[46] Surprisingly, the autonomously folded large amino-terminal ECD module of CaSR seems necessary for forward trafficking of the receptor. A receptor variant devoid of the ECD poorly expresses at the cell surface and presence of ECD structure in these CaSR variants improves cell-surface transport of the receptor, suggesting that the ECD plays an important role in plasma membrane transport activity. The secretion of ECD variant is inhibited greatly by mutating cysteines responsible for disulfide-linked dimerization and elimination of four N-glycosylation sites, indicating that disulfide-linked dimerization and addition of complex carbohydrates on the ECD in the ER lumen are critical for ER exit and transport. A simplistic explanation is that these mutations affect the correct folding of the CaSR ECD and thus the ER quality control checkpoints prevent ER exit of this secreted ECD. These results also raise interesting questions regarding a role of CaSR amino-terminal ECD in a receptor-mediated mechanism for ER export. CaSR may have to display a specific assembly by disulfide-linked dimer formation and N-linked glycosylation of the ECD to serve as a signal for incorporation of fully assembled CaSR cargo into the COPII vesicles. Unfortunately, the roles of the amino-termini motifs in regulating GPCR export trafficking have been much less investigated and remain controversial. A novel ER export signal consisting of a proline residue at the +2 position from the signal peptide cleavage site in nucleobindin-1, a Golgi-localized soluble protein, has been reported.[54] A Tyr-Ser motif within the amino-terminus of α_2-AR has been demonstrated to modulate trafficking of the receptor.[55] Because amino-terminal domains of secreted proteins cannot rely on a direct interaction with the COPII coat for topological reasons, selective export of some soluble luminal cargos requires specific transmembrane cargo receptors, such as ER-Golgi intermediate compartment (ERGIC)53, p24 family of proteins, to mediate the interaction of cargo in the ER lumen with COPII components in the cytosol.[51,55] Interestingly, a cargo receptor family member p24A transmembrane emp24 domain trafficking protein 2 is identified in yeast-2 hybrid screening and shown to interact only with immaturely glycosylated forms of CaSR.[56] P24A protein, which is predominantly localized in the early secretory pathway and cycles between the ER, the ERGIC, and *cis*-Golgi membranes, may provide stability of the intracellular forms of CaSR, thereby increasing plasma membrane level of the receptor. The CaSR C-terminal tail distal to Thr868 is shown necessary for interaction with p24A, but this interaction seems not to involve the ECD portion of the CaSR. Thus, it remains to be determined whether the CaSR ECD requires other p24 family transmembrane cargo receptor interaction and specific signal motifs for association with the cargo receptor to be packaged into COPII vesicles.

Many GPCRs destined for Golgi and plasma membrane exit the ER via vesicular or tubular structures. Rab small GTP-binding proteins, the largest branch of the Ras-related superfamily, are involved in almost every step of this vesicle-mediated protein transport including targeting, tethering, and fusion of transport vesicles with the appropriate acceptor membrane. Several Rab proteins have been documented also in internalization and degradation of GPCRs.[57,59] Rab1 is involved in vesicular transport and assembly of the fusion complex for delivery to Golgi membranes. Vesicular transport from the ER of many cell-surface proteins involves their concentration in ER-derived COPII-coated vesicles. After the assembly of COPII-coated vesicles from the ER, Rab1 regulates the anterograde vesicular trafficking of the cargo proteins from the ER to the *cis*-Golgi network and possibly to intra-Golgi transport. Of the GPCRs, the angiotensin II type 1 receptor (AT1R), and β_2-AR, α_{1A}-AR use conventional Rab1-dependent transport pathway in HEK293 cells, whereas, α_{2B}-AR undergoes Rab1-independent export in these cells.[58] Rab1, Rab2, and Rab6 are shown to be involved in both anterograde and retrograde transport of AT1R, α_{1A}-AR, and α_{1B}-AR.[50,59] The role of Rab1 in vesicular transport from ER to Golgi to regulate CaSR transport to the cell surface has been investigated.[52,53] These data demonstrate that transient expression of the dominant negative Rab1N124I mutant and Rab1 siRNA in HeLa and COS-7 cells and in HEK293 cell type inhibited cell-surface expression of the full-length and C-tail-truncated CaSR variants, and endogenous Rab1 is limiting for CaSR transport from the ER to the cell surface in this expression paradigm. CaSR transport in HEK293 cells is Rab1-dependent similar to AT1R, β_2-AR, and α-$_{1A}$AR and unlike the α_{2B}-AR, which uses a nonconventional Rab1-independent pathway in HEK293 cells.[58] The results identify a Rab1-dependent mechanism regulating the transit from the ER to the Golgi that enables supply of CaSR to the cell surface.

Forward trafficking of many GPCRs from the ER to the cell surface may also require another family of chaperone proteins, receptor-activity-modifying proteins (RAMPs).[60] RAMPS (RAMP1, RAMP2, and RAMP 3) are single transmembrane spanning GPCR accessory proteins that function to modify the expression and pharmacology of calcitonin (CT) receptor and calcitonin-like receptor.[61] RAMPs appear to allosterically influence the structure of CT family receptors allowing for their terminal glycosylation in the ER, thereby facilitating their expression at the cell surface. RAMPs not only regulate the pharmacology of the CT receptor family but also the intracellular trafficking and posttranslational modification of the receptors, essential for the regulation of signal transduction mechanisms of these receptors. Reviewed in Ref. 62, in heterologous cell system (HEK-293 and COS-7 cells) expressing CaSR, the receptor is not transported to the plasma membrane unless either RAMP1 or RAMP3 is coexpressed with the receptor. By associating with RAMP1 and RAMP3, partially glycosylated CaSR can bypass the ER retention and traffic to Golgi apparatus for terminal glycosylation before trafficking to the plasma membrane.

Table 1 Roles of Known Intracellular Protein Interacting with CaSR in Receptor Trafficking

Protein	Reported Effect on CaR	References
ER → Golgi		
Sar1	Regulates transport, ↑PM expression	52,56
P24A	ER exit, stability, ↑PM targeting	56
Rab1	CaR ECD anterograde transport	53
RAMPs	ER to PM forward trafficking	62
14-3-3	ER retention, regulates transport	34,63,64
CaM	Anterograde trafficking?? Endocytosis?	32
Dorfin	Regulates ubiquitination, degradation?	65
Plasma membrane		
PKC	Phosphorylates C-tail of CaR, desensitization, Ca_i^{2+} oscillations	29,86
GRKs	↑Endocytosis	66,67
β-arrestins	Minor effect, constitutive recycling of receptor?	66,67
Rab7, Rab11a	↑Endocytosis	68,87
FLNA	Direct interaction, PM microdomain targeting?	69,88,90
RhoA		69,88,90
Integrins	Cellular adhesion? Cell migration?	29,89
AMSH	↓PM expression	68

No other direct interaction between RAMPs and the CaSR or reports suggesting roles of RAMPs in CaSR signaling in endogenous tissues have been documented thus far. Also, interaction of RAMPs with other family-C/3 GPCRs have not been reported, so generalization of the mechanisms by which RAMPs may regulate CaSR and possibly other family-C/3 GPCRs signaling, pharmacology, and trafficking are not known. Table 1 provides a list of known protein modulators that affect trafficking of the CaSR.

4. UNIQUE REGULATION OF CaSR AGONIST-DRIVEN INSERTIONAL SIGNALING AT THE CELL SURFACE

As noted earlier, the CaSR is a novel GPCR that must function in the chronic presence of calcium and continuously monitors small changes in Ca_o^{2+}. In particular, the CaSR expressed in different cellular compartments is required to sense changes in the concentrations of other metabolites such as L-amino acids as well as modalities such as pH and ionic strength. In the case of Ca_o^{2+} sensing, the receptor provides responses to persistent deviations in Ca_o^{2+} level, which can last for several hours or more.[35] In this respect, it differs significantly from many other GPCRs, such as β-ARs, which are rapidly desensitized within minutes following ligand binding (see review in Ref. 70). Resistance to desensitization requires the maintenance of a functional receptor pool at the cell surface and persistent coupling of the receptor to its heterotrimeric G-proteins and downstream signaling pathways. It is unclear though what regulates the activation of CaSR at the plasma membrane in the constant presence of agonist and how functional desensitization is regulated. Some studies indicate that CaSR on the cell surface undergoes only minor functional desensitization when exposed continuously to elevated extracellular calcium level and in the presence of positive allosteric modulators.[66] This resistance to desensitization is likely important to enable sufficient level of CaSR at the cell surface to continuously monitor and maintain constant Ca_o^{2+} level. Interestingly, the CaSR is localized to both the plasma membrane and intracellular compartments (such as Golgi apparatus, ER) in recombinant as well as native CaSR-expressing cells.[41,71,72] CaSR agonists and positive allosteric modulators promote the forward trafficking and glycosylation of intracellular receptors to the plasma membrane and in the steady-state level of the receptor at the cell surface,[63,73] a phenomenon that has been termed agonist-driven insertional signaling (ADIS). This is directly linked to membrane-localized receptor signaling events, as evidenced by attenuation of ADIS by inactivating CaSR mutations.[63,74] Strikingly, however, the small-molecule negative modulator (calcilytic) NPS 2143 also rescues cell-surface expression of loss-of-expression receptor mutants in HEK293 cells.[75] NPS 2143 negatively modulates CaSR-mediated signaling and positively modulates receptor trafficking to the cell surface and offers an example of biased allosteric modulation, with a complete reversal in cooperativity (positive vs. negative) between pathways (trafficking vs. acute signaling). The net increase in plasma membrane CaSR results mostly from an increase in anterograde trafficking of the receptor through the secretory pathways and a constant rate of endocytosis (Fig. 2, these rates are designated as k_1, k_2, and k_3) (see Ref 76, for details). Minimal functional desensitization of CaSR by phosphorylation and endocytosis of CaSR is possibly balanced by continuous supply of newly formed CaSR from the secretory pathway at the cell surface. The G-protein mediated signaling that induces this anterograde trafficking of the receptor has been shown to be blocked by a $G_{\beta\gamma}$ inhibitor galleon with effectors, by brefeldin A, an inhibitor of ARF1 that blocks forward trafficking, and by tunicamycin, an inhibitor that blocks N-linked glycosylation of newly synthesized receptor.[63]

The unusual ADIS phenomenon raises many questions about the nature of signaling events that may trigger and contribute in diverse cell types and the activation of CaSR, which drives its own trafficking to the plasma membrane. The underlying signaling mechanism and its G-protein requirements have not yet been determined. Characterization of ADIS mechanism has been performed mostly in heterologously expressed receptor in HEK293 cells and reproduced only in human endothelial cells with N-terminal pH-sensitive green fluorescent protein-tagged CaSR. Some results are worth mentioning here to point out probable events that may trigger and regulate ADIS. Increase in Ca_o^{2+} is noted to increase the net plasma membrane CaSR level, which is inhibited by gallein. Second, although it is shown that elevated intracellular calcium is likely involved in CaSR activation, this intracellular calcium release is inhibited by tunicamycin, implying that signaling by endogenous CaSR probably requires ongoing CaSR biosynthesis and exit of the carbohydrate-modified receptor from the ER (reviewed in Ref 76). Phosphorylation of the C-tail PKC sites may also contribute in the release of CaSR from the ER, since a CaSR mutant S889A receptor exhibits enhanced binding to 14-3-3 proteins and reduces optimal ADIS response. Interaction of 14-3-3 proteins with the proximal CaSR C-tail via an arginine-rich domain causes partial ER retention.[34,64,77] Flanking Ser889 phosphorylation site mutation prevents 14-3-3-proteins binding and increases CaSR plasma membrane expression. So, ADIS has potential therapeutic utility for disorders of calcium homeostasis in which CaSR expression is impaired.

5. CELL-SURFACE COMPARTMENTALIZATION AND CYTOSKELETAL ALTERATION OF CaSR SIGNALING

Association of the CaSR with other proteins has been postulated to influence its function. In some cells, such as parathyroid chief cells and cardiac myocytes, the CaSR localizes to plasma membrane invaginations known as caveolae and caveolin-enriched membrane fractions purified from bovine parathyroid cells are enriched for CaSR and for a variety of signaling proteins including heterotrimeric G-proteins and PKC family members.[10,78,79] The CaSR has been coimmunoprecipitated with membrane proteins caveolins, caveolin-1 in bovine and human parathyroid glands and caveolin-3 from mouse cardiac myocytes. Direct interaction of CaSR and caveolin has not been confirmed by other experimental methods. So, it is unclear whether CaSR traffics to the plasma membrane with caveolins after incorporating into caveolae in the Golgi. Another possibility is that accumulation of signaling complex in plasma membrane might partition CaSR into caveolae and thus interaction with caveolins is dependent on signaling mechanisms in cell-specific manner in different cell types. A direct interaction of CaSR with an adaptor protein, including actin-binding protein filamin A has been reported in CaSR-transfected HEK293 cells and in bovine parathyroid cells.[80] The intracellular domain of the CaSR (amino acids 907–997) binds directly to filamin A, and cotreatment of a peptide that blocks CaSR binding to filamin A attenuates CaSR-mediated ERK activation in CaSR-HEK cells. It is speculated that filamin A acts as an intracellular signaling complex integrating CaSR-mediated signals and facilitates the receptor-mediated MAP kinase signaling pathway.[81] Surprisingly, filamin A also increases the total cellular CaSR level and speculated to prevent proteosomal degradation.[82] In keratinocytes, disruption of the CaSR–filamin interaction prevents Ca_o^{2+}-induced Rho activation and E-cadherin-mediated cell–cell adhesion and thereby attenuates keratinocyte differentiation.[69] Interactions with filamin A and caveolins indicate that CaSR signaling likely does not take place randomly throughout the cell but is compartmentalized and organized, where both filamin A and caveolin-1/3 act as scaffolds that bind signaling components and other cellular components (cytoskeleton) to facilitate the interaction of the receptor with its various signaling pathways.

6. ENDOCYTOSIS AND RECYCLING OF CaSR

The functional impact of GPCR signaling depends not only on ligand-dependent initiation of signaling but also on the timing and rate of termination of signaling events. Thus, control of desensitization is critical to receptor function with impacts on the duration of cellular responses, ligand selectivity, and even the selection of specific signaling pathways. There is limited information regarding the mechanisms of agonist-stimulated desensitization and internalization of cell-surface-expressed functional CaSR. A common feature among GPCR signaling system is desensitization, i.e., a loss of responsiveness of a receptor when continuously exposed to the agonist. G-protein-coupled receptor kinases (GRKs) and arrestins are important regulators of GPCR desensitization.[83] Upon agonist or hormone binding, many GPCRs are rapidly phosphorylated by a GRK resulting in binding of arrestin, which uncouples the receptor from G-proteins and initiates GPCR endocytosis by dynamin-mediated clathrin-dependent endocytic mechanisms (Fig. 3). Once internalized, GPCRs are dephosphorylated and subsequently recycled back to the cell surface where they can again respond to agonists. Constitutive, agonist-independent mechanism of receptor internalization has also been reported for some GPCRs that are β-arrestin and dynamin-independent but is also mediated by clathrin-coated vesicles. For many GPCRs, GRK-mediated phosphorylation on its own is insufficient to mediate the desensitization of many GPCRs. Instead, the recruitment of β-arrestin proteins to agonist-activated and GRK-phosphorylated GPCRs facilitates the uncoupling of the receptor from heterotrimeric G-proteins. However, for $G\alpha_{q/11}$-coupled GPCRs, phosphorylation-independent desensitization can be achieved as the consequence of the displacement of $G\alpha_{q/11}$ from the receptor complex by GRK2 and GRK3. Thus, GRKs regulate GPCR desensitization by both phosphorylation-dependent and -independent mechanisms. The endocytosis of many GPCRs appears to be mediated by the same mechanism that is required for GPCR desensitization. GRK-mediated phosphorylation promotes the binding of β-arrestin, which function as endocytic adaptor proteins that facilitate the targeting of receptors for clathrin-mediated endocytosis. Thus, β-arrestins play an essential role in recruiting proteins that are not only

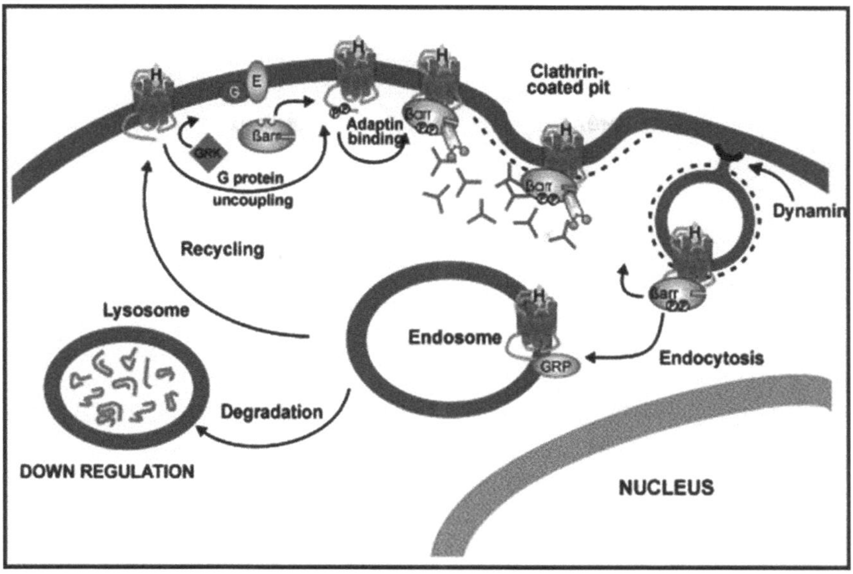

Figure 3 Schematic representation of GPCR desensitization and recycling following agonist activation. βarr, β-arrestin; E, effector enzyme; G, heterotrimeric G protein; GRK, G-protein-coupled receptor kinase; GRP, G-protein-coupled receptor phosphatase; H, hormone; P, phosphate group. See text for details. *Reproduced with copyright permission from Elsevier Press: Authors—Dhami GK, Ferguson SS. Book—Physiology and Therapeutics; Volume 111, Issue 1, July 2006, Pages 260–271; Regulation of metabotropic glutamate receptor signaling, desensitization and endocytosis.* (See the color plate.)

essential for the internalization of GPCRs but also for the regulation of the endocytic machinery.

The processes of CaSR desensitization and internalization are currently poorly understood. The CaSR is phosphorylated by PKC as well as by GRK2 and GRK4 and also shown to bind β-arrestins. CaSR functional desensitization is facilitated by binding of GRK2 to $G\alpha_{q/11}$ subunit, but kinase activity of GRK2 is not required for agonist-dependent accumulation of inositol phosphates.[66] GRK2 also rapidly terminates a subset of CaSR signaling pathways dependent upon Gq without requiring receptor phosphorylation, β-arrestin recruitment, or subsequent receptor internalization. In another study, a significant reduction of CaSR signaling by GRK2, GRK4, and β-arrestin have been noticed, which the investigators attribute to both phosphorylation-dependent and -independent mechanisms.[67] Interestingly, this study reveals that CaSR undergoes only a minor agonist-stimulated internalization and binding of β-arrestin seems to be PKC-dependent but not GRK-dependent. The constitutive internalization and recycling of CaSR are less clear as different research groups have obtained different results and the reason for these differences are unclear. A number of reports suggest that CaSR endocytosis is rapid, constitutive, and facilitated by Rab7 and Rab11a.[68,84] Recycling of the CaSR has been noted in some studies[68,84] but not in others,[63] suggesting that the constitutive receptor internalization may require a different endocytic machinery. The CaSR endocytosis is essential for the transactivation of epidermal growth factor receptor that leads to the MAP kinase signaling cascade and links receptor signaling to PTH-related peptide secretion via a Rab11a-dependent and associated molecule with the SH3 domain of STAM (AMSH)-sensitive mechanism. It is proposed that this internalization and downregulation might be important regulatory mechanisms for rapid and efficient control of CaSR cell-surface expression and for its signaling activities. Because all these studies were performed in heterologously expressed cells and contributions of anterograde trafficking or recycling from early endosomes under physiological conditions have not been simultaneously tested, cell-type specific differences in signaling dynamics are unclear. It appears that CaSR internalizes by multiple mechanisms depending upon the nature of the trigger for internalization as well as the cell or tissue environment where it is expressed.

7. PROTEOSOMAL AND LYSOSOMAL DEGRADATION PATHWAYS

The principal mechanism underlying downregulation of GPCR degradation is a multistep process often involving endocytosis and subsequent delivery of the receptor to lysosomes for degradation.[85] Little is known about the molecular mechanisms involved in sorting GPCRs to lysosomes. Once internalized, receptors are often targeted to specialized endosomal compartments, dephosphorylated, and recycled back to the cell surface or targeted to lysosomes for degradation (Fig. 3). In addition to the lysosomes, degradation of receptor protein is often accompanied by proteosomal pathway. The CaSR and other GPCRs including human opioid receptor subtypes, rhodopsin, and follicle-stimulating hormone receptor have been shown to bind ubiquitin and undergo ubiquitin-targeted proteosomal degradation.[65,85] The cytoskeletal actin-binding protein filamin A facilitates the CaSR-mediated MAPK signaling pathway and increases the total cellular CaSR level possibly by preventing proteosomal degradation.[82,85] Also, CaSR ubiquitination and degradation are linked to the activity of E3 ubiquitin ligase, also known as Dorfin.[82] The CaSR is ubiquitinated by E3 ligase Dorfin and immunoprecipitation of Dorfin pulls down the immature high-mannose forms of the CaSR and the ATPase VCP (p97) interacts with CaSR in this context. Thus, ubiquitination is possibly happening for the quality control during the synthesis of the CaSR in the ER and in contrary to ubiquitination of other GPCRs and/or β-arrestin at the plasma membrane that regulates both endocytosis and recycling.

Another protein, a deubiquitinating enzyme-specific for K63-linkages known as the enzyme AMSH, reduces CaSR expression.[68] The role of AMSH is to mediate deubiquitination of proteins targeted for lysosomal degradation from endosomes in the endosomal sorting complexes required for transport pathway. The idea that the CaSR targeted for multivesicular body (MVP)/lysosomal degradation is ubiquitinated is supported by the inverse relationship between AMSH and CaSR expression levels. Further work is needed to identify the E3-ligase that ubiquitinates the cell-surface CaSR. The C-tail of CaSR also contains an unusually long cellular sorting signal encoded by 51 amino acid residues containing a proline, glutamine, serine and threonine (PEST)-like motif and several glutamine (Q) repeats, which specify an endosomal–lysosomal degradation pathway required for downregulation of hCaSR protein contributing importantly to the regulation of cell-surface expression level of the receptor.[84] The relative contributions of the PEST-like sequence and ubiquitination to CaSR lysosomal degradation remain to be determined.

8. CONCLUSION

The CaSR functions in the chronic presence of ligand, and recent studies indicate that the life cycle of CaSR is complex and depends upon novel mechanisms regulating its cellular trafficking. A large pool of preplasma membrane localization has been confirmed using ^{35}S-cysteine-labeled receptor in pulse chase experiments. The study showed slow maturation of CaSR in the biogenesis pathway by carbohydrate modification in the Golgi and less than 50% newly formed receptors are converted to mature glycosylated forms after 24 h.[77] The CaSR is also stable with less than 50% decrease in labeled receptor after 24 h. This indicates that a significant pool of CaSR is stored intracellularly and can be mobilized to the plasma membrane upon signal initiation. Minimal functional desensitization of CaSR by phosphorylation and endocytosis of CaSR is possibly balanced by continuous supply of newly formed CaSR from the secretory pathway at the cell surface. Anterograde transport of CaSR utilizes proteins common in secretory pathways (Sar1, Rab1, and ARF-small GTP-binding proteins) and other interacting scaffolding protein partners (p24A, 14-3-3, and CaM), some of which can be regulated by phosphorylation also (by PKC and GRK). An endosomal–lysosomal degradation pathway is responsible for down-regulation of CaSR protein, possibly contributing to regulating of cell-surface expression level of the receptor. Only limited information is available in the knowledge of cell-specific regulation of CaSR trafficking in plasma membrane microdomains specialized signaling in apical and basolateral cell membranes or by concentrating large signaling complexes in caveolei.

REFERENCES

1. Ulloa-Aguirre A, Conn PM. Targeting of G protein-coupled receptors to the plasma membrane in health and disease. *Front Biosci (Landmark Ed)*. 2009;14:973–994.
2. Wu G. Regulation of post-Golgi traffic of G protein-coupled receptors. *Subcell Biochem*. 2012;63:83–95.
3. Kang DS, Tian X, Benovic JL. Role of β-arrestins and arrestin domain-containing proteins in G protein-coupled receptor trafficking. *Curr Opin Cell Biol*. 2014;27:63–71.
4. Brauner-Osborne H, Wellendorph P, Jensen AA. Structure, pharmacology and therapeutic prospects of family C G-protein coupled receptors. *Curr Drug Targets*. 2007;8:169–184.
5. Brown EM, MacLeod RJ. Extracellular calcium sensing and extracellular calcium signaling. *Physiol Rev*. 2001;81:239–297.
6. Tfelt-Hansen J, Brown EM. The calcium-sensing receptor in normal physiology and pathophysiology: a review. *Crit Rev Clin Lab Sci*. 2005;42:35–70.
7. Quinn SJ, Ye CP, Diaz R, et al. The Ca2+-sensing receptor: a target for polyamines. *Am J Physiol*. 1997;273(4 Pt 1):C1315–C1323.
8. Conigrave AD, Hampson DR. Broad-spectrum amino acid-sensing class C G-protein coupled receptors: molecular mechanisms, physiological significance and options for drug development. *Pharmacol Ther*. 2010;127(3):252–260.
9. Ho C, Conner DA, Pollak MR, et al. A mouse model of human familial hypocalciuric hypercalcemia and neonatal severe hyperparathyroidism. *Nat Genet*. 1995;11:389–394.
10. Chakravarti B, Chattopadhyay N, Brown EM. Signaling through the extracellular calcium-sensing receptor (CaSR). *Adv Exp Med Biol*. 2012;740:103–142.
11. Alfadda TI, Saleh AM, Houillier P, Geibel JP. Calcium-sensing receptor 20 years later. *Am J Physiol Cell Physiol*. 2014;307:C221–C231.
12. Herberger AL, Loretz CA. Vertebrate extracellular calcium-sensing receptor evolution: selection in relation to life history and habitat. *Comp Biochem Physiol Part D Genomics Proteomics*. 2013;8:86–94.
13. Houillier P, Nicolet-Barousse L, Maruani G, Paillard M. What keeps serum calcium level stable? *Joint Bone Spine*. 2003;70(6):407–413.
14. Magno AL, Ward BK, Ratajczak T. The calcium-sensing receptor: a molecular perspective. *Endocr Rev*. 2011;32(1):3–30.
15. Nearing J, Betka M, Quinn S, et al. Polyvalent cation receptor proteins (CaRs) are salinity sensors in fish. *Proc Natl Acad Sci U S A*. 2002;99:9231–9236.
16. Hubbard PC, Canario AV. Evidence that olfactory sensitivities to calcium and sodium are mediated by different mechanisms in the goldfish *Carassius auratus*. *Neurosci Lett*. 2007;414:90–93.

17. Silve C, Petrel C, Leroy C, et al. Delineating a Ca^{2+} binding pocket within the venus flytrap module of the human calcium sensing receptor. *J Biol Chem*. 2005;280(45): 37917–37923.
18. Huang Y, Zhou Y, Yang W, et al. Identification and dissection of Ca(2+) -binding sites in the extracellular domain of Ca(2+) -sensing receptor. *J Biol Chem*. 2007; 282:19000–19010.
19. Huang Y, Zhou Y, Castiblanco A, Yang W, Brown EM, Yang JJ. Multiple Ca(2+)-binding sites in the extracellular domain of the Ca(2+)-sensing receptor corresponding to cooperative Ca(2+) response. *Biochemistry*. 2009;48(2):388–398.
20. Nemeth EF. Calcimimetic and calcilytic drugs: just for parathyroid cells? *Cell Calcium*. 2004;35:28–39.
21. Nemeth EF. Allosteric modulators of the extracellular calcium receptor. *Drug Discov Today Technol*. 2013;10(2):277–284.
22. Miedlich SU, Gama L, Seuwen K, Wolf RM, Breitwieser GE. Homology modeling of the transmembrane domain of the human calcium sensing receptor and localization of an allosteric binding site. *J Biol Chem*. 2004;279(8):7254–7263.
23. Jensen AA, Bräuner-Osborne H. Allosteric modulation of the calcium-sensing receptor. *Curr Neuropharmacol*. 2007;5(3):180–186.
24. Petrel C, Kessler A, Dauban P, Dodd RH, Rognan D, Ruat M. Positive and negative allosteric modulators of the Ca^{2+}-sensing receptor interact within overlapping but not identical binding sites in the transmembrane domain. *J Biol Chem*. 2004;279(18): 18990–18997.
25. Ray K, Tisdale J, Dodd RH, Dauban P, Ruat M, Northup JK. Calindol, a positive allosteric modulator of the human Ca(2+) receptor, activates an extracellular ligand-binding domain-deleted rhodopsin-like seven-transmembrane structure in the absence of Ca(2+). *J Biol Chem*. 2005;280(44):37013–37020.
26. Ray K, Northup JK. Evidence for distinct cation and calcimimetic compound (NPS 568) recognition domains in the transmembrane regions of the human Ca^{2+} receptor. *J Biol Chem*. 2002;277(21):18908–18913.
27. Mun HC, Franks AH, Culverston EL, Krapcho K, Nemeth EF, Conigrave AD. The Venus Fly Trap domain of the extracellular Ca^{2+}-sensing receptor is required for L-amino acid sensing. *J Biol Chem*. 2004;279(50):51739–51744.
28. Ray K, Fan GF, Goldsmith PK, Spiegel AM. The carboxyl terminus of the human calcium receptor. Requirements for cell-surface expression and signal transduction. *J Biol Chem*. 1997;272:31355–31361.
29. Bai M, Trivedi S, Lane CR, Yang Y, Quinn SJ, Brown EM. Protein kinase C phosphorylation of threonine at position 888 in Ca_o^{2+} -sensing receptor (CaR) inhibits coupling to Ca^{2+} store release. *J Biol Chem*. 1998;273:21267–21275.
30. Jiang YF, Zhang Z, Kifor O, Lane CR, Quinn SJ, Bai M. Protein kinase C (PKC) phosphorylation of the Ca_o^{2+} -sensing receptor (CaR) modulates functional interaction of G proteins with the CaR cytoplasmic tail. *J Biol Chem*. 2002;277:50543–50549.
31. Pdasheva S, D'Souza Li L, Canaff L, Cole DE, Hendy GN. CASRdb: calcium-sensing receptor locus-specific database for mutations causing familial (benign) hypocalciuric hypercalcemia, neonatal severe hyperparathyroidism, and autosomal dominant hypocalcemia. *Hum Mutat*. 2004;24:107–111.
32. Huang Y, Zhou Y, Wong HC, et al. Calmodulin regulates Ca^{2+}-sensing receptor-mediated Ca^{2+} signaling and its cell surface expression. *J Biol Chem*. 2010; 285: 35919–35931.
33. Lienhardt A, Garabedian MG, Bai M, et al. A large homozygous or heterozygous in-frame deletion within the calcium-sensing receptor's carboxylterminal cytoplasmic tail that causes autosomal dominant hypocalcemia. *J Clin Endocrinol Metab*. 2000;85: 1695–1702.

34. Stepanchick A, McKenna J, McGovern O, Huang Y, Breitweiser GE. Calcium sensing receptor mutations implicated in pancreatitis and idiopathic epilepsy syndrome disrupt an arginine-rich retention motif. *Cell Physiol Biochem*. 2010;26:363–374.
35. Gama L, Breitwieser GE. A carboxyl-terminal domain controls the cooperativity for extracellular Ca^{2+} activation of the human calcium sensing receptor. A study with receptor-green fluorescent protein fusions. *J Biol Chem*. 1998;273:29712–29718.
36. Chang W, Pratt S, Chen TH, Bourguignon L, Shoback D. Amino acids in the cytoplasmic C terminus of the parathyroid Ca^{2+}-sensing receptor mediate efficient cell-surface expression and phospholipase C activation. *J Biol Chem*. 2001;276:44129–44136.
37. Simonds WF, James-Newton LA, Agarwal SK, et al. Familial isolated hyperparathyroidism: clinical and genetic characteristics of 36 kindreds. *Medicine (Baltimore)*. 2002;81:1–26.
38. Nissen PH, Christensen SE, Heickendorff L, Brixen K, Mosekilde L. Molecular genetic analysis of the calcium sensing receptor gene in patients clinically suspected to have familial hypocalciuric hypercalcemia: phenotypic variation and mutation spectrum in a Danish population. *J Clin Endocrinol Metab*. 2007;92:4373–4379.
39. Ray K, Hauschild BC, Steinbach PJ, Goldsmith PK, Hauache O, Spiegel AM. Identification of the cysteine residues in the amino-terminal extracellular domain of the human Ca(2+) receptor critical for dimerization. Implications for function of monomeric Ca(2+) receptor. *J Biol Chem*. 1999;274(39):27642–27650.
40. Ray K, Clapp P, Goldsmith PK, Spiegel AM. Identification of the sites of N-linked glycosylation on the human calcium receptor and assessment of their role in cell surface expression and signal transduction. *J Biol Chem*. 1998;273:34558–34567.
41. Pidasheva S, Grant M, Canaff L, Ercan O, Kumar U, Hendy GN. Calcium-sensing receptor dimerizes in the endoplasmic reticulum: biochemical and biophysical characterization of CASR mutants retained intracellularly. *Hum Mol Genet*. 2006;15(14):2200–2209.
42. Fan GF, Ray K, Zhao XM, Goldsmith PK, Spiegel AM. Mutational analysis of the cysteines in the extracellular domain of the human Ca^{2+} receptor: effects on cell surface expression, dimerization and signal transduction. *FEBS Lett*. 1998;436(3):353–356.
43. Ray K, Hauschild BC. Cys-140 is critical for metabotropic glutamate receptor-1 dimerization. *J Biol Chem*. 2000;275(44):34245–34251.
44. Romano C, Miller JK, Hyrc K, et al. Covalent and noncovalent interactions mediate metabotropic glutamate receptor mGlu5 dimerization. *Mol Pharmacol*. 2001;59(1):46–53.
45. Magalhaes AC, Dunn H, Ferguson SS. Regulation of GPCR activity, trafficking and localization by GPCR-interacting proteins. *Br J Pharmacol*. 2012;165(6):1717–1736.
46. Sato K, Nakano A. Mechanisms of COPII vesicle formation and protein sorting. *FEBS Lett*. 2007;581:2076–2082.
47. Yoo JS, Moyer BD, Bannykh S, Yoo HM, Riordan JR, Balch RE. Non-conventional trafficking of the cystic fibrosis transmembrane conductance regulator through the early secretory pathway. *J Biol Chem*. 2002;277:11401–11409.
48. Zheng H, McKay J, Buss JE. H-Ras does not need COPI- or COPII-dependent vesicular transport to reach plasma membrane. *J Biol Chem*. 2007;282:25760–25768.
49. Dong C, Zhou F, Fugetta EK, Filipeanu CM, Wu G. Endoplasmic reticulum export of adrenergic and angiotensin II receptors is differentially regulated by Sar1 GTPase. *Cell Signal*. 2008;6:1035–1043.
50. Dong C, Filipeanu CM, Duvernay MT, Wu G. Regulation of G-protein-coupled receptor export trafficking. *Biochim Biophys Acta*. 2007;1768:853–870.
51. Barlowe C. Signals for COPII-dependent export from the ER: what's the ticket out? *Trends Cell Biol*. 2003;13:295–299.
52. Zhuang X, Chowdhury S, Northup JK, Ray K. Sar-1 dependent trafficking of human calcium receptor to the cell surface. *Biochem Biophys Res Commun*. 2010;396: 874–880.

53. Zhuang X, Adipietro KA, Datta S, Northup JK, Ray K. Rab-1 small GTP-binding protein regulates cell surface trafficking of the human calcium-sensing receptor. *Endocrinology*. 2010;151:5114–5123.
54. Tsukumo Y, Tsukahara S, Saito S, Tsuruo T, Tomida A. A novel ER export signal: proline at the +2 position from the signal cleavage site. *J Biol Chem*. 2009;284: 27500–27510.
55. Dong C, Wu G. Regulation of anterograde transport of alpha2-adrenergic receptors by the N-termini at multiple intracellular compartments. *J Biol Chem*. 2006;281: 38543–38554.
56. Stepanchick A, Breitwieser GE. The cargo receptor p24A facilitates calcium sensing receptor maturation and stabilization in the early secretory pathway. *Biochem Biophys Res Commun*. 2010;395(1):136–140.
57. Martinez O, Goud B. Rab proteins. *Biochim Biophys Acta*. 1998;1404(1–2):101–112.
58. Wu G, Zhao G, He Y. Distinct pathways for the trafficking of angiotensin II and adrenergic receptors from the endoplasmic reticulum to the cell surface: Rab1-independent transport of a G protein-coupled receptor. *J Biol Chem*. 2003;278(47): 47062–47069.
59. Dong C, Wu G. Regulation of anterograde transport of adrenergic and angiotensin II receptors by Rab2 and Rab6 GTPases. *Cell Signal*. 2007;19(11):2388–2399.
60. Bomberger JM, Parameswaran N, Spielman WS. Regulation of GPCR trafficking by RAMPs. *Adv Exp Med Biol*. 2012;744:25–37.
61. McLatchie LM, Fraser NJ, Main MJ, et al. RAMPs regulate the transport and ligand specificity of the calcitonin-receptor-like receptor. *Nature*. 1998;393(6683): 333–339.
62. Bouschet T, Martin S, Henley JM. Regulation of calcium sensing receptor trafficking by RAMPs. *Adv Exp Med Biol*. 2012;744:39–48.
63. Grant MP, Stepanchick A, Cavanaugh A, Breitwieser GE. Agonist-driven maturation and plasma membrane insertion of calcium-sensing receptors dynamically control signal amplitude. *Sci Signal*. 2011;4(200):ra78.
64. Arulpragasam A, Magno AL, Ingley E, et al. The adaptor protein 14-3-3 binds to the calcium-sensing receptor and attenuates receptor-mediated Rho kinase signalling. *Biochem J*. 2012;441(3):995–1006.
65. Huang Y, Niwa J, Sobue G, Breitwieser GE. Calcium-sensing receptor ubiquitination and degradation mediated by the E3 ubiquitin ligase dorfin. *J Biol Chem*. 2006;281(17):11610–11617.
66. Lorenz S, Frenzel R, Paschke R, Breitwieser GE, Miedlich SU. Functional desensitization of the extracellular calcium-sensing receptor is regulated via distinct mechanisms: role of G protein-coupled receptor kinases, protein kinase C and beta-arrestins. *Endocrinology*. 2007;148:2398–2404.
67. Pi M, Oakley RH, Gesty-Palmer D, et al. Beta-arrestin and G protein receptor kinase-mediated calcium-sensing receptor desensitization. *Mol Endocrinol*. 2005;19: 1078–1087.
68. Reyes-Ibarra AP, Garcia-Regalado A, Ramirez-Rangel I, et al. Calcium-sensing receptor endocytosis links extracellular calcium signaling to parathyroid hormone-related peptide secretion via a Rab11a-dependent and AMSH-sensitive mechanism. *Mol Endocrinol*. 2007;21:1394–1407.
69. Tu C, Chang W, Bikle D. The calcium-sensing receptor-dependent regulation of cell-cell adhesion and keratinocyte differentiation requires Rho and filamin A. *J Invest Dermatol*. 2011;131:1119–1128.
70. Charlton S. Agonist efficacy and receptor desensitization: from partial truths to a fuller picture. *Br J Pharmacol*. 2009;158:165–168.
71. Rodriguez L, Tu C, Cheng Z, et al. Expression and functional assessment of an alternatively spliced extracellular Ca^{2+}-sensing receptor in growth plate chondrocytes. *Endocrinology*. 2005;146(12):5294–5303.

72. Bonomini M, Giardinelli A, Morabito C, et al. Calcimimetic R-568 and its enantiomer S-568 increase nitric oxide release in human endothelial cells. *PLoS One*. 2012;7(1):e30682.
73. McCormick WD, Atkinson-Dell R, Campion KL, Mun HC, Conigrave AD, Ward DT. Increased receptor stimulation elicits differential calcium-sensing receptor(T888) dephosphorylation. *J Biol Chem*. 2010;285(19):14170–14177.
74. Grant MP, Stepanchick A, Breitwieser GE. Calcium signaling regulates trafficking of familial hypocalciuric hypercalcemia (FHH) mutants of the calcium sensing receptor. *Mol Endocrinol*. 2012;26(12):2081–2091.
75. Leach K, Wen A, Cook AE, Sexton PM, Conigrave AD, Christopoulos A. Impact of clinically relevant mutations on the pharmacoregulation and signaling bias of the calcium-sensing receptor by positive and negative allosteric modulators. *Endocrinology*. 2013;154(3):1105–1116.
76. Breitweiser G. The calcium sensing receptor life cycle; trafficking, cell surface expression, and degradation. *Best Practice & Research Clinical Endocrinology & Metabolism*. 2013;27:303–313.
77. Cavanaugh A, McKenna J, Stepanchick A, Breitwieser GE. Calcium-sensing receptor biosynthesis includes a cotranslational conformational checkpoint and endoplasmic reticulum retention. *J Biol Chem*. 2010;285(26):19854–19864.
78. Kifor O, Diaz R, Butters R, Kifor I, Brown EM. The calcium-sensing receptor is localized in caveolin-rich plasma membrane domains of bovine parathyroid cells. *J Biol Chem*. 1998;273(34):21708–21713.
79. Sowa G. Caveolae, caveolins, cavins, and endothelial cell function: new insights. *Front Physiol*. 2012;2:120.
80. Hjälm G, MacLeod RJ, Kifor O, Chattopadhyay N, Brown EM. Filamin-A binds to the carboxyl-terminal tail of the calcium-sensing receptor, an interaction that participates in CaR-mediated activation of mitogen-activated protein kinase. *J Biol Chem*. 2001; 276(37):34880–34887.
81. Huang C, Miller RT. The calcium-sensing receptor and its interacting proteins. *J Cell Mol Med*. 2007;11:923–934.
82. Zhang M, Breitwieser GE. High affinity interaction with filamin A protects against calcium-sensing receptor degradation. *J Biol Chem*. 2005;280:11140–11146.
83. Walther C, Ferguson SS. Arrestins: role in the desensitization, sequestration, and vesicular trafficking of G protein-coupled receptors. *Prog Mol Biol Transl Sci*. 2013; 118:93–113.
84. Zhuang X, Northup JK, Ray K. Large putative PEST-like sequence motif at the carboxyl tail of human calcium receptor directs lysosomal degradation and regulates cell surface receptor level. *J Biol Chem*. 2012;287(6):4165–4176.
85. Marchese A, Paing MM, Temple BR, Trejo J. G protein-coupled receptor sorting to endosomes and lysosomes. *Annu Rev Pharmacol Toxicol*. 2008;48:601–629.
86. Young SH, Rozengurt E. Amino acids and Ca2+ stimulate different patterns of Ca2+ oscillations through the Ca2+-sensing receptor. *Am J Physiol Cell Physiol*. 2002;282(6): C1414–C1422.
87. Holstein OM, Berg KA, leeb-lundberg lM, et al. Calcium-sensing receptor-mediated ERKl/2 activation requires Galphi2 coupling and dynamin-independent receptor internalization. *J Biol Chem*. 2004;279:10060–10069.
88. Rey O, Young SH, Yuan J, et al. Amino acid-stimulated Ca2∼ oscillations produced by the Ca2∼-sensing receptor are mediated by a phospholipase C/inositoll,4, S-trisphosphate-independent pathway that requires G12, Rho, filamin-A, and the actin cytoskeleton. *J Biol Chem*. 2005;280:22875–22882.
89. Tharmalingam S, Daulat AM, Antflick JE, et al. Calcium-sensing receptor modulates cell adhesion and migration via integrins. *J Biol Chem*. 2011;286:40922–40933.
90. Pi M, Spurney RF, Tu Q, et al. Calcium-sensing receptor activation of rho involves filamin and rhoguanine nucleotide exchange factor. *Endocrinology*. 2002;143: 3830–3838.

CHAPTER SEVEN

Trafficking of β-Adrenergic Receptors: Implications in Intracellular Receptor Signaling

Qin Fu*, Yang K. Xiang†,[1]
*Department of Pharmacology, School of Basic Medicine, Tongji Medical College, Huazhong University of Science and Technology, Wuhan, PR China
†Department of Pharmacology, University of California, Davis California, USA
[1]Corresponding author: e-mail address: ykxiang@ucdavis.edu

Contents

Progress in Molecular Biology and Translational Science, Volume 132
ISSN 1877-1173
http://dx.doi.org/10.1016/bs.pmbts.2015.03.008

Abstract

β-Adrenergic receptors (βARs), prototypical G-protein-coupled receptors, play a pivotal role in regulating neuronal and cardiovascular responses to catecholamines during stress. Agonist-induced receptor endocytosis is traditionally considered as a primary mechanism to turn off the receptor signaling (or receptor desensitization). However, recent progress suggests that intracellular trafficking of βAR presents a mean to translocate receptor signaling machinery to intracellular organelles/compartments while terminating the signaling at the cell surface. Moreover, the apparent multidimensionality of ligand efficacy in space and time in a cell has forecasted exciting pathophysiological implications, which are just beginning to be explored. As we begin to understand how these pathways impact downstream cellular programs, this will have significant implications for a number of pathophysiological conditions in heart and other systems, that in turn open up new therapeutic opportunities.

1. INTRODUCTION

G-protein-coupled receptors (GPCRs) constitute the largest family of membrane-bound receptors, initiate diverse signal processes like neurotransmission, metabolism, cell growth, and immune response, and are targets for many clinically used drugs.[1] β-Adrenergic receptors (βARs), a family of prototypical GPCRs, play a pivotal role in regulating cardiovascular response to catecholamines during stress.[2,3] Over the past three decades, a wealth of studies have revealed extensively the signaling properties of βARs. Much work has been focused on illustrating the ways in which βARs regulate discrete effector molecules including adenylyl cyclases, phosphodiesterases (PDEs), phospholipases, and ion channels in animal hearts. Still further work has shed light on molecular mechanisms by which βAR signaling is regulated and has led to the discovery of additional proteins including G-protein receptor kinases (GRKs)[4,5] and β-arrestins,[6] which respectively, phosphorylate agonist-activated βAR and bind phosphorylated βAR. The binding of β-arrestin to phosphorylated βAR is postulated as a mean to physically disrupt receptor/G-protein interaction, thereby leading to desensitization of receptor-mediated G-protein activation. Comparatively, less work has focused on βAR trafficking, much of it is related to mechanisms regulating endocytosis of βAR from the cell surface, including the role of β-arrestin in facilitating βAR endocytosis. Mechanistic understanding of βAR sorting in the endocytic pathway is only beginning to emerge. This chapter highlights our current understanding of intracellular distribution and trafficking of βAR with emphasis on recent progress made on βAR signaling in subcellular compartments/organelles such as endosomes. We focus on the role of receptor distribution and trafficking that might be involved in βAR signaling and their possible pathophysiological and pharmacological implications in heart.

2. CELL SURFACE βAR DISTRIBUTION AND SIGNALING

Several studies indicate that proteins involved in adrenergic signaling (e.g., Gβγ, $G\alpha_s$, adenylate cyclase, and βAR) colocalize within same microdomains, possibly because they contain an "address" for that specific domain.[7] The cytoplasmic tail of βAR is a potential region that likely contains an address site.[7] Indeed, scaffold proteins containing PDZ domains and membrane structural protein caveolins have been shown to associate with the C-termini of βAR.[8,9] Thus, cellular distribution and trafficking is precisely controlled by regulation of the receptor and its association with a variety of scaffold/structural proteins.[8] Such arrangement is essential for a specific cellular action under receptor stimulation.

2.1 β_1 Adrenergic Receptor and β_2 Adrenergic Receptor Distribution in Cardiomyocytes

Stimulation of βAR by the endogenous agonist noradrenaline and adrenaline represents the major mechanism to increase cardiac chronotropy and inotropy. β_1 adrenergic receptor (β_1AR) and β_2 adrenergic receptor (β_2AR) are found on the surface of cardiac muscle cells (cardiomyocytes), coupling primarily to G_s to promote production of the common second messenger cyclic adenosine monophosphate (cAMP),[2] whereas coupling of β_2AR to G_i has been described in several animal species and in failing human cardiomyocytes. The second messenger cAMP then leads to activation of exchange protein directly activated by cAMP and protein kinase A (PKA). The latter phosphorylates key regulators of cardiac excitation/contraction machinery, including L-type Ca^{2+} channel, phospholamban, ryanodine receptor, and troponin T and I. However, selective stimulation of these two receptor subtypes leads to distinct physiological and pathophysiological responses. Stimulation of β_1AR, but not β_2AR, by moderate overexpression[10] induces hypertrophy and promotes cardiomyocyte apoptosis.[11,12] In contrast, activation of β_2AR has an anti apoptotic effect in both rat and mouse adult cardiac myocytes, which is mediated by βγ subunits of G_i.[13] Upon local receptor stimulation, β_1AR-mediated cAMP signaling propagated throughout the entire cell, whereas β_2AR-cAMP responses were locally confined.[14] Yet, these findings cannot be sufficiently explained by differential coupling of β_1AR and β_2AR to G_s and G_i proteins. Differences between β_1- and β_2AR signaling have also been attributed to compartmentation of local signaling events, such as formation of signalosomes[15] and localized control of cAMP degradation through PDEs.[16] These studies raised the possibility that the precise distribution of the β_1- and β_2ARs is related to highly organized sarcomeric structure of cardiomyocytes and its potential functional implications.

Indeed, lipid raft or caveolae have been identified as membrane subdomains that compartmentalize βAR signaling pathways.[17] Lipid rafts are generally used to refer to the small, mobile membrane subdomains dispersed over the cell surface of mammalian cells. Lipid rafts are enriched in cholesterol and sphingolipids and often contain associated proteins such as caveolins, flotillins, and stomatins, which may serve as scaffolds for signaling complexes.[18] Caveolae are a type of lipid raft distinguished by the presence of caveolins, a family of $\approx$20 kDa cholesterol-binding proteins, which line the internal surface of caveolae and promote their typical flask-like shape.[19] Thus, caveolae may act as a scaffold promoting interaction of specific signaling molecules for local actions. A number of studies have demonstrated that caveolae are enriched in components of signal transduction cascades, including G proteins, GPCRs, and effector molecules.[17] Immunofluorescence microscopy demonstrated colocalization of βAR with caveolin, indicating a nonrandom distribution of βAR in the plasma membrane. Using polyhistidine-tagged recombinant proteins, βAR were copurified with caveolin, suggesting that they were physically bound. These results also suggest that caveolae may act as platforms to nucleate, regulate, or propagate βAR-dependent cAMP signals. However, studies that examined the subcellular localization of βAR and adenylate cyclase in adult cardiomyocytes showed that cholesterol depletion with cyclodextrin augments agonist-stimulated cAMP accumulation, indicating that caveolae function as negative regulators of cAMP accumulation. The inhibitory interaction between caveolae and the cAMP signaling pathway as well as domain-specific differences in the stoichiometry of individual elements in βAR signaling cascades represent important modifiers of cAMP-dependent signaling.[20]

2.2 β_2AR Signaling Components Localize in Lipid Raft Caveolin-3-Associated Microdomains

Studies identify that caveolae contains all of β_2AR and only a fraction of β_1AR, whereas the remainder of the plasma membrane contains only

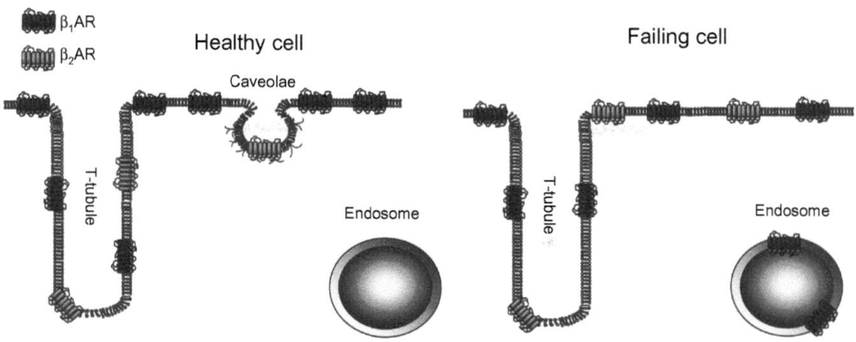

Figure 1 Distribution of βAR in heart failure. In normal cardiac myocytes, β_1AR is distributed between the T-tubules and the crest membrane and β_2AR in T-tubules and caveolae membrane. In failing cardiac myocytes, β_1AR is relocated to endosomes, whereas β_2AR is not confined in caveolae due to reduction of numbers of caveolae. (See the color plate.)

β_1AR (Fig. 1). Overall, the vast majority of the β_1AR population is excluded from caveolae.[20,21] Differential modes for β_1- and β_2AR regulation of adenylyl cyclase in cardiomyocytes are consistent with spatial distribution of receptor signaling components in microdomains of the plasma membrane.[20] Caveolae from quiescent rat ventricular cardiomyocytes are also highly enriched in Gα_i, PKA RIIα subunit, caveolin-3, and flotillins (caveolin functional homologues). In contrary, m_2 muscarinic cholinergic receptors, Gα_s, and cardiac type V/VI adenylyl cyclase distribute between caveolae and other cell fractions, whereas PKA RIα subunit, GRK2, and clathrin are largely excluded from caveolae. β_2AR is localized to caveolae in cardiomyocytes and cardiac fibroblasts (with markedly different β_2AR expression levels), indicating that the fidelity of β_2AR targeting to caveolae is maintained over a physiologic range of β_2AR expression. These differences in the spatial distribution of β_1- and β_2ARs as well as their downstream signaling proteins in cardiomyocytes might contribute to apparent compartmentation of βAR subtype-specific responses.

Both β_1- and β_2ARs mediate increases in cardiac contraction upon agonist stimulation through the G_s-adenylyl cyclase-cAMP-dependent PKA pathway.[22] Activated β_2AR will couple to G_s for a stimulatory effect on cAMP/PKA activities for increasing cardiac contraction at low concentrations of catecholamines. However, compared to β_1AR, in the presence of high concentrations of catecholamines, activated β_2AR will switch from G_s to G_i, which inhibits adenylyl cyclase to reduce cardiac contraction and initiates anti apoptotic and cell growth signaling.[23–26] Stimulation of β_1AR leads to a PKA-dependent increase in contraction rate.[21] β_2AR is confined to caveolae and induces a biphasic effect, which includes an initial PKA-independent increase in contraction rate followed by a sustained decrease in contraction rate that can be blocked by G_i inhibitor pertussis toxin, suggesting sequential β_2AR coupling to G_s followed by G_i.[21] The β_2AR-stimulated increase in myocyte contraction rate is enhanced and markedly prolonged by filipin (an agent that disrupts lipid rafts such as caveolae and significantly reduces coimmunoprecipitation of β_2AR and caveolin 3, and comigration of β_2AR and caveolin 3-enriched membranes). In contrast, filipin has no effect on β_1AR signaling. These observations suggest that β_2AR are normally restricted to caveolae in myocyte membranes and that localization is essential for physiologic signaling of this receptor subtype.[21] Disruption of caveolar structures in cardiomyocytes selectively enhances and prolongs the increase in myocyte contraction rate mediated by β_2AR activation, but has no effect on β_1AR signaling.[21]

Accordingly, several laboratories have demonstrated membrane partitioning of ion channels serving as βAR effectors. It has been reported that both sodium channels and $G\alpha_s$ are associated with caveolar membranes. Sodium channels present within the cardiomyocytes caveolar membrane are functionally capable of mediating the PKA-independent isoproterenol-induced increase in sodium current in heart.[27] Moreover, shaker-like potassium channels reside in caveolae and are regulated by local lipid microenvironment based on protein-lipid interactions.[28] Caveolae have been implicated in control of local sarcoplasmic reticulum calcium release events, thereby playing a role in triggering calcium sparks (transient and spatially localized elevations in $[Ca^{2+}]_i$).[29] Alterations in molecular assembly and ultrastructure of caveolae may lead to pathophysiological changes in Ca^{2+} signaling. Studies using cell-attached patch-clamp technique show that β_2AR signaling can modulate L-type Ca^{2+} channel activity in distinct subcellular microdomains in cardiomyocytes.[30] Stimulation of β_1AR or β_2AR in the patch membrane, by adding agonist into patch pipette, activates L-type calcium channels in the patch. But when agonist is applied to the membrane outside the patch pipette, only β_1AR stimulation activates the channels in the patch. Thus, β_1AR signal globally and activate L-type calcium currents throughout the cell, whereas β_2AR signaling is localized to the cell membrane.[30] Furthermore, in cardiac muscle, Ca^{2+} sparks are induced by membrane potential-dependent entry of Ca^{2+} through L-type Ca^{2+} channels at transverse tubules. These results suggest that channel localization to caveolae might be as yet another generalized mechanism to regulate βAR signaling in excitable tissues. Thus, caveolae may be intimately involved in cardiovascular diseases by regulating βAR signaling.

2.3 βAR Redistribution in Heart Failure

Heart failure is a common clinical syndrome that results from virtually all forms of cardiac diseases and it is consistently characterized by extensive abnormalities in the βAR system. Considerable evidence supports the concept that chronic increase in circulating catecholamine levels is largely responsible for the βAR abnormalities found in failing hearts.[31] Agonist-induced receptor dysfunction begins with βAR phosphorylation by GRK2, followed by β-arrestin binding that may sterically interdicts further G protein coupling and initiates the process of receptor internalization.[31] Once internalized, receptors are targeted to specialized intracellular compartments, where they can be dephosphorylated and recycled to the plasma membrane (early endosomes) or sent to degradation pathway (late endosomes).[31] According to this traditional paradigm, a combination of increased rates of β_1AR lysosomal degradation and reduced receptor transcript determines selective β_1AR downregulation at the plasma membrane under conditions of heart failure.[32] However, Rockman laboratory has shown that chronic catecholamine stimulation and heart failure lead to a loss of βAR on the plasma membrane and a redistribution of receptors to endosomal compartments (Fig. 1). Seven-day isoproterenol administration in wild-type mice induced desensitization of βAR and their redistribution from the plasma membrane to early and late endosomes.[33] An emerging concept points out that internalization of βAR is a pathological process *per se*, because it directly activates maladaptive signaling pathways in a G-protein-independent fashion. Therefore, strategies that prevent this redistribution may exert a beneficial effect in heart failure.

In pigs with pacing-induced heart failure, βAR redistribution to intracellular compartments was accompanied by a significant increase in membrane-targeted phosphoinositide 3-kinase (PI3K) activity.[33] Rockman laboratory has recently shown that efficient βAR internalization requires recruitment of PI3K to agonist-stimulated βAR.[34,35] This process depends on the cytosolic association of PI3K with GRK2 through the helical domain of PI3K, also known as the phosphoinositide kinase domain.[34,35] Importantly, targeted PI3K inhibition prevents β_1AR sequestration into endosomal compartments despite chronic agonist stimulation and reverses β_1AR abnormalities in a pig model of heart failure.[33] Consistently, failing human hearts displayed a marked increase in GRK2-associated PI3K activity that was attributed exclusively to enhanced activity of the PI3Kγ-isoform.[36] Increased GRK2-coupled PI3K activity in failing hearts were associated with downregulation of β_1AR from the plasma membrane and enhanced sequestration into early and late endosomes when compared with unmatched nonfailing controls. Importantly, left ventricular assist device (LVAD) support reversed PI3K activation, normalized the levels of agonist-responsive βAR at the plasma membrane, and depleted βAR from endosome compartments without changing the total number of receptors (sum of plasma membrane and early and late endosome receptors).[36] These results suggest a novel paradigm in which human βAR undergoes a process of intracellular sequestration that is dynamically reversed after LVAD support. These findings indicate that heart failure state is associated with a maladaptive redistribution of βAR away from the plasma membrane. This is an important concept and a shift from the current understanding wherein it is thought that receptors targeted for degradation are trafficked to late endosomes.[37] The dynamic sequestration of βAR into endosomal compartments may represent an important mechanism to regulate adrenergic responsiveness in failing human heart that can be counteracted through a strategy that targets formation of GRK2/PI3K complex. Thus, targeting the GRK2/PI3K complex with molecular interventions or LVAD support represents a novel approach to restore βAR function in heart failure. This would prevent βAR redistribute into intracellular pools, preserving their plasma membrane levels and restoring their capability to properly signal without interfering with GRK2 phosphorylation of activated receptors[35] or with other PI3K downstream signaling pathways.

The spatial localization of β_2AR in caveolae/lipid rafts and the compartmentation of their signaling are also thought to play a critical role in cardiac physiology and development of heart failure.[13,38,39] In both human and murine heart failure, left ventricle dysfunction is associated with selective decrease in expression of caveolin 3, and reduced caveolae on the membrane[40]. Using the SICM–FRET technique, Gorelik laboratory has identified that in normal cardiomyocytes, β_2AR is exclusively localized to the T-tubules, whereas β_1AR are present in both the cell crests and the T-tubules.[41] The interaction of β_2AR with lipid rafts might be responsible for T-tubule-selective localization of this receptor.[39] The investigators found that β_1AR-cAMP signals were detectable in both cell crests and T-tubular regions, whereas β_2AR showed locally confined cAMP signals in the T-tubules. In heart failure, β_2AR redistributed from the T-tubules to the cell crest (Fig. 1). Membrane cholesterol depletion by methyl-β-cyclodextrin did not cause any loss of T-tubules but induced β_2AR redistribution and propagation of β_2AR-cAMP signals from the crest of the cell.[41] Redistribution of β_2AR from the T-tubules to the cell crest in failing cardiomyocytes and the loss of proper PKA localization, observed also in human heart failure,[42] result in uncoupling of β_2AR from the localized pools of PKA that are responsible for the compartmentation of the β_2AR-cAMP signaling. Thus, in failing cells, activation of β_2AR leads to cell-wide cAMP signal propagation, similar to the patterns observed for β_1AR. Upon redistribution of the receptor, β_2AR signaling may lose normal cardioprotective properties and acquire the characteristics of the β_1AR response, thus contributing to the heart failure phenotype.[41] These findings suggest that the interaction of β_2AR with cholesterol-rich membrane domains is important for normal β_2AR localization and signal compartmentation.

3. MOLECULAR MACHINERY FOR βAR ENDOCYTOSIS

The distinct cell signaling processes and endocytic membrane trafficking are intimately and bidirectionally linked in animal cells. The activation of receptors or downstream effectors often stimulates receptor endocytosis. Endocytosis is increasingly understood to play crucial roles in many signaling pathways, from determining which signaling components are activated, to how the signal is subsequently transduced and/or terminated. Endocytosis of GPCRs regulates the long-term sensitivity of cells to their specific ligands simply by controlling the number of receptors available for activation in the plasma membrane, and functions as a homeostatic regulatory loop to prevent excessive ligand-induced activation of downstream effectors.

Entering the endocytic pathway removes the receptors from the surface; subsequently, receptors are sorted into various endosomal compartments. The structural and molecular machinery utilized can serve to define distinct endocytic pathways, including clathrin-dependent,[43] caveolae-dependent,[44] and clathrin/caveolae-independent pathways (Fig. 2). Whether a receptor-ligand complex is internalized via a clathrin-dependent or a clathrin-independent endocytic route, the internalized cargo is trafficked into endosomes, where it is sorted either back to the cell surface or into other compartments (multivesicular bodies and lysosomes) for degradation. Although these pathways are spatially distinct, they utilize a subset of overlapping machinery and in some cases may merge within the cell, suggesting that the various endocytic pathways are highly integrated. The receptor complex trafficking through specific endocytic compartments has profound effects on its signaling output.

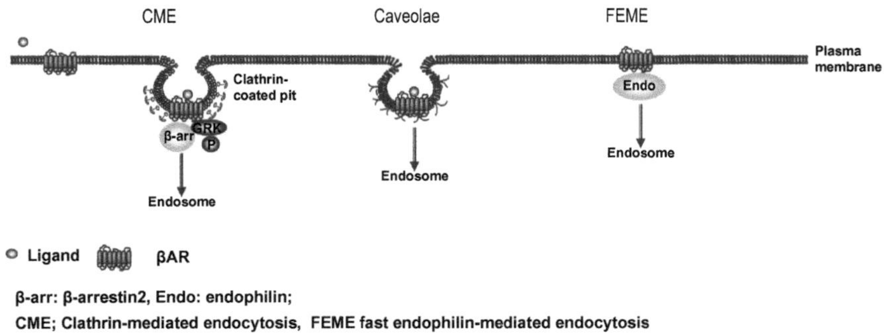

Figure 2 βAR endocytosis machinery. Three major endocytic machineries involved βAR endocytosis. β_1AR utilizes clathrin-dependent and -independent (FEME and caveolae-mediated) endocytosis. In contrast, β_2AR undergoes endocytosis primarily through clathrin-dependent pathway. (See the color plate.)

3.1 Clathrin-Dependent βAR Endocytosis

Clathrin-mediated endocytosis (CME) is an extremely important endocytic mechanism. β_2AR, like many other GPCRs, undergo rapid endocytosis through clathrin-coated pits (CCPs), and subsequently form clathrin-coated vesicles (CCVs).[45] Agonist binding to β_2AR leads to receptor activation and G-protein coupling. GRKs then phosphorylate agonist-activated β_2AR, initiating arrestin recruitment. β_2AR/arrestin complexes are then targeted to CCPs, where arrestin forms a multicomponent complex with clathrin, AP-2, and phosphoinositides, resulting in β_2AR internalized in CCPs in a dynamin-dependent fashion and then proceeded to tubular-vesicular early endosomes.[46] There, β_2AR is subsequently sorted to either recycling endosomes, which traffic receptors back to the plasma membrane, or multivesicular late endosomes, which traffic receptors to lysosomes for degradation. Overexpression of β-arrestin can rescue a β_2AR sequestration-defective mutant.[47] In addition, overexpression of dominant negative forms of β-arrestin or other endocytic proteins related to the clathrin pathway, such as dynamin, inhibit β_2AR internalization.[47] Thus, β-arrestin plays pivotal roles as adapters and scaffolds in the process of β_2AR endocytosis through CCVs.

Most GPCRs use arrestin as an adapter for CCP targeting. However, there are a few notable examples in which GPCRs directly bind to AP-2 via C-tail motifs.[46] Internalization motifs that target GPCRs to localized CCP zones[48] include the classical dileucine- or tyrosine-based (YXXφ, where φ is a residue with a bulky hydrophobic side chain) motifs described for sorting of non-GPCRs. In general, these motifs mediate direct binding to AP-2 complexes[49] in a lipid-dependent fashion[50] and in turn promote binding to clathrin endocytic components at the plasma membrane.

3.2 Clathrin-Independent βAR Endocytosis

Although CME is an extremely important endocytic mechanism, accounting for a large proportion of endocytic events, an ever expanding array of cargoes has been shown to undergo endocytosis in a clathrin-independent manner.[51,52] Much of the budding that occurs from the plasma membrane does not appear to require the formation of clathrin coat, which is readily observable around CCPs and CCVs. Indeed, cells devoid of CME events are capable of endocytosis. Clathrin-dependent and -independent endocytosis are expected to be roughly equivalent modes of internalization in fibroblastic cells, and the relative proportions of each will presumably differ in other cell types owing to their adaptions for specific function. Clathrin-independent endocytosis itself has been further dissected into seemingly distinct pathways, based on the reliance of these pathways on certain proteins and lipids, their differential drug sensitivities and their abilities to internalize particular cargoes.[53]

The Bin/amphiphysin/Rvs (BAR) domain-containing protein endophilin is an endocytic protein that recruits dynamin and synaptojanin,[54] and the disruption of which has a profound effect on endocytosis.[55] The latest study presents evidence that endophilin functions at the nexus of a clathrin-independent, dynamin-dependent pathway of β_1AR endocytosis, operating from distinct regions of the cells, on a different timescale to CME.[56] Building an endocytic vesicle requires cargo recruitment adaptors, membrane curvature effectors and a membrane scission machinery.[57] Endophilin has all these characteristics in one protein, thus explaining its central role in fast endophilin-mediated endocytosis (FEME) carrier formation. Its SH3 domain binds to cargo β_1AR, its BAR domain induces membrane curvature, and by insertion of its multiple amphipathic helices, it can support membrane scission[41], aided by recruitment of dynamin.[58] Interestingly, endophilin binds directly to a proline-rich motif present in the third intracellular loop of β_1AR but not β_2AR.[59] Mechanistically, β_1AR needs to be activated cargoes, and PtdIns(3,4)P_2, produced from the dephosphorylation of PtdIns(3,4,5)P_3 by SHIP phosphatases, mediates the engagement of lamellipodia, which in turn recruits endophilin at the plasma membrane.[56] These findings reveal that endophilin is not simply a peripheral component of CME but principally marks and controls a distinct endocytic pathway.

3.3 Caveolin-Dependent βAR Endocytosis

Although very different structurally, both caveolae and CCVs serve as scaffolds that integrate signal-transduction complexes, providing microdomains for cross talk between specific signaling molecules.[60] The apparent association of β_2AR with caveolin-enriched membranes is disrupted following agonist-induced internalization,[20] presumably as activated β_2AR undergoes clathrin-dependent endocytosis.[61] In contrast, β_1AR partitioning between caveolae membranes and noncaveolar cell surface membranes is not detectably altered by treatment with βAR agonist isoproterenol. Evidence suggests that the pathway selected for β_1AR internalization in fibroblasts is primarily determined by the kinase that phosphorylates the receptor. GRK-mediated phosphorylation directs internalization through a CCP pathway,[62] whereas PKA-mediated phosphorylation directs internalization via a caveolae pathway.[63] Importantly, at maximally efficacious concentration of agonist, β_1AR endocytosis occurs via both CCPs and caveolae. Each pathway contributes approximately to half of the observed response, and the two pathways are additive. Thus, endocytosis through CCPs cannot compensate for loss of the caveolar pathway, and vice versa. However, PKA-mediated phosphorylation still undergoes significant internalization at low agonist concentrations, suggesting that the clathrin-independent mechanism of β_1AR internalization does contribute to receptor endocytosis over a wide range of agonist concentrations.[63] These data contrast dramatically with findings in cardiomyocytes, in which β_1AR display minimal endocytosis under acute stimulation with agonist.[64] These data are also in contrast to those obtained using PKA and GRK mutants of β_2AR. Although both PKA and GRK phosphorylation contribute to desensitization of β_2AR,[65] PKA phosphorylation does not play a significant role in endocytosis of this receptor.[65]

4. POSTTRANSLATIONAL MODIFICATIONS OF βAR IN TRAFFICKING AND SIGNALING

The seven-membrane spanning characteristic of βAR gives rise to several structural features. Each protein has an N-terminal extracellular domain; the seven-transmembrane helices, which also define three extracellular and three intracellular loops; and an intracellular carboxyl-terminal domain.

To this topographical organization, posttranslational modifications add an additional level of complexity. βARs are also extensively phosphorylated by several kinases. The sites of phosphorylation have been mapped mainly to the carboxyl tail and the third intracellular loop, and have been linked to regulatory processes, such as desensitization and internalization.[66,67] While GPCR signaling is typically regulated by desensitization and endocytosis mediated by phosphorylation and β-arrestins, it can also be modulated by ubiquitination. Ubiquitination is emerging as an important regulatory process that may have unique roles in governing GPCR trafficking and signaling. In addition, βARs are subject to covalent modification with fatty acid palmitate. Here, we will focus on these posttranslational modifications, the regulation and the roles in βAR endosome trafficking events.

4.1 Agonist-Dependent Phosphorylation in βAR Trafficking and Signaling

Reversible posttranslational modification or protein interaction occurring in the plasma membrane represents a key principle by which GPCR trafficking itineraries are specified and regulated. Phosphorylation of GPCRs has long been known to influence receptor function and trafficking.[68] It may initiate sorting by "tagging" receptors for a subsequent trafficking fate.[69] Phosphorylation of βAR in the plasma membrane can also influence later sorting events, either by initiating additional posttranslational modification or by controlling receptor interaction with downstream sorting proteins. Phosphorylation of the receptor by second messenger (cAMP) activated kinases (like PKA or PKC) and GRKs promotes functional uncoupling of activated βAR from its cognate G proteins.[70] In this context, PKA or PKC activated as a consequence of βAR stimulation can phosphorylate βAR to reduce G-protein coupling.[71] PKA/PKC can also phosphorylate βAR independent of ligand occupancy or activation status.[23,71] While PKC-mediated phosphorylation contributes to internalization of β_2AR, and it does not play a role in the internalization of β_1AR.[62,63] In comparison, phosphorylation of βAR by GRKs results in recruitment of β-arrestin to the agonist-occupied receptor complex, which not only sterically hinders G-protein coupling but also prepares the receptor toward internalization.[72] β-arrestin falls off the βAR complex after internalization.[72] Internalized βAR is directed to recycling endosomes, wherein they are dephosphorylated and recycled back to the plasma membrane as naive receptors ready for new stimulation (resensitized) or trafficked to lysosomes for degradation.[37,73]

Early studies suggest that PKA phosphorylation of β_2AR enhances the receptor affinity to G_i while reducing affinity to G_s[23,74], thus mediating switch of coupling from G_s to G_i,[23,74] presumably operating in a feedback after receptor activation. Wang *et al.* have shown that β_2AR/G_i coupling is also dependent on receptor internalization and recycling.[75] Meanwhile, direct inhibition of GRK2 prevents β_2AR/G_i coupling in mouse cardiac myocytes,[75] supporting a role of the β_2AR phosphorylation by GRK in receptor/G_i coupling. Moreover, Liu *et al.* show that the receptor coupling to G_i proteins is differentially regulated by the PKA- and GRK-mediated phosphorylation of activated β_2AR.[24] At both low and high concentrations of agonist, activated β_2AR undergoes the PKA-mediated phosphorylation. In comparison, only high concentrations of agonist induce the GRK-mediated phosphorylation of β_2AR for subsequent internalization, which is also necessary for sufficient receptor coupling to G_i proteins.[21,75] One possibility is that the GRK-phosphorylated receptors undergo internalization to promote the access of the receptor to G_i protein. Another possibility is that the GRK-mediated phosphorylation of β_2AR directly enhances the binding affinity of the receptor to G_i protein.[24] These studies link together various components of β_2AR, including receptor phosphorylation, receptor trafficking, and differential receptor/G-protein coupling in cardiac cells, which may allow the activation of β_2AR signaling pathway to function as either a stimulatory or protective mechanism for cardiac cells under different levels of stress. Meanwhile, scaffold proteins containing PDZ domains have been shown to associate with the C-termini of βAR,[8] and these interactions can be disrupted through phosphorylation by GRK5.[76] Thus, cellular signaling and trafficking are also controlled by phosphorylation-dependent regulation of the receptor and its association with scaffold proteins.[8] In neonatal cardiac myocytes, the PDZ motif at the C-terminus of β_1AR is responsible for its limited internalization and that mutation of this domain increases internalization to levels similar to those observed with β_2AR.[63] Therefore, differential phosphorylation of β_1AR plays a critical role in determining its internalization pathway, indicating that the site of phosphorylation may serve as molecular address that directs receptor internalization.

4.2 Signaling Cross Talk Prompts β_2AR Phosphorylation-Mediated Internalization and G_i Coupling

β_2AR internalization is generally considered to be an agonist-dependent phenomenon; however, recent studies suggest that β_2AR can undergo endocytosis via signaling cross talk between β_2AR and other membrane receptors even in the absence of βAR agonist.[77] Tumor necrosis factor α (TNFα) is upregulated in conditions of cardiac stress and failure. Vasudevan *et al.* have reported that βAR dysfunction is independent of sympathetic overdrive in conditions of inflammation.[78] The observed βAR dysfunction is associated with selective upregulation of GRK2 in two proinflammatory mouse models of heart failure (Myo-Tg and TNFα-Tg). Studies from TNFR1 or TNFR2 knockout mice show that TNFR2 preferentially recruits GRK2-mediating β_2AR phosphorylation. Furthermore, *in vitro* and *in vivo* studies show that TNFα alone is sufficient to induce βAR dysfunction and TNFR2 preferentially recruits GRK2-mediating β_2AR phosphorylation. Importantly, TNFα-induced βAR internalization is agonist independent because βAR phosphorylation is observed despite the presence of the βAR antagonist propranolol. Moreover, cardiac ablation of GRK2 (GRK2 del) is able to normalize the reduction in myocyte contractility following pretreatment with TNFα. Therefore, these findings have identified that TNFα-induced agonist-independent βAR internalization is mediated by GRK2, uncovering a cross talk between TNFR2 and βAR function, providing the underpinnings of inflammation-mediated cardiac dysfunction.[78]

The neurohormone arginine vasopressin (AVP) is elevated in patients with heart failure, and there is a direct relationship between plasma levels of AVP and disease severity and mortality. Tilley *et al.* have reported that AVP acutely inhibits βAR-mediated cardiac contractility via a GRK-dependent and Gq protein-independent mechanism.[79] The AVP-induced and GRK-mediated inhibition of β_1AR signaling occurs even in the absence of Gq protein activity. With AVP pretreatment, they only observe β_1AR signaling when all possible C-terminal GRK phosphorylation sites on V1AR are mutated to alanines. Although the mutant V1AR construct reveals a required role for GRK-dependent regulation of βAR activity after acute V1AR stimulation, the precise mechanism of this effect remains to be defined. Because GRKs are primarily known for their role in phosphorylation-mediated GPCR endocytosis, a likely molecular explanation exists in which V1AR stimulation results in enhanced association of active GRK at the plasma membrane, which enables more rapid GRK-mediated phosphorylation of βAR on subsequent catecholamine stimulation, thereby enhancing the kinetics of βAR-G_s protein uncoupling and receptor endocytosis for reduced signal output. These results may explain the increased mortality observed in patients with acute heart failure and elevated AVP levels, and provide support for the potential use of a V1AR antagonist in treatment of these patients.[79]

Early studies indicate reciprocal regulation between βAR and insulin receptor (IR) signaling pathways.[80–82] Stimulation of IR promotes phosphorylation of βAR at threonine/serine and tyrosine residues, leading to receptor endocytosis.[83] More recently, evidence suggests the existence of a complex consisting of IR and β_2AR in heart, and stimulation of IR resulted in the reduction of association of the β_2AR–IR complex.[84] This is consistent with the study of BRET, which provides further evidence for direct interaction between β_2AR and IR.[85] Activation of IR with insulin induces PKA and GRK2 phosphorylation of β_2AR and also prompts β_2AR internalization by recruiting GRK2 to β_2AR.[77,84] Internalization of β_2AR selectively promotes G_i coupling to attenuate cAMP/PKA signaling, which inhibits contractile response in isolated neonatal and adult cardiomyocytes and in Langendorff perfused hearts.[84] These findings underscore the critical role of IR in β_2AR trafficking and signaling. It also opens a question whether other types of receptors regulate β_2AR trafficking. Further studies on functional cross talk between different receptors will be of translational significance, which may provide a potential general mechanism to understand cross talk between other classes of receptors and βAR regulatory systems in cardiac diseases.

4.3 Palmitoylation is Required for β-Arrestin 2-Mediated β_2AR Internalization

Palmitoylation is the covalent attachment of fatty acids, such as palmitic acid, to cysteine and less frequently to serine and threonine residues of proteins, which are typically membrane proteins. The precise function of palmitoylation depends on the particular protein being modified. Palmitoylation enhances the hydrophobicity of proteins and contributes to their membrane association. Palmitoylation also appears to play a significant role in subcellular trafficking of proteins between membrane compartments,[86,87] as well as in modulating protein–protein interactions.[88] β_2AR was the first hormone-binding GPCR for which palmitoylation was demonstrated[89]; this modification is a general feature of the GPCR superfamily. By site-directed mutagenesis of β_2AR, Cys341 in the carboxyl tail has been identified as the primary site of palmitoylation. Mutation of Cys341 to glycine results in a nonpalmitoylated form of the receptor that exhibits a drastically reduced ability to mediate isoproterenol stimulation of adenylyl cyclase. The functional impairment of this mutated β_2AR is also reflected in a markedly reduced ability to form a guanyl nucleotide-sensitive high-affinity state for agonists, characteristic of wild-type receptor. These results indicate that posttranslational modification of β_2AR by palmitate may play a crucial role in the normal coupling of the receptor to adenylyl cyclase signal transduction system.

Moreover, palmitoylation is required for β_2AR signaling formation. β_2AR is known to bind to AKAP79 (AKAP5) and AKAP250 (AKAP12),[90] and both AKAP79 and mAKAP have been shown to scaffold adenylyl cyclase to facilitate signaling transduction efficiency and specificity.[91] On the contrary, β_2AR binds to PDE4 enzymes in an arrestin-dependent manner,[92] which shapes spatiotemporal distribution of intracellular cAMP. Mutation of palmitoylation site on β_2AR (C341A) displays a reduced receptor interaction with β-arrestin 2 and PDE4D enzymes for cAMP degradation. This reduced binding of PDE4D is due to a reduced association of β-arrestin 2 with the mutant β_2AR,[93] similar to the effect of palmitoylation on the recruitment of arrestin to activated vasopressin receptor.[94] The mutant β_2AR-C341A displays normal GRK phosphorylation but an increased and prolonged PKA phosphorylation under agonist stimulation. Both GRK phosphorylation and palmitoylation are necessary for recruitment of arrestin for clathrin-dependent internalization of the activated β_2AR.[95] Despite the inability of the mutant β_2AR-C341A to recruit β-arrestin 2, the receptor is able to undergo agonist-induced and PKA-dependent internalization via a caveolae-dependent pathway. Thus, the mutant β_2AR-C341A undergoes internalization without recruiting PDE4D, which alters subcellular distribution of PDE4D and contributes to a higher and more sustained intracellular cAMP accumulation under agonist stimulation.

4.4 Ubiquitination of β-Arrestin 2, but Not β_2AR, Involves in β_2AR Endocytosis

Ubiquitination is another class of posttranslational modification that is important in regulation of various aspects of receptor signaling and trafficking.[96] Ubiquitin is a 76 amino acid polypeptide that is typically attached to proteins through the formation of an isopeptide bond between the carboxyl terminus of ubiquitin and the ε-amino group of lysine side chains on target proteins.[97,98] Ubiquitination of β_2AR on either the third intracellular loop or C-tail lysine residues is induced by agonist activation and is required for lysosomal degradation.[99] A β_2AR mutant lacking lysine residues, which was not ubiquitinated, was internalized normally but was degraded ineffectively.[100] Although prevention of receptor ubiquitination has been shown to have little effect on the endocytosis of a number of GPCRs including β_2AR,[96,100,101] there is evidence for significant "indirect" regulation via ubiquitination of β-arrestin.[96] Agonist stimulation of endogenous or transfected β_2AR led to rapid ubiquitination of both the receptors and β-arrestin.[100] Agonist-promoted ubiquitination of β_2AR was observed in wild-type cells, and to a lesser extent in β-arrestin1 knockout cells, but not in β-arrestin 2 knockout cells, suggesting an obligatory role of the β-arrestin 2 isoform in ubiquitination of β_2AR.[100] Upon agonist stimulation, β-arrestin 2 interacts with the E3 ligase Mdm2 [murine double minute oncogene encodes this protein] and this interaction is required for efficient ubiquitination of β-arrestin 2. Either RNAi-mediated depletion of Mdm2 or overexpression of a catalytically inactive mutant version inhibits endocytosis of β_2AR.[100] β_2AR ubiquitination, however, is unaffected in Mdm2-null cells, and a dominant negative Mdm2 (which hinders both β-arrestin ubiquitination and β_2AR internalization) is unable to inhibit β_2AR ubiquitination, allowing the receptor degradation to proceed normally. Thus, β-arrestin may bind and recruit other E3 ligases that can act on the receptor.[100] Indeed, β-arrestin 2 interacts with the HECT (homologous to E6AP carboxyl terminus)-domain E3 ubiquitin ligase neural precursor development downregulated protein 4 (Nedd4-1), which mediates ubiquitination of β_2AR.[102] siRNA that downregulates Nedd4 expression inhibits β_2AR ubiquitination and lysosomal degradation, and the interaction between β_2AR and Nedd4-1 is dependent upon the presence of β-arrestin 2.[102] This indicates that the fate of β_2AR in the lysosomal compartments is dependent upon β-arrestin 2-mediated recruitment of Nedd4 to the activated receptor and Nedd4-catalyzed ubiquitination.[102] Additionally, Nedd4-1 may be recruited to β_2AR independent of β-arrestin 2 through a mechanism mediated by the arrestin domain-containing protein arrestin domain-containing (ARRDC3).[103] ARRDC3 interacts with NEDD4 through two conserved PPXY motifs and recruits NEDD4 to the activated receptor. ARRDC3 also interacts and co-localizes with activated β_2AR. Knockdown of ARRDC3 expression abolishes the association between NEDD4 and β_2AR, thus attenuating agonist-induced ubiquitination and lysosomal sorting of β_2AR.[103] These findings delineate an adapter role of β-arrestin in mediating ubiquitination of β_2AR and indicate that ubiquitination of the receptor and of β-arrestin have distinct and obligatory roles in the trafficking and degradation of this prototypic GPCR.[100]

Similar to phosphorylation, ubiquitination is often transient and may be removed from proteins by various ubiquitin-specific proteases (USPs) or deubiquitinating enzymes (DUBs).[104] Two DUBs USP20 and USP33 have been shown to mediate deubiquitination of β_2AR.[105] DUBs remove ubiquitin, thereby opposing ubiquitin-dependent lysosomal targeting of β_2AR, while concomitantly promoting receptor recycling from the late-endosomal compartments as well as resensitization of recycled receptors at the cell surface.[105] A recent study examined the subcellular distribution of USP33 in cells and found no evidence that it was localized to the endocytic pathway, but rather localized to the Golgi apparatus suggesting a function in the secretory pathway.[106] It is possible that USP33 regulates β_2AR endocytic trafficking *in trans*, whereby Golgi or ER sites may be in close proximity to the endocytic pathway and thereby facilitate trafficking.[106]

Many questions regarding how β_2AR ubiquitination occurs in intact cells and the physiological consequences at the whole animal level remain unknown.[96] The physiological consequences of particular GPCRs ubiquitination/deubiquitination reactions influencing receptor traffic remain essentially undefined in intact animals. This is clearly an important avenue for future study, and represents an exciting frontier for both physiological and membrane trafficking research. Further, considering the high level of diversity and specificity that is already evident from cell-based studies of ubiquitin-dependent regulation of mammalian GPCRs, GPCRs ubiquitination/deubiquitination reactions could represent promising new targets for therapeutic drug development.[96]

5. REGULATION OF ENDOSOME βAR SIGNALING

It is increasingly clear that GPCRs are physically separated into divergent pathways after endocytosis, which can exert effects on signal transduction. Agonist-induced endocytosis of GPCRs was initially recognized as a phenomenon coinciding with rapid desensitization of G-protein-mediated cellular responses.[107] It was traditionally thought that receptor-mediated activation of cognate heterotrimeric G proteins is restricted to the plasma membrane and the endosome-associated receptor pool is functionally inactive with regard to canonical second messenger signaling. Evidences supporting this traditional view are based on analytical methods that provide limited or no subcellular resolution.[108] Recent studies reveal that diverse GPCRs do not always follow this conventional paradigm. It has been subsequently proposed that GPCRs activation elicits a discrete form of persistent G-protein signaling,[109–113] or that internalized GPCRs can indeed contribute to acute G-protein-mediated response.[114] Thus, GPCRs mediate G-protein signaling not only from the plasma membrane but also from endosomal membranes. In supporting the concept of endosome-based G-protein signaling, heterotrimeric G proteins and adenylyl cyclases can be observed in endosome as well as at the plasma membrane.[109,111,114] Such internal signaling to G proteins has been observed for the GPCRs of parathyroid hormone,[109] thyroid-stimulating hormone,[111,112] and dopamine.[114] However, a fundamental problem in interpreting all of the studies summarized above is their reliance on temporal correlation, together with possible complications of off-target or pleiotropic effects of endocytic inhibitors.[115] It remains unknown if endosome-localized GPCRs are even present in an active form.

5.1 Endosome G-Protein-Dependent Signaling

von Zastrow laboratory generated a biosensor of activated β_2AR based on a conformation-specific single-domain camelid antibody Nb80 nanobody, which selectively binds agonist-occupied β_2AR because it mimics the cognate G protein (Gs) subunit in its nucleotide-free form.[116] Nb80 is able to effectively detect the activated receptor conformation without force activation in the absence of agonist when present at a low concentration.[116] After application of agonist, endosome recruitment of NB80-GFP was visible several minutes after recruitment to the plasma membrane and occurred as a discrete second phase, after the delivery of receptors to endosome devoid of bound nanobody.[115,116] Moreover, Irannejad and coworkers developed a distinct biosensor based on another nanobody, Nb37, which specifically recognizes the guanine-nucleotide-free form of Gα_s representing the catalytic intermediate of G protein activation to directly investigate the subcellular location of G protein activation.[116] Agonist application initiated two phases of Nb37-GFP recruitment, first to the plasma membrane and then to endosome ~1 min after receptor arrival. These findings provide photographic and video evidence pointing to two temporally and spatially separated waves of β_2AR signaling through G$_s$,[117] first from the plasma membrane before receptors are internalized and then from endosomes after ligand-induced endocytosis. Both waves of signaling led to accumulation of cAMP.[117] This is arguably the strongest evidence supporting that activated β_2AR and conformational activation of cognate G proteins, can indeed occur in endosome. Nanobody-based biosensors suggest that similar protein conformational states accompany β_2AR and Gs activation in endosome and the plasma membrane (Fig. 3). The simplest interpretation is that some β_2AR can activate G proteins in endosomes by a similar (or the same) mechanism as in the plasma membrane.[115]

A major fascinating question raised by the discovery of endosome-based signaling is whether there is any functional importance to the spatial separation of cAMP production sites apart from its temporal impact through prolonging the cellular response. The downstream manifestations of these spatiotemporal effects remain largely unexplored but, considering how many physiological processes depend on signal timing, they are likely widespread.[109,111,117] Temporal effects may also be important to mediating the therapeutic or toxic actions of drugs, particularly high-affinity compounds that remain associated with target GPCRs for long periods of time.[115] A subsequent study from von Zastrow laboratory profiled global changes in gene expression in response to β_2AR activation and found that endocytosis is required for the full repertoire of downstream cAMP-dependent transcriptional control.[118] They described an orthogonal optogenetic approach to definitively establish that the location of cAMP production is

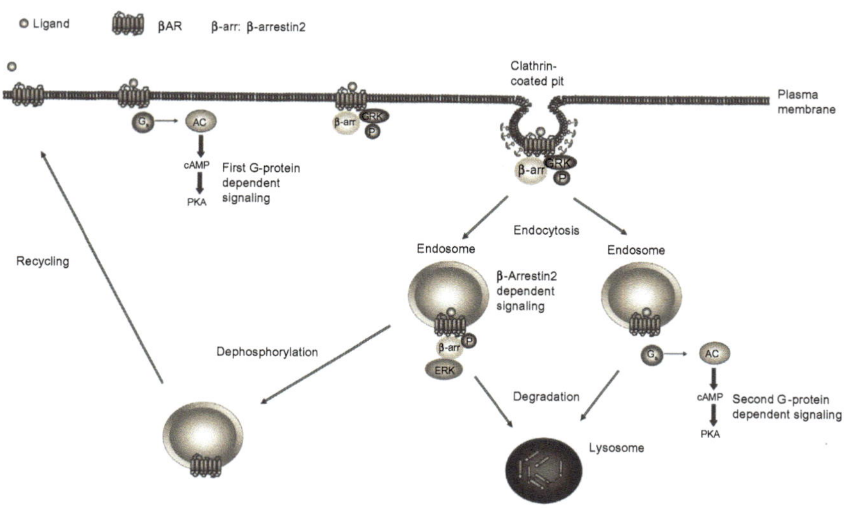

Figure 3 Endosome βAR signaling. Agonist stimulation promotes βAR endocytosis, and the receptor undergoes subsequent sorting for either recycling or degradation. After agonist stimulation, βAR can signal to G proteins at the cell surface and at the endosome after agonist-induced endocytosis. In addition, β_1AR can signal through β-arrestin pathways. (See the color plate.)

indeed a critical variable determining the transcriptional response.[118] Endocytic inhibitors were found to reduce the magnitude of β_2AR-elicited induction of a large repertoire of cAMP-dependent genes, including PCK1, the gene-encoding phosphoenolpyruvate carboxykinase 1 that determines the rate of gluconeogenesis. These results suggest that β_2AR-G protein activation in endosomes confers a discrete type of spatial control over the specificity of downstream signaling, likely by increasing the efficiency of cAMP-dependent phosphorylation of CREB.[115] These findings reveal that endosomes function as flexible signal delivery vehicles that physically move, in response to receptor activation, the site of receptor-elicited cAMP production away from the plasma membrane and in proximity to the nucleus for efficient downstream control of CREB-dependent transcription.[118]

These results, however, are limited to a relatively undifferentiated cell model, and it will be interesting in future studies to investigate the signaling consequences of endocytosis in native cell types or tissues.[118] How particular the mechanisms of endosome-based signaling are terminated is another interesting question.[115] Recent evidence suggests that endocytosis machinery may participate in the pathogenesis of cardiac diseases. β_1AR undergoes prolonged endocytosis (>4 h) after isoproterenol stimulation and that activation of the endocytosis machinery, instead of turning off the signal through the receptor, positively mediates βAR-induced cardiac hypertrophy.[119] Concanavalin A treatment not only inhibits endocytosis of β_1AR but also blocks isoproterenol-induced cardiac hypertrophy.[119] Furthermore, inhibition of endocytosis of βAR[120,121] also prevents isoproterenol-induced increases in ANF transcription and myocyte hypertrophy.[122,123] These results argue that endocytosis of β_1AR itself or activation of the endocytosis machinery is required for isoproterenol-induced cardiac hypertrophy. Several pathophysiological conditions in heart elevate sympathetic nervous activity; the dynamic sequestration of β_1AR into endosome compartments may represent an important mechanism to regulate adrenergic responsiveness in failing human heart.[36] Mechanistically, internalization of βAR plays a critical role in activation of downstream MAP kinases and Akt, which may in turn mediate cardiac hypertrophy in response to βAR stimulation.[119] Thus, the β_1AR endocytosis machinery may be an important target for treatment of heart failure, because inhibition of βAR endocytosis may selectively uncouple cardiac hypertrophy while preserving coupling between βAR and cardiac contractility. Accordingly, targeted PI3K inhibition prevents β_1AR sequestration into endosome compartments in animal hearts and reverses β_1AR abnormalities in a large-animal model of heart failure.[33]

5.2 Endosome G-Protein-Independent Signaling

Early studies, based on the effects of endocytic inhibitors, suggested that β_2AR initiates G-protein-dependent activation of adenylyl cyclase specifically from the plasma membrane and G-protein-independent activation of MAP kinase signaling specifically from endosomes (Fig. 3).[124,125] Shortly, it was found that between these two waves, β_2AR bound to β-arrestin and then clustered in CCPs, where it is thought to trigger a wave of nonconventional signals, including ERK activation.[126] Terrillon and Bouvier then showed, using a clever chemical strategy, that plasma membrane recruitment of arrestin is sufficient to activate MAP kinase signaling.[127] These latter observations are in line with general observation that β_2AR (like many other GPCRs) associate with arrestins primarily in the plasma membrane, but not strongly in endosomes. However, there is a subset of GPCRs that do robustly recruit arrestin to endosome as well as the plasma membrane, apparently because they remain persistently phosphorylated after endocytosis.[128] For several of these GPCRs, endosome recruitment of MAP kinase components has also been demonstrated and is thought to contribute to localized cellular responses.[129,130]

Although these studies have not directly established the occurrence of endosome signaling, it has an implication in the treatment of heart failure. For instance, β-arrestin bias confers positive effects, whereas G-protein-dependent signaling may cause side effects.[131] An example of beneficial β-arrestin-dependent effects is provided by the β-blocker carvedilol, a β-arrestin-biased ligand, acting at both β_1AR and β_2AR subtypes. Carvedilol stimulates epidermal growth factor receptor transactivation and ERK phosphorylation in an arrestin-dependent manner.[132,133] Interestingly, chronic βAR coupling to G_s is thought to be cardiotoxic,[134] whereas epidermal growth factor receptor transactivation has been reported to confer cardioprotection.[133] Together, these observations suggest that carvedilol, which acts as an antagonist of G-protein signaling and simultaneously engages cardioprotective β-arrestin signaling, might provide an added therapeutic benefit in treatment of heart failure compared with other antagonists that block all βAR signaling.

5.3 Endosome Recycling of βAR

Endocytosis is an endocytic mechanism in which specific molecules are ingested into the cell, which allows the interactions between the cell and its environment to be precisely regulated. In addition to its role in mediating rapid desensitization, endocytosis of certain GPCRs is thought to play a major role in mediating the distinct process of receptor resensitization.[135] Endocytosis is an essential way to rebalance the receptor spatial signaling in cells for acute stress response. Endocytosis of membrane receptors, removing them from the surface where they are able to interact with extracellular cues, regulates the long-term sensitivity of cells to their specific ligands. Endocytosis brings receptors in close proximity to an endosome-associated phosphatase, which mediates dephosphorylation of receptors previously phosphorylated (hence "desensitized") at the cell surface. Internalized receptors are resensitized via dephosphorylation in the early endosomes by protein phosphatase 2A (PP2A)[136] prior to recycling back to the plasma membrane.[31] PP2A, a serine–threonine phosphatase, is regulated by endogenously inhibitor proteins called the inhibitors of PP2A (I1- and I2PP2A).[137] Agonist stimulation can lead to PI3K activation, which phosphorylates I2PP2A thereby inhibiting PP2A-mediated dephosphorylation of the receptor at the plasma membrane, thus driving the system towards internalization of the receptor.[138] Upon dephosphorylation, receptors are then recycled back to the plasma membrane in a "resensitized" state, which is fully functional to mediate subsequent rounds of signal transduction upon reexposure to agonist.[122] For some GPCRs, slow recycling to the plasma membrane is attributed to the dissociation kinetics with β-arrestin. Transient association facilitates rapid sorting while strong association (mediated by GPCR phosphorylation and β-arrestin ubiquitination) leads to trafficking to the perinuclear compartment and slow recycling kinetics.[123] A recurring theme in understanding membrane trafficking is that the pattern or timing of activity, defined by protein-protein interactions or posttranslational modifications, are important determinants of receptor fate.[139] Thus, the pattern of receptor phosphorylation can determine the sorting fate. By altering the pattern of receptor phosphorylation, a cell can alter its recycling kinetics, which in turn provides a mechanism to alter the kinetics of resensitization and tissue responsiveness.

The identity of the kinases involved in GPCRs phosphorylation and postendocytic sorting includes the GRKs. D2 dopamine receptor requires GRK2 and GRK3 phosphorylation, not for rapid desensitization or β-arrestin binding, but as an important determinant of receptor recycling.[140] This agrees with prior data on β_2AR, where a GRK5 phosphorylated serine in the receptor C-tail recycling sequence (DSLL) is necessary for receptor targeting to this pathway.[76] Recently, the cAMP-dependent PKA has been shown to be involved in unprecedented features of regulated recycling for β_2AR.[141] With total-internal reflection (TIRF) microscopy, GPCR recycling (measured by membrane insertion events) was observed to be very rapid, occurring within 3–5 min of agonist stimulation.[141] In neurons, β_2AR insertion events differed in their time to laterally diffuse in the membrane, termed as transient or persistent events.[142] β_2AR is a $G\alpha_s$-coupled receptor that activates PKA, the authors identified that chemical inhibition of PKA increases the frequency of transiently localized recycling events.[142] The site of PKA action is at a PKA consensus site in the receptor C-tail, distinct from the distal recycling sequence. Mutation of the PKA site also increased frequency of recycling events. These findings illustrate the fine-tuning capacity of these pathways and also allude to a signal compartmentalization role for persistent receptor insertion events.[141,142] Moreover, membrane scaffolding proteins such as A kinase anchoring proteins (AKAPs) can form complexes with β_2AR to fine-tune receptor signaling. AKAP5 and AKAP12 not only provide a platform for compartmentalizing signaling molecules such as PKA and c-Src with β_2AR but are also involved in regulating recycling and resensitization, and even recently, signaling to the MAP kinase pathway.[143] In a similar fashion, the signalsome of β_1AR, which contains SAP97 and AKAP5, also promotes PKA-mediated phosphorylation of the receptor at the third intracellular loop necessary for receptor recycling to the cell surface.[144]

GPCRs recycling is also a targeted process occurring via a "sequence-directed" mechanism.[69,145] These so-called recycling sequences are highly diverse and interact specifically with distinct cytoplasmic sorting proteins,[139] suggesting a combinatorial mechanism controlling endocytic regulatory profile of individual GPCRs in complex mammalian cells.[69] A well-defined class of recycling sequences is PSD-95/Discs-large/ZO-1 (PDZ) domain binding motifs (also called PDZ ligands) that are usually located at the carboxyl-terminal end of different GPCR tails.[146,147] Studies of β_2AR show that efficient recycling of the receptor requires a short C-terminal PDZ motif (DSLL).[76] Fusion of this motif to the cytoplasmic tail of delta opioid receptor, a distinct mammalian GPCR that normally traffics to lysosomes after endocytosis, is sufficient to reroute receptors into the recycling pathway.[148] The β_2AR PDZ motif can bind to a family of PDZ proteins named sodium–hydrogen exchange regulatory factor-1 (NHERF1) or ezrin-binding phosphoprotein of 50 kDa (EBP50). NHERF/EBP50 family proteins mediate indirect connectivity of cognate motif-bearing integral membrane proteins to actin filaments,[149] which is sufficient to promote plasma membrane recycling of GPCRs.[150] However, the major PDZ protein essential for efficient recycling of the wild-type

β_2AR is sorting nexin 27 (SNX27),[151] which may function at a step prior to those mediated by NHERF/EBP50 proteins. In comparison, β_1AR possesses a distinct type 1 PDZ motif (ESKV) that binds a largely non-overlapping spectrum of PDZ proteins.[152] Of these, SAP97 was shown to be required for efficient recycling of internalized receptors to the plasma membrane and, consequently, to promote functional recovery of cellular signaling following agonist-induced desensitization. Further, SAP97 was shown to bind AKAP79 and thereby linked β_1AR in an organized "receptosome" complex.[144] Thus, for β_1AR, the same PDZ protein interaction mediates discrete signaling and trafficking functions of its PDZ motif.

5.4 Endosome Sorting for βAR Degradation

Chronic GPCRs stimulation can lead to rerouting of GPCRs from recycling to degradation pathway as part of mechanism of receptor downregulation. Such reprogramming of the trafficking fate has significant therapeutic implications as it contributes to the phenomenon of tachyphylaxis or drug tolerance.[153] Most GPCRs undergo endocytosis in response to activation, yet their subsequent sorting in endosomes is variable, creating variable regulation of their activity during prolonged or repeated stimulation. There is also evidence that core endocytic machinery could regulate this trafficking event. The early endosome-localized adaptor protein Hrs has been identified as such a protein for recycling of β_2AR, μ-opioid receptor and calcitonin receptor-like receptor.[101,154] Of note, Hrs-dependent recycling of all these receptors is mediated via the N-terminal VHS domain of Hrs, and Hrs may not directly bind to the receptor cargo.[101] Hrs is also considered to be a scaffolding protein at the early endosome membrane that promotes ESCRT-dependent sorting of GPCRs to lysosomes. Together, Hrs plays a pivotal role in controlling diverse GPCRs sorting fates, and the precise mechanisms involved in different sorting fates remains to be examined.

GPCRs targeting to lysosomes leads to downregulate cellular responses mediated by the receptor.[155] The best-characterized pathway-mediating proteolytic downregulation of GPCRs involves endocytosis of receptors followed by membrane trafficking to lysosomes. Additional proteolytic machinery, such as proteasomes or cell-associated endoproteases, are also implicated in mediating downregulation of certain GPCRs. GPCRs may be targeted to lysosomes after initial endocytosis by CCPs or may follow a distinct membrane pathway involving alternate mechanisms of endocytosis.[156,157] Furthermore, distinct GPCRs differ in their sorting between divergent membrane pathways when coexpressed in the same cells.[158,159] Recent studies have also identified cytoplasmic sequences present in certain GPCRs that promote sorting of internalized receptors to lysosomes.[160] Together with the sequences identified for promoting or preventing rapid recycling of receptors,[76,161,162] there are multiple biochemical mechanisms which distinguish the postendocytic sorting of specific GPCRs and play a critical role in determining the precise functional consequences of agonist-induced endocytosis. Recycling of β_2AR back to the plasma membrane promotes functional resensitization of receptors, whereas sorting of internalized receptors to lysosomes promotes downregulation of receptors and long-term desensitization of receptor-mediated signal transduction.[70] Thus, the sorting of internalized β_2AR between recycling endosomes and lysosomes is responsible for opposite effects on signal transduction,[70] and may be fundamental in the physiological regulation of signal transduction.

In conclusion, in its simplest form, endocytosis can act to downregulate the levels of a receptor at the cell surface to inhibit further signaling. However, it is becoming increasingly clear that endocytosis can modulate signaling in a number of other manners. Compartmentalization of the plasma membrane, for example in cholesterol-rich lipid microdomains, serves to cluster receptors and signaling components in specific constellations to accomplish particular signaling aims.[163] Endocytosis also can direct signaling-active endosomes to different intracellular compartments for modification of the signal or degradation or recycling of the signaling components.

6. CONCLUSION AND REMARKS

The endocytic sorting machinery, by specifically regulating the number and membrane localization of GPCRs, can profoundly affect cellular responses to natural ligands as well as pharmacological agents. Thus, endocytic trafficking of GPCRs, in addition to mediating acute and chronic regulation of the strength of "classical" G protein-linked signaling from the plasma membrane, can promote receptor signaling via altogether distinct effector pathways.[69] For example, β_2AR can activate G-protein-dependent signaling pathways as well as signal via β-arrestin-dependent pathways. The majority of known β_2AR agonists exhibit relative efficacies for β-arrestin-associated activities (β-arrestin membrane translocation and β_2AR internalization) parallel to the efficacies for G protein-dependent signaling (cAMP generation). However, three βAR ligands display a marked bias toward β-arrestin signaling; these ligands stimulate greater β-arrestin-dependent receptor activities relative to their efficacy for G-protein-dependent activity.[164] This selective signal activation is referred as "ligand bias."[165] β-arrestin through its scaffolding function provides a platform for formation of multifunctional signaling cascades, which can initiate a variety of cellular responses through MAPKs, Src tyrosine kinase, nuclear factor κB and PI3K.[166,167] The ERK1/2 signaling pathway is known to be cardioprotective, in part due to inhibition of cardiomyocyte apoptosis that results from ischemia/reperfusion injury or oxidative stress.[168,169] βAR blocking agents are routinely used in treatment of heart failure, and so a broader understanding of whether their capacity to activate (or *not* activate) particular MAPK or other non-cAMP pathways correlates with their clinical efficacy will be of immense value.[170] It is also reasonable to hypothesize that βAR antagonists with a similar ability to inhibit βAR-G-protein activation and with the simultaneous capacity to stimulate β-arrestin-dependent signaling pathways may have additional salutary effects to those already recognized for β-blockers.

As the importance of ligand-directed signaling becomes more fully appreciated, we envisage that both existing and novel βAR agonists and antagonists will be subject to screening for their interaction with multiple signaling pathways. In the case of existing drugs, it may be possible to determine activity profiles that correlate positively or negatively with clinical efficacy as has been done for the series of antipsychotic drugs acting at dopamine D2 receptor.[171] The ability to predict therapeutic benefit for newly developed drugs will depend largely on the power of this profiling, and it will be interesting to see whether profiling can be done in recombinant systems with high receptor abundance or whether it must be augmented by the use of primary human cell systems expressing endogenous receptors. Indeed, a number of compounds that specifically activate β-arrestin signaling, but not G-protein coupling, have been reported.[164 165] There are also reports of compounds that can induce differential GPCRs sorting, rerouting receptors from the recycling pathway utilized by the native ligand and inducing receptor ubiquitination and downregulation.[172] This highlights the potential value of GPCRs trafficking assays as tools in drug discovery programs. Ultimately, the application of broader screening methods to drug development will need to be validated by *post hoc* clinical trials and long-term monitoring of clinical outcomes.[170] The remarkable diversity of signaling effects of GPCRs endocytic trafficking, together with unexpected differences among individual ligands in their ability to promote various trafficking events, increasingly challenge conventional concepts of quantitative pharmacology based on a single dimension of agonist efficacy.[69] These complex trafficking systems provide avenues for development of pathway-specific compounds, such as interest in ligands that could have high specificity to particular signaling or trafficking pathways (also termed biased-agonists, or ligand-directed signaling) and thus have minimal side effects.[165,173]

In conclusion, endocytic sorting of GPCRs plays a critical role in determining cellular signaling patterns, beyond the traditional paradigm of GPCRs desensitization by receptor phosphorylation. The cellular fate of a receptor can be determined at multiple points in the endocytic pathway, and via receptor association with many different cytoplasmic proteins. Considering the long-recognized importance of GPCR–G-protein signal termination at the plasma membrane,[68,70] this will remain as a critical future direction toward investigating broader implications of endomembrane G-protein activation. Another interesting direction is to investigate the importance of the endosome-based signaling *in vivo*, which might help to develop therapeutically useful compounds affecting endosome-based signaling by GPCRs.[69] Moreover, the apparent multidimensionality of ligand efficacy has exciting physiological implications, which are just beginning to be explored. The study of endocytic mechanisms not only provides a basis for new therapeutic opportunities but also hopefully inspires new techniques to deliver drugs to specific intracellular locations. As we begin to understand more of how these endosome βAR pathways impact downstream cellular programs, we expect better therapies for an increasing number of pathophysiological cardiac conditions with minimal undesired effects of chronic or repeated drug exposure, such as tachyphylaxis and tolerance intervention.

ACKNOWLEDGMENTS

This study was supported by National Natural Science Foundation of China grants 81473212 and 81102438, and a Central Authorities of an Institution of Higher Learning of Scientific Research Special Fund of China 2014QN031 to QF, a NIH grant RO1 HL082846, an AHA established investigator grant 12EIA8410007, and a National Natural Science Foundation of China grant 81428022 to YKX. YKX is a Shanghai Eastern Scholar.

REFERENCES

1. Pierce KL, Premont RT, Lefkowitz RJ. Seven-transmembrane receptors. *Nat Rev Mol Cell Biol.* 2002;3(9):639–650.
2. Xiang Y, Kobilka BK. Myocyte adrenoceptor signaling pathways. *Science.* 2003; 300(5625):1530–1532.
3. Xiao RP, Zhu W, Zheng M, et al. Subtype-specific alpha1- and beta-adrenoceptor signaling in the heart. *Trends Pharmacol Sci.* 2006;27(6):330–337.

4. Benovic JL, Mayor Jr F, Staniszewski C, Lefkowitz RJ, Caron MG. Purification and characterization of the beta-adrenergic receptor kinase. *J Biol Chem.* 1987; 262(19):9026–9032.
5. Pitcher JA, Freedman NJ, Lefkowitz RJ. G protein-coupled receptor kinases. *Annu Rev Biochem.* 1998;67:653–692.
6. Lohse MJ, Benovic JL, Codina J, Caron MG, Lefkowitz RJ. beta-Arrestin: a protein that regulates beta-adrenergic receptor function. *Science.* 1990;248(4962):1547–1550.
7. Anderson RG, Jacobson K. A role for lipid shells in targeting proteins to caveolae, rafts, and other lipid domains. *Science.* 2002;296(5574):1821–1825.
8. Hall RA, Lefkowitz RJ. Regulation of G protein-coupled receptor signaling by scaffold proteins. *Circ Res.* 2002;91(8):672–680.
9. Hu LA, Tang Y, Miller WE, et al. beta 1-adrenergic receptor association with PSD-95. Inhibition of receptor internalization and facilitation of beta 1-adrenergic receptor interaction with N-methyl-D-aspartate receptors. *J Biol Chem.* 2000;275(49):38659–38666.
10. Engelhardt S, Hein L, Wiesmann F, Lohse MJ. Progressive hypertrophy and heart failure in beta1-adrenergic receptor transgenic mice. *Proc Natl Acad Sci USA.* 1999;96(12):7059–7064.
11. Communal C, Singh K, Sawyer DB, Colucci WS. Opposing effects of beta(1)- and beta(2)-adrenergic receptors on cardiac myocyte apoptosis: role of a pertussis toxin-sensitive G protein. *Circulation.* 1999;100(22):2210–2212.
12. Zhu WZ, Zheng M, Koch WJ, Lefkowitz RJ, Kobilka BK, Xiao RP. Dual modulation of cell survival and cell death by beta(2)-adrenergic signaling in adult mouse cardiac myocytes. *Proc Natl Acad Sci USA.* 2001;98(4):1607–1612.
13. Fischmeister R, Castro LR, Abi-Gerges A, et al. Compartmentation of cyclic nucleotide signaling in the heart: the role of cyclic nucleotide phosphodiesterases. *Circ Res.* 2006;99(8):816–828.
14. Nikolaev VO, Bunemann M, Schmitteckert E, Lohse MJ, Engelhardt S. Cyclic AMP imaging in adult cardiac myocytes reveals far-reaching beta1-adrenergic but locally confined beta2-adrenergic receptor-mediated signaling. *Circ Res.* 2006;99(10):1084–1091.
15. Insel PA, Head BP, Ostrom RS, et al. Caveolae and lipid rafts: G protein-coupled receptor signaling microdomains in cardiac myocytes. *Ann NY Acad Sci.* 2005;1047:166–172.
16. Soderling SH, Beavo JA. Regulation of cAMP and cGMP signaling: new phosphodiesterases and new functions. *Curr Opin Cell Biol.* 2000;12(2):174–179.
17. Okamoto T, Schlegel A, Scherer PE, Lisanti MP. Caveolins, a family of scaffolding proteins for organizing "preassembled signaling complexes" at the plasma membrane. *J Biol Chem.* 1998;273(10):5419–5422.
18. Galbiati F, Razani B, Lisanti MP. Emerging themes in lipid rafts and caveolae. *Cell.* 2001;106(4):403–411.
19. Cohen AW, Hnasko R, Schubert W, Lisanti MP. Role of caveolae and caveolins in health and disease. *Physiol Rev.* 2004;84(4):1341–1379.
20. Rybin VO, Xu X, Lisanti MP, Steinberg SF. Differential targeting of beta-adrenergic receptor subtypes and adenylyl cyclase to cardiomyocyte caveolae. A mechanism to functionally regulate the cAMP signaling pathway. *J Biol Chem.* 2000; 275(52):41447–41457.
21. Xiang Y, Rybin VO, Steinberg SF, Kobilka B. Caveolar localization dictates physiologic signaling of beta 2-adrenoceptors in neonatal cardiac myocytes. *J Biol Chem.* 2002;277(37):34280–34286.
22. Lefkowitz RJ. Seven transmembrane receptors: something old, something new. *Acta Physiol (Oxf).* 2007;190(1):9–19.

23. Daaka Y, Luttrell LM, Lefkowitz RJ. Switching of the coupling of the beta2-adrenergic receptor to different G proteins by protein kinase A. *Nature*. 1997;390(6655):88–91.
24. Liu R, Ramani B, Soto D, De Arcangelis V, Xiang Y. Agonist dose-dependent phosphorylation by protein kinase A and G protein-coupled receptor kinase regulates beta2 adrenoceptor coupling to G(i) proteins in cardiomyocytes. *J Biol Chem*. 2009; 284(47):32279–32287.
25. Xiao RP. Beta-adrenergic signaling in the heart: dual coupling of the beta2-adrenergic receptor to G(s) and G(i) proteins. *Sci STKE*. 2001;2001(104):re15.
26. Singh K, Xiao L, Remondino A, Sawyer DB, Colucci WS. Adrenergic regulation of cardiac myocyte apoptosis. *J Cell Physiol*. 2001;189(3):257–265.
27. Yarbrough TL, Lu T, Lee HC, Shibata EF. Localization of cardiac sodium channels in caveolin-rich membrane domains: regulation of sodium current amplitude. *Circ Res*. 2002;90(4):443–449.
28. Martens JR, Navarro-Polanco R, Coppock EA, et al. Differential targeting of Shaker-like potassium channels to lipid rafts. *J Biol Chem*. 2000;275(11):7443–7446.
29. Lohn M, Furstenau M, Sagach V, et al. Ignition of calcium sparks in arterial and cardiac muscle through caveolae. *Circ Res*. 2000;87(11):1034–1039.
30. Chen-Izu Y, Xiao RP, Izu LT, et al. G(i)-dependent localization of beta(2)-adrenergic receptor signaling to L-type Ca(2+) channels. *Biophys J*. 2000;79(5):2547–2556.
31. Rockman HA, Koch WJ, Lefkowitz RJ. Seven-transmembrane-spanning receptors and heart function. *Nature*. 2002;415(6868):206–212.
32. Bristow MR, Ginsburg R, Umans V, et al. Beta 1- and beta 2-adrenergic-receptor subpopulations in nonfailing and failing human ventricular myocardium: coupling of both receptor subtypes to muscle contraction and selective beta 1-receptor down-regulation in heart failure. *Circ Res*. 1986;59(3):297–309.
33. Perrino C, Naga Prasad SV, Schroder JN, Hata JA, Milano C, Rockman HA. Restoration of beta-adrenergic receptor signaling and contractile function in heart failure by disruption of the betaARK1/phosphoinositide 3-kinase complex. *Circulation*. 2005;111(20):2579–2587.
34. Naga Prasad SV, Barak LS, Rapacciuolo A, Caron MG, Rockman HA. Agonist-dependent recruitment of phosphoinositide 3-kinase to the membrane by beta-adrenergic receptor kinase 1. A role in receptor sequestration. *J Biol Chem*. 2001;276(22):18953–18959.
35. Naga Prasad SV, Laporte SA, Chamberlain D, Caron MG, Barak L, Rockman HA. Phosphoinositide 3-kinase regulates beta2-adrenergic receptor endocytosis by AP-2 recruitment to the receptor/beta-arrestin complex. *J Cell Biol*. 2002; 158(3):563–575.
36. Perrino C, Schroder JN, Lima B, et al. Dynamic regulation of phosphoinositide 3-kinase-gamma activity and beta-adrenergic receptor trafficking in end-stage human heart failure. *Circulation*. 2007;116(22):2571–2579.
37. Drake MT, Shenoy SK, Lefkowitz RJ. Trafficking of G protein-coupled receptors. *Circ Res*. 2006;99(6):570–582.
38. Zaccolo M, Pozzan T. Discrete microdomains with high concentration of cAMP in stimulated rat neonatal cardiac myocytes. *Science*. 2002;295(5560):1711–1715.
39. Head BP, Patel HH, Roth DM, et al. G-protein-coupled receptor signaling components localize in both sarcolemmal and intracellular caveolin-3-associated microdomains in adult cardiac myocytes. *J Biol Chem*. 2005;280(35):31036–31044.
40. Feiner EC, Chung P, Jasmin JF, et al. Left ventricular dysfunction in murine models of heart failure and in failing human heart is associated with a selective decrease in the expression of caveolin-3. *J Card Fail*. 2011;17(3):253–263.
41. Nikolaev VO, Moshkov A, Lyon AR, et al. Beta2-adrenergic receptor redistribution in heart failure changes cAMP compartmentation. *Science*. 2010;327(5973):1653–1657.

42. Zakhary DR, Moravec CS, Bond M. Regulation of PKA binding to AKAPs in the heart: alterations in human heart failure. *Circulation*. 2000;101(12):1459–1464.
43. Schmid SL. Clathrin-coated vesicle formation and protein sorting: an integrated process. *Annu Rev Biochem*. 1997;66:511–548.
44. Chini B, Parenti M. G-protein coupled receptors in lipid rafts and caveolae: how, when and why do they go there? *J Mol Endocrinol*. 2004;32(2):325–338.
45. Sorkin A, von Zastrow M. Endocytosis and signalling: intertwining molecular networks. *Nat Rev Mol Cell Biol*. 2009;10(9):609–622.
46. Moore CA, Milano SK, Benovic JL. Regulation of receptor trafficking by GRKs and arrestins. *Annu Rev Physiol*. 2007;69:451–482.
47. Ferguson SS, Downey 3rd WE, Colapietro AM, Barak LS, Menard L, Caron MG. Role of beta-arrestin in mediating agonist-promoted G protein-coupled receptor internalization. *Science*. 1996;271(5247):363–366.
48. Santini F, Gaidarov I, Keen JH. G protein-coupled receptor/arrestin3 modulation of the endocytic machinery. *J Cell Biol*. 2002;156(4):665–676.
49. Ohno H, Stewart J, Fournier MC, et al. Interaction of tyrosine-based sorting signals with clathrin-associated proteins. *Science*. 1995;269(5232):1872–1875.
50. Honing S, Ricotta D, Krauss M, et al. Phosphatidylinositol-(4,5)-bisphosphate regulates sorting signal recognition by the clathrin-associated adaptor complex AP2. *Mol Cell*. 2005;18(5):519–531.
51. Doherty GJ, McMahon HT. Mechanisms of endocytosis. *Annu Rev Biochem*. 2009;78:857–902.
52. Kirkham M, Parton RG. Clathrin-independent endocytosis: new insights into caveolae and non-caveolar lipid raft carriers. *Biochim Biophys Acta*. 2005;1746(3):349–363.
53. Mayor S, Pagano RE. Pathways of clathrin-independent endocytosis. *Nat Rev Mol Cell Biol*. 2007;8(8):603–612.
54. Ringstad N, Nemoto Y, De Camilli P. The SH3p4/Sh3p8/SH3p13 protein family: binding partners for synaptojanin and dynamin via a Grb2-like Src homology 3 domain. *Proc Natl Acad Sci USA*. 1997;94(16):8569–8574.
55. Milosevic I, Giovedi S, Lou X, et al. Recruitment of endophilin to clathrin-coated pit necks is required for efficient vesicle uncoating after fission. *Neuron*. 2011;72(4):587–601.
56. Boucrot E, Ferreira AP, Almeida-Souza L, et al. Endophilin marks and controls a clathrin-independent endocytic pathway. *Nature*. 2015;517(7535):460–465.
57. McMahon HT, Boucrot E. Molecular mechanism and physiological functions of clathrin-mediated endocytosis. *Nat Rev Mol Cell Biol*. 2011;12(8):517–533.
58. Meinecke M, Boucrot E, Camdere G, Hon WC, Mittal R, McMahon HT. Cooperative recruitment of dynamin and BIN/amphiphysin/Rvs (BAR) domain-containing proteins leads to GTP-dependent membrane scission. *J Biol Chem*. 2013; 288(9):6651–6661.
59. Tang Y, Hu LA, Miller WE, et al. Identification of the endophilins (SH3p4/p8/p13) as novel binding partners for the beta1-adrenergic receptor. *Proc Natl Acad Sci USA*. 1999;96(22):12559–12564.
60. Marx J. Caveolae: a once-elusive structure gets some respect. *Science*. 2001; 294(5548):1862–1865.
61. Steinberg SF, Brunton LL. Compartmentation of G protein-coupled signaling pathways in cardiac myocytes. *Annu Rev Pharmacol Toxicol*. 2001;41:751–773.
62. Lin F, Wang H, Malbon CC. Gravin-mediated formation of signaling complexes in beta 2-adrenergic receptor desensitization and resensitization. *J Biol Chem*. 2000;275(25):19025–19034.
63. Rapacciuolo A, Suvarna S, Barki-Harrington L, et al. Protein kinase A and G protein-coupled receptor kinase phosphorylation mediates beta-1 adrenergic receptor endocytosis through different pathways. *J Biol Chem*. 2003;278(37):35403–35411.

64. Xiang Y, Devic E, Kobilka B. The PDZ binding motif of the beta 1 adrenergic receptor modulates receptor trafficking and signaling in cardiac myocytes. *J Biol Chem.* 2002;277(37):33783–33790.
65. Ferguson SS, Menard L, Barak LS, Koch WJ, Colapietro AM, Caron MG. Role of phosphorylation in agonist-promoted beta 2-adrenergic receptor sequestration. Rescue of a sequestration-defective mutant receptor by beta ARK1. *J Biol Chem.* 1995; 270(42):24782–24789.
66. Ferguson SS, Zhang J, Barak LS, Caron MG. Molecular mechanisms of G protein-coupled receptor desensitization and resensitization. *Life Sci.* 1998;62(17–18): 1561–1565.
67. Tsao PI, von Zastrow M. Diversity and specificity in the regulated endocytic membrane trafficking of G-protein-coupled receptors. *Pharmacol Ther.* 2001; 89(2):139–147.
68. Krupnick JG, Benovic JL. The role of receptor kinases and arrestins in G protein-coupled receptor regulation. *Annu Rev Pharmacol Toxicol.* 1998;38:289–319.
69. Hanyaloglu AC, von Zastrow M. Regulation of GPCRs by endocytic membrane trafficking and its potential implications. *Annu Rev Pharmacol Toxicol.* 2008;48:537–568.
70. Lefkowitz RJ, Pitcher J, Krueger K, Daaka Y. Mechanisms of beta-adrenergic receptor desensitization and resensitization. *Adv Pharmacol.* 1998;42:416–420.
71. Benovic JL, Pike LJ, Cerione RA, et al. Phosphorylation of the mammalian beta-adrenergic receptor by cyclic AMP-dependent protein kinase. Regulation of the rate of receptor phosphorylation and dephosphorylation by agonist occupancy and effects on coupling of the receptor to the stimulatory guanine nucleotide regulatory protein. *J Biol Chem.* 1985;260(11):7094–7101.
72. Lefkowitz RJ. G protein-coupled receptors. III New roles for receptor kinases and beta-arrestins in receptor signaling and desensitization. *J Biol Chem.* 1998; 273(30):18677–18680.
73. Magalhaes AC, Dunn H, Ferguson SS. Regulation of GPCR activity, trafficking and localization by GPCR-interacting proteins. *Br J Pharmacol.* 2012;165(6):1717–1736.
74. Zamah AM, Delahunty M, Luttrell LM, Lefkowitz RJ. Protein kinase A-mediated phosphorylation of the beta 2-adrenergic receptor regulates its coupling to Gs and Gi. Demonstration in a reconstituted system. *J Biol Chem.* 2002;277(34):31249–31256.
75. Wang Y, De Arcangelis V, Gao X, Ramani B, Jung YS, Xiang Y. Norepinephrine- and epinephrine-induced distinct beta2-adrenoceptor signaling is dictated by GRK2 phosphorylation in cardiomyocytes. *J Biol Chem.* 2008;283(4):1799–1807.
76. Cao TT, Deacon HW, Reczek D, Bretscher A, von Zastrow M. A kinase-regulated PDZ-domain interaction controls endocytic sorting of the beta2-adrenergic receptor. *Nature.* 1999;401(6750):286–290.
77. Fu Q, Xu B, Parikh D, Cervantes D, Xiang YK. Insulin induces IRS2-dependent and GRK2-mediated beta2AR internalization to attenuate betaAR signaling in cardiomyocytes. *Cell Signal.* 2015;27(3):707–715.
78. Vasudevan NT, Mohan ML, Gupta MK, et al. Gbetagamma-independent recruitment of G-protein coupled receptor kinase 2 drives tumor necrosis factor alpha-induced cardiac beta-adrenergic receptor dysfunction. *Circulation.* 2013;128(4):377–387.
79. Tilley DG, Zhu W, Myers VD, et al. beta-adrenergic receptor-mediated cardiac contractility is inhibited via vasopressin type 1A-receptor-dependent signaling. *Circulation.* 2014;130(20):1800–1811.
80. Doronin S, Wang Hy HY, Malbon CC. Insulin stimulates phosphorylation of the beta 2-adrenergic receptor by the insulin receptor, creating a potent feedback inhibitor of its tyrosine kinase. *J Biol Chem.* 2002;277(12):10698–10703.
81. Hadcock JR, Port JD, Gelman MS, Malbon CC. Cross-talk between tyrosine kinase and G-protein-linked receptors. Phosphorylation of beta 2-adrenergic receptors in response to insulin. *J Biol Chem.* 1992;267(36):26017–26022.

82. Shumay E, Song X, Wang HY, Malbon CC. pp60Src mediates insulin-stimulated sequestration of the beta(2)-adrenergic receptor: insulin stimulates pp60Src phosphorylation and activation. *Mol Biol Cell.* 2002;13(11):3943–3954.
83. Karoor V, Wang L, Wang HY, Malbon CC. Insulin stimulates sequestration of beta-adrenergic receptors and enhanced association of beta-adrenergic receptors with Grb2 via tyrosine 350. *J Biol Chem.* 1998;273(49):33035–33041.
84. Fu Q, Xu B, Liu Y, et al. Insulin inhibits cardiac contractility by inducing a Gi-biased beta2-adrenergic signaling in hearts. *Diabetes.* 2014;63(8):2676–2689.
85. Mandic M, Drinovec L, Glisic S, Veljkovic N, Nohr J, Vrecl M. Demonstration of a direct interaction between beta2-adrenergic receptor and insulin receptor by BRET and bioinformatics. *PLoS One.* 2014;9(11):e112664.
86. Moffett S, Mouillac B, Bonin H, Bouvier M. Altered phosphorylation and desensitization patterns of a human beta 2-adrenergic receptor lacking the palmitoylated Cys341. *EMBO J.* 1993;12(1):349–356.
87. Rocks O, Peyker A, Kahms M, et al. An acylation cycle regulates localization and activity of palmitoylated Ras isoforms. *Science.* 2005;307(5716):1746–1752.
88. Torrecilla I, Tobin AB. Co-ordinated covalent modification of G-protein coupled receptors. *Curr Pharm Des.* 2006;12(14):1797–1808.
89. O'Dowd BF, Hnatowich M, Caron MG, Lefkowitz RJ, Bouvier M. Palmitoylation of the human beta 2-adrenergic receptor. Mutation of Cys341 in the carboxyl tail leads to an uncoupled nonpalmitoylated form of the receptor. *J Biol Chem.* 1989; 264(13):7564–7569.
90. Malbon CC. A-kinase anchoring proteins: trafficking in G-protein-coupled receptors and the proteins that regulate receptor biology. *Curr Opin Drug Discov Devel.* 2007;10(5):573–579.
91. Dessauer CW. Adenylyl cyclase-A-kinase anchoring protein complexes: the next dimension in cAMP signaling. *Mol Pharmacol.* 2009;76(5):935–941.
92. De Arcangelis V, Liu R, Soto D, Xiang Y. Differential association of phosphodiesterase 4D isoforms with beta2-adrenoceptor in cardiac myocytes. *J Biol Chem.* 2009;284(49):33824–33832.
93. Liu R, Wang D, Shi Q, Fu Q, Hizon S, Xiang YK. Palmitoylation regulates intracellular trafficking of beta2 adrenergic receptor/arrestin/phosphodiesterase 4D complexes in cardiomyocytes. *PLoS One.* 2012;7(8):e42658.
94. Halls ML, Cooper DM. Sub-picomolar relaxin signalling by a pre-assembled RXFP1, AKAP79, AC2, beta-arrestin 2, PDE4D3 complex. *EMBO J.* 2010;29(16):2772–2787.
95. Vaughan DJ, Millman EE, Godines V, et al. Role of the G protein-coupled receptor kinase site serine cluster in beta2-adrenergic receptor internalization, desensitization, and beta-arrestin translocation. *J Biol Chem.* 2006;281(11):7684–7692.
96. Hislop JN, von Zastrow M. Role of ubiquitination in endocytic trafficking of G-protein-coupled receptors. *Traffic.* 2011;12(2):137–148.
97. Hershko A, Ciechanover A. The ubiquitin system. *Annu Rev Biochem.* 1998;67:425–479.
98. Pickart CM, Eddins MJ. Ubiquitin: structures, functions, mechanisms. *Biochim Biophys Acta.* 2004;1695(1–3):55–72.
99. Xiao K, Shenoy SK. Beta2-adrenergic receptor lysosomal trafficking is regulated by ubiquitination of lysyl residues in two distinct receptor domains. *J Biol Chem.* 2011;286(14):12785–12795.
100. Shenoy SK, McDonald PH, Kohout TA, Lefkowitz RJ. Regulation of receptor fate by ubiquitination of activated beta 2-adrenergic receptor and beta-arrestin. *Science.* 2001;294(5545):1307–1313.
101. Hanyaloglu AC, McCullagh E, von Zastrow M. Essential role of Hrs in a recycling mechanism mediating functional resensitization of cell signaling. *EMBO J.* 2005;24(13):2265–2283.

102. Shenoy SK, Xiao K, Venkataramanan V, Snyder PM, Freedman NJ, Weissman AM. Nedd4 mediates agonist-dependent ubiquitination, lysosomal targeting, and degradation of the beta2-adrenergic receptor. *J Biol Chem.* 2008;283(32):22166–22176.
103. Nabhan JF, Pan H, Lu Q. Arrestin domain-containing protein 3 recruits the NEDD4 E3 ligase to mediate ubiquitination of the beta2-adrenergic receptor. *EMBO Rep.* 2010;11(8):605–611.
104. Nijman SM, Luna-Vargas MP, Velds A, et al. A genomic and functional inventory of deubiquitinating enzymes. *Cell.* 2005;123(5):773–786.
105. Berthouze M, Venkataramanan V, Li Y, Shenoy SK. The deubiquitinases USP33 and USP20 coordinate beta2 adrenergic receptor recycling and resensitization. *EMBO J.* 2009;28(12):1684–1696.
106. Thorne C, Eccles RL, Coulson JM, Urbe S, Clague MJ. Isoform-specific localization of the deubiquitinase USP33 to the Golgi apparatus. *Traffic.* 2011; 12(11):1563–1574.
107. Waldo GL, Northup JK, Perkins JP, Harden TK. Characterization of an altered membrane form of the beta-adrenergic receptor produced during agonist-induced desensitization. *J Biol Chem.* 1983;258(22):13900–13908.
108. Lohse MJ, Nuber S, Hoffmann C. Fluorescence/bioluminescence resonance energy transfer techniques to study G-protein-coupled receptor activation and signaling. *Pharmacol Rev.* 2012;64(2):299–336.
109. Ferrandon S, Feinstein TN, Castro M, et al. Sustained cyclic AMP production by parathyroid hormone receptor endocytosis. *Nat Chem Biol.* 2009;5(10):734–742.
110. Feinstein TN, Wehbi VL, Ardura JA, et al. Retromer terminates the generation of cAMP by internalized PTH receptors. *Nat Chem Biol.* 2011;7(5):278–284.
111. Calebiro D, Nikolaev VO, Gagliani MC, et al. Persistent cAMP-signals triggered by internalized G-protein-coupled receptors. *PLoS Biol.* 2009;7(8):e1000172.
112. Werthmann RC, Volpe S, Lohse MJ, Calebiro D. Persistent cAMP signaling by internalized TSH receptors occurs in thyroid but not in HEK293 cells. *FASEB J.* 2012;26(5):2043–2048.
113. Calebiro D, Nikolaev VO, Lohse MJ. Imaging of persistent cAMP signaling by internalized G protein-coupled receptors. *J Mol Endocrinol.* 2010;45(1):1–8.
114. Kotowski SJ, Hopf FW, Seif T, Bonci A, von Zastrow M. Endocytosis promotes rapid dopaminergic signaling. *Neuron.* 2011;71(2):278–290.
115. Tsvetanova NG, Irannejad R, von Zastrow M. GPCR signaling via heterotrimeric G proteins from endosomes. *J Biol Chem.* 2015;290:6689–6696.
116. Irannejad R, Tomshine JC, Tomshine JR, et al. Conformational biosensors reveal GPCR signalling from endosomes. *Nature.* 2013;495(7442):534–538.
117. Lohse MJ, Calebiro D. Cell biology: receptor signals come in waves. *Nature.* 2013;495(7442):457–458.
118. Tsvetanova NG, von Zastrow M. Spatial encoding of cyclic AMP signaling specificity by GPCR endocytosis. *Nat Chem Biol.* 2014;10(12):1061–1065.
119. Morisco C, Marrone C, Galeotti J, et al. Endocytosis machinery is required for beta1-adrenergic receptor-induced hypertrophy in neonatal rat cardiac myocytes. *Cardiovasc Res.* 2008;78(1):36–44.
120. Lin FT, Krueger KM, Kendall HE, et al. Clathrin-mediated endocytosis of the beta-adrenergic receptor is regulated by phosphorylation/dephosphorylation of beta-arrestin1. *J Biol Chem.* 1997;272(49):31051–31057.
121. Koch WJ, Rockman HA, Samama P, et al. Cardiac function in mice overexpressing the beta-adrenergic receptor kinase or a beta ARK inhibitor. *Science.* 1995; 268(5215):1350–1353.
122. Pippig S, Andexinger S, Lohse MJ. Sequestration and recycling of beta 2-adrenergic receptors permit receptor resensitization. *Mol Pharmacol.* 1995;47(4):666–676.

123. Shenoy SK, Barak LS, Xiao K, et al. Ubiquitination of beta-arrestin links seven-transmembrane receptor endocytosis and ERK activation. *J Biol Chem*. 2007; 282(40):29549–29562.
124. Daaka Y, Luttrell LM, Ahn S, et al. Essential role for G protein-coupled receptor endocytosis in the activation of mitogen-activated protein kinase. *J Biol Chem*. 1998;273(2):685–688.
125. Irannejad R, von Zastrow M. GPCR signaling along the endocytic pathway. *Curr Opin Cell Biol*. 2014;27:109–116.
126. Luttrell LM, Ferguson SS, Daaka Y, et al. Beta-arrestin-dependent formation of beta2 adrenergic receptor-Src protein kinase complexes. *Science*. 1999;283(5402):655–661.
127. Terrillon S, Bouvier M. Receptor activity-independent recruitment of betaarrestin2 reveals specific signalling modes. *EMBO J*. 2004;23(20):3950–3961.
128. Oakley RH, Laporte SA, Holt JA, Barak LS, Caron MG. Association of beta-arrestin with G protein-coupled receptors during clathrin-mediated endocytosis dictates the profile of receptor resensitization. *J Biol Chem*. 1999;274(45):32248–32257.
129. DeFea KA, Zalevsky J, Thoma MS, Dery O, Mullins RD, Bunnett NW. beta-arrestin-dependent endocytosis of proteinase-activated receptor 2 is required for intracellular targeting of activated ERK1/2. *J Cell Biol*. 2000;148(6):1267–1281.
130. DeFea KA, Vaughn ZD, O'Bryan EM, Nishijima D, Dery O, Bunnett NW. The proliferative and anti apoptotic effects of substance P are facilitated by formation of a beta-arrestin-dependent scaffolding complex. *Proc Natl Acad Sci USA*. 2000; 97(20):11086–11091.
131. Reiter E, Ahn S, Shukla AK, Lefkowitz RJ. Molecular mechanism of beta-arrestin-biased agonism at seven-transmembrane receptors. *Annu Rev Pharmacol Toxicol*. 2012;52:179–197.
132. Wisler JW, DeWire SM, Whalen EJ, et al. A unique mechanism of beta-blocker action: carvedilol stimulates beta-arrestin signaling. *Proc Natl Acad Sci USA*. 2007;104(42):16657–16662.
133. Noma T, Lemaire A, Naga Prasad SV, et al. Beta-arrestin-mediated beta1-adrenergic receptor transactivation of the EGFR confers cardioprotection. *J Clin Invest*. 2007;117(9):2445–2458.
134. Lohse MJ, Engelhardt S, Eschenhagen T. What is the role of beta-adrenergic signaling in heart failure? *Circ Res*. 2003;93(10):896–906.
135. Pippig S, Andexinger S, Daniel K, et al. Overexpression of beta-arrestin and beta-adrenergic receptor kinase augment desensitization of beta 2-adrenergic receptors. *J Biol Chem*. 1993;268(5):3201–3208.
136. Krueger KM, Daaka Y, Pitcher JA, Lefkowitz RJ. The role of sequestration in G protein-coupled receptor resensitization. Regulation of beta2-adrenergic receptor dephosphorylation by vesicular acidification. *J Biol Chem*. 1997;272(1):5–8.
137. Li M, Damuni Z. I1PP2A and I2PP2A. Two potent protein phosphatase 2A-specific inhibitor proteins. *Methods Mol Biol*. 1998;93:59–66.
138. Vasudevan NT, Mohan ML, Gupta MK, Hussain AK, Naga Prasad SV. Inhibition of protein phosphatase 2A activity by PI3Kgamma regulates beta-adrenergic receptor function. *Mol Cell*. 2011;41(6):636–648.
139. Jean-Alphonse F, Hanyaloglu AC. Regulation of GPCR signal networks via membrane trafficking. *Mol Cell Endocrinol*. 2011;331(2):205–214.
140. Namkung Y, Dipace C, Urizar E, Javitch JA, Sibley DR. G protein-coupled receptor kinase-2 constitutively regulates D2 dopamine receptor expression and signaling independently of receptor phosphorylation. *J Biol Chem*. 2009;284(49):34103–34115.
141. Yudowski GA, Puthenveedu MA, Henry AG, von Zastrow M. Cargo-mediated regulation of a rapid Rab4-dependent recycling pathway. *Mol Biol Cell*. 2009; 20(11):2774–2784.

142. Yudowski GA, Puthenveedu MA, von Zastrow M. Distinct modes of regulated receptor insertion to the somatodendritic plasma membrane. *Nat Neurosci.* 2006; 9(5):622–627.
143. Tao J, Malbon CC. G-protein-coupled receptor-associated A-kinase anchoring proteins AKAP5 and AKAP12: differential signaling to MAPK and GPCR recycling. *J Mol Signal.* 2008;3:19.
144. Gardner LA, Naren AP, Bahouth SW. Assembly of an SAP97-AKAP79-cAMP-dependent protein kinase scaffold at the type 1 PSD-95/DLG/ZO1 motif of the human beta(1)-adrenergic receptor generates a receptosome involved in receptor recycling and networking. *J Biol Chem.* 2007;282(7):5085–5099.
145. Marchese A, Paing MM, Temple BR, Trejo J. G protein-coupled receptor sorting to endosomes and lysosomes. *Annu Rev Pharmacol Toxicol.* 2008;48:601–629.
146. Bockaert J, Dumuis A, Fagni L, Marin P. GPCR-GIP networks: a first step in the discovery of new therapeutic drugs? *Curr Opin Drug Discov Devel.* 2004;7(5):649–657.
147. Gage RM, Matveeva EA, Whiteheart SW, von Zastrow M. Type I PDZ ligands are sufficient to promote rapid recycling of G Protein-coupled receptors independent of binding to N-ethylmaleimide-sensitive factor. *J Biol Chem.* 2005;280(5):3305–3313.
148. Gage RM, Kim KA, Cao TT, von Zastrow M. A transplantable sorting signal that is sufficient to mediate rapid recycling of G protein-coupled receptors. *J Biol Chem.* 2001;276(48):44712–44720.
149. Bretscher A. Regulation of cortical structure by the ezrin-radixin-moesin protein family. *Curr Opin Cell Biol.* 1999;11(1):109–116.
150. Lauffer BE, Chen S, Melero C, Kortemme T, von Zastrow M, Vargas GA. Engineered protein connectivity to actin mimics PDZ-dependent recycling of G protein-coupled receptors but not its regulation by Hrs. *J Biol Chem.* 2009;284(4):2448–2458.
151. Lauffer BE, Melero C, Temkin P, et al. SNX27 mediates PDZ-directed sorting from endosomes to the plasma membrane. *J Cell Biol.* 2010;190(4):565–574.
152. He J, Bellini M, Inuzuka H, et al. Proteomic analysis of beta1-adrenergic receptor interactions with PDZ scaffold proteins. *J Biol Chem.* 2006;281(5):2820–2827.
153. von Zastrow M, Svingos A, Haberstock-Debic H, Evans C. Regulated endocytosis of opioid receptors: cellular mechanisms and proposed roles in physiological adaptation to opiate drugs. *Curr Opin Neurobiol.* 2003;13(3):348–353.
154. Hasdemir B, Bunnett NW, Cottrell GS. Hepatocyte growth factor-regulated tyrosine kinase substrate (HRS) mediates post-endocytic trafficking of protease-activated receptor 2 and calcitonin receptor-like receptor. *J Biol Chem.* 2007; 282(40):29646–29657.
155. von Zastrow M. Mechanisms regulating membrane trafficking of G protein-coupled receptors in the endocytic pathway. *Life Sci.* 2003;74(2–3):217–224.
156. Tsao P, Cao T, von Zastrow M. Role of endocytosis in mediating downregulation of G-protein-coupled receptors. *Trends Pharmacol Sci.* 2001;22(2):91–96.
157. Marchese A, Chen C, Kim YM, Benovic JL. The ins and outs of G protein-coupled receptor trafficking. *Trends Biochem Sci.* 2003;28(7):369–376.
158. Dale LB, Bhattacharya M, Seachrist JL, Anborgh PH, Ferguson SS. Agonist-stimulated and tonic internalization of metabotropic glutamate receptor 1a in human embryonic kidney 293 cells: agonist-stimulated endocytosis is beta-arrestin1 isoform-specific. *Mol Pharmacol.* 2001;60(6):1243–1253.
159. Shapiro MJ, Coughlin SR. Separate signals for agonist-independent and agonist-triggered trafficking of protease-activated receptor 1. *J Biol Chem.* 1998;273(44): 29009–29014.
160. Parnot C, Miserey-Lenkei S, Bardin S, Corvol P, Clauser E. Lessons from constitutively active mutants of G protein-coupled receptors. *Trends Endocrinol Metab.* 2002;13(8):336–343.

161. Barak LS, Oakley RH, Laporte SA, Caron MG. Constitutive arrestin-mediated desensitization of a human vasopressin receptor mutant associated with nephrogenic diabetes insipidus. *Proc Natl Acad Sci USA*. 2001;98(1):93–98.
162. Whistler JL, Gerber BO, Meng EC, Baranski TJ, von Zastrow M, Bourne HR. Constitutive activation and endocytosis of the complement factor 5a receptor: evidence for multiple activated conformations of a G protein-coupled receptor. *Traffic*. 2002;3(12):866–877.
163. Andersson ER. The role of endocytosis in activating and regulating signal transduction. *Cell Mol Life Sci*. 2012;69(11):1755–1771.
164. Drake MT, Violin JD, Whalen EJ, Wisler JW, Shenoy SK, Lefkowitz RJ. beta-arrestin-biased agonism at the beta2-adrenergic receptor. *J Biol Chem*. 2008;283(9):5669–5676.
165. Violin JD, Lefkowitz RJ. Beta-arrestin-biased ligands at seven-transmembrane receptors. *Trends Pharmacol Sci*. 2007;28(8):416–422.
166. Rajagopal S, Rajagopal K, Lefkowitz RJ. Teaching old receptors new tricks: biasing seven-transmembrane receptors. *Nat Rev Drug Discov*. 2010;9(5):373–386.
167. Shenoy SK, Drake MT, Nelson CD, et al. beta-arrestin-dependent, G protein-independent ERK1/2 activation by the beta2 adrenergic receptor. *J Biol Chem*. 2006;281(2):1261–1273.
168. Yue TL, Wang C, Gu JL, et al. Inhibition of extracellular signal-regulated kinase enhances ischemia/reoxygenation-induced apoptosis in cultured cardiac myocytes and exaggerates reperfusion injury in isolated perfused heart. *Circ Res*. 2000;86(6):692–699.
169. Lips DJ, Bueno OF, Wilkins BJ, et al. MEK1-ERK2 signaling pathway protects myocardium from ischemic injury in vivo. *Circulation*. 2004;109(16):1938–1941.
170. Evans BA, Sato M, Sarwar M, Hutchinson DS, Summers RJ. Ligand-directed signalling at beta-adrenoceptors. *Br J Pharmacol*. 2010;159(5):1022–1038.
171. Masri B, Salahpour A, Didriksen M, et al. Antagonism of dopamine D2 receptor/beta-arrestin 2 interaction is a common property of clinically effective antipsychotics. *Proc Natl Acad Sci USA*. 2008;105(36):13656–13661.
172. Gonzalez-Cabrera PJ, Hla T, Rosen H. Mapping pathways downstream of sphingosine 1-phosphate subtype 1 by differential chemical perturbation and proteomics. *J Biol Chem*. 2007;282(10):7254–7264.
173. Kenakin T. Collateral efficacy in drug discovery: taking advantage of the good (allosteric) nature of 7TM receptors. *Trends Pharmacol Sci*. 2007;28(8):407–415.

CHAPTER EIGHT

Postendocytic Sorting of Adrenergic and Opioid Receptors: New Mechanisms and Functions

Shanna L. Bowman, Manojkumar A. Puthenveedu[1]

Department of Biological Sciences, Carnegie Mellon University, Pittsburgh, Pennsylvania, USA
[1]Corresponding author: e-mail address: map3@andrew.cmu.edu

Contents

Abstract

The endocytic pathway tightly regulates the activity of G protein-coupled receptors (GPCRs). Much of our understanding of this relationship between GPCR endocytic trafficking and signaling comes from studies done on catecholamine and opioid receptors. After ligand-induced endocytosis, a key sorting step in the endosome determines whether receptors are recycled back to the cell surface, leading to recovery of signaling, or are degraded in the lysosome, leading to desensitization. Recycling of GPCRs, unlike that of many other proteins, is an active process driven by specific sequences on the receptor and proteins that interact with this sequence. Recent data suggest that sequence-dependent recycling plays complex roles in regulating both the timing and location of GPCR signaling. This chapter will describe our current understanding of the mechanisms regulating GPCR sorting in the endosome and discuss emerging ideas on their role in GPCR signaling, focusing on adrenergic and opioid receptors as prototypes.

Progress in Molecular Biology and Translational Science, Volume 132
ISSN 1877-1173
http://dx.doi.org/10.1016/bs.pmbts.2015.03.005

1. INTRODUCTION: THE ENDOSOME AS A SORTING STATION FOR INTERNALIZED GPCRs

Endocytic trafficking is a fundamental cellular process that regulates GPCR function. GPCR activation on the cell surface results in their removal from the cell surface by clathrin-mediated endocytosis.[1–3] This was first recognized by studies measuring ligand-induced desensitization of signaling, which required receptor phosphorylation, binding of beta-arrestins, and endocytosis. Endocytosed GPCRs are transported to the endosome, where a critical sorting step determines the further fate of GPCRs.[4–6] They may be either recycled to the cell surface or degraded in the lysosome. An established consequence of this sorting is that it directly controls the number of receptors on the cell surface, causing either recovery of sensitivity of the cell to the signal or downregulation of signaling, respectively.[7,8] Recent evidence, however, suggests that this postendocytic sorting might have much more complex roles in regulating the function of many GPCRs, including the prototypic adrenergic and opioid receptors.

The early endosome serves as the primary sorting station for most internalized proteins including GPCRs. How endosomes sort receptors, considering the amount of membrane cargo that traffic through these dynamic organelles, is a fundamental question with many implications. There are four main pathways for internalized proteins out of the endosome (Fig. 1). Proteins may be recycled by bulk membrane flow, sorted to the

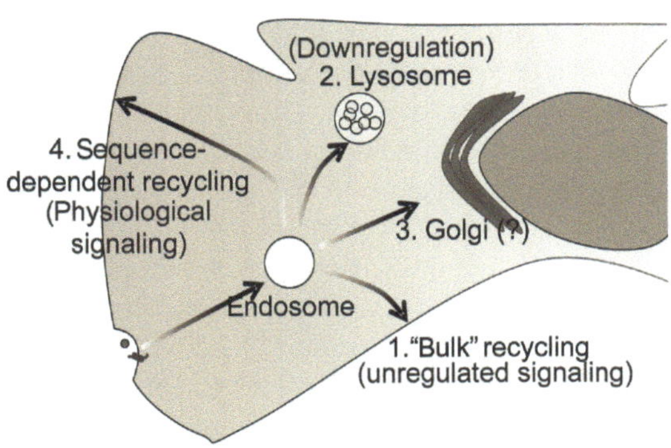

Figure 1 Potential pathways for endocytosed GPCRs from the endosome. Sequence-dependent recycling is the physiological scenario for recovery of signaling, while sorting to the lysosome leads to degradation and downregulation. Bulk recycling pathway, taken normally by nutrient receptors but not GPCRs, potentially leads to unregulated signaling for GPCRs. While some proteins can travel to the Golgi apparatus, this is not established as a common route for GPCRs. (See the color plate.)

lysosome to be degraded, recycled via a specialized regulated "sequence-dependent" recycling pathway, or transported to the Golgi apparatus.

Many proteins, like the transferrin receptor (TfR), are recycled back to the cell surface as part of bulk membrane flow, apparently without any specific requirements. This "bulk recycling" was first described over 20 years ago by Maxfield and colleagues, who labeled and tracked the membrane and fluid phase compartments of endosomes.[9,10] They observed that the endosome extruded narrow tubules with a large surface area (i.e., membranes) but low volume (i.e., fluids). Continued fission and recycling of these tubules provided a geometric basis for how nutrient receptors could be recycled as part of bulk membrane flow, simply because they are membrane proteins, leaving their soluble ligands in the lumen of the endosome to be eventually degraded in the lysosome.[9–11] Several membrane-modifying proteins responsible for membrane tubulation have been identified. Such geometric sorting likely occurs extensively along the endocytic pathway, iteratively recycling receptors from the early, recycling, and the late endosomes.

Considering that recycling is thought to be a "default" fate for membrane proteins, it is interesting that GPCRs are not commonly sorted via bulk recycling. Many GPCRs, like the delta opioid receptor (DOR), and signaling receptors, like the EGF receptor, are degraded in the lysosome.[12,13] Degradation of these proteins takes advantage of geometric sorting. These proteins are packaged into vesicles that bud off into the interior of the endosomes, essentially partitioning the protein-containing vesicles into the fluid phase of these endosomes, which eventually mature into or fuse with lysosomes.[14–16] For GPCRs that recycle, this is a regulated process that requires specific protein sequences and interactions.[17–30] Mutation of these sequences directs GPCRs to the lysosome, reiterating that their recycling is not simply a function of them being membrane proteins.[17,22,25] This chapter will discuss our current understanding of GPCR sorting between the degradative and recycling pathways, using adrenergic and opioid receptors as model receptors.

2. MECHANISMS OF SEQUENCE-DEPENDENT GPCR SORTING

2.1 GPCRs Are Sorted into Lysosomes by Multiple Mechanisms

The first conceptual step in GPCR sorting to the lysosome is to segregate receptors away from bulk recycling, by packaging them into intraluminal vesicles (ILVs). The process of generating vesicles that pinch off to the interior of the vesicles, called involution, is a topologically highly interesting process, as this involves the generation of vesicles away from the cytoplasm, unlike with the "classical" coat-mediated vesicle transport processes like endocytosis. Most of our understanding of protein sorting to lysosomes comes from studies done in yeast. Genetic and biochemical studies of protein sorting to the vacuole, a compartment homologous to the mammalian lysosome, have identified the ESCRT proteins as the primary protein machinery responsible for this process. The current views on the mechanisms of how ESCRT proteins induce negative curvature and generate ILVs are discussed in structural detail in several recent authoritative reviews.[31–34]

The main sorting signal that targets proteins to the lysosome is the addition of ubiquitin to cytoplasmic lysine residues of proteins. Early evidence for the importance of ubiquitination in GPCR trafficking came from studies on the yeast GPCR Ste2, for which ubiquitination promoted transport to the vacuole. Since then, ubiquitination has been shown to promote lysosomal targeting of many membrane proteins in mammalian cells, including the EGF receptor.[13,35] In the case of many of these proteins, ubiquitin interacts with the ubiquitin-interacting motif (UIM) of Hrs, an endosomal protein often termed ESCRT-0. Hrs then transfers these proteins to Tsg101, an ESCRT-I component that also has a UIM. This localizes the cargo in ESCRT domains, which allows them to be incorporated in ILVs. While many GPCRs are known to be ubiquitinated, the role of these interactions in its lysosomal targeting has been shown only for CXCR4.[36] The role of ubiquitination in GPCR trafficking is discussed much more extensively elsewhere in this book.

Receptor ubiquitination or ESCRTs, however, are not required for sorting all GPCRs to lysosomes. A mutant of the DOR lacking all cytoplasmic lysines is endocytosed and trafficked to the lysosome.[12] This still requires Hrs and Tsg101, suggesting ubiquitin-independent binding of GPCRs with ESCRT.[37] The protease-activated receptor PAR1, on the other hand, can be sorted to the lysosome independent of both ubiquitination and ESCRTs.[38] Instead, this sorting depends on the endosomal protein sorting nexin-1 (SNX1).[39] Additionally, members of the family of GPCR-associated sorting proteins (GASP1 and GASP2) and Beclin-2 have been implicated in degradation of DOR.[40,41] While both SNX1 and GASP proteins interact with several GPCRs, including DOR, the exact mechanism by which they sort GPCRs into the lysosome is still not well understood.[42] Our current understanding is that these proteins may play a role in down-regulation of receptors under conditions of chronic agonist stimulation.

2.2 GPCRs Contain Diverse Recycling Sequences

It has been appreciated for over 30 years that adrenergic receptors recycle to the surface, based on the reappearance of activity after agonist-induced desensitization.[2,43,44] Over the years, however, it has become clear that the simple model of bulk recycling cannot explain GPCR recycling, mostly from research on the beta adrenergic receptors. The recycling of the beta-2 adrenergic receptor (B2AR) depends on a specific sequence on its C-terminus that interacts with postsynaptic density 95/disc large/zonula occludins-1 (PDZ)-domain-containing proteins.[17] Similar sequences that conform to classical type I PDZ-ligand sequences have been identified on several GPCRs, including the related beta-1 adrenergic receptor (B1AR) and the kappa opioid receptor (KOR), over the past decade.[19,22,29] For many of these receptors, these sequences are required for the recycling of these receptors and are also sufficient, as transplanting these sequences onto a non-recycling receptor like the DOR allows the receptor to recycle.[45] Alpha adrenergic receptors also contain PDZ ligands, although they are more diverse, and their role in recycling is less well understood.[46] One critical feature of these PDZ-ligand sequences is that they need to be on the C-terminal tail of the receptors. The terminal peptide binds in an antiparallel fashion in a hydrophobic cleft formed by a beta strand, a loop, and an alpha helix in the PDZ protein, with the free carboxyl group occupying a hydrophobic pocket.[47] In addition, sequence comparisons have identified many internal PDZ ligands on GPCRs that might be involved in receptor recycling.[48,49]

How do PDZ-ligand sequences mediate adrenergic receptor recycling? For the B2AR, a complement of proteins that bind the PDZ ligand, including NHERF-1, NHERF-2, PDZK1, and MAGI-3, was identified soon after the identification of the sequence.[17,50] Similarly, several proteins, such as PIST, MAGI-2, MAGI-3, PSD-95, and SAP97, that bind the PDZ-ligand sequence of the B1AR have also been identified.[19,20,29] It is interesting that these two related receptors bind distinct complements of PDZ proteins. SAP97 has been implicated as the main protein that mediates B1AR recycling, by directly interacting with the PDZ ligand and recruiting the A-kinase-anchoring protein-79 (AKAP-79) which phosphorylates the B1AR on a residue that contributes to recycling.[29] Similarly, B2AR also interacts with AKAP-12 (Gravin), which might recruit c-Src to this complex.[51]

However, while these protein interactions have been delineated, we have only recently started to understand how GPCRs are recycled by these interactions. Some of the key breakthroughs came from recent advances in live cell imaging that allow for direct visualization of receptor sorting and recycling in the endosome. These studies showed that the recycling of the B2AR is mediated by specific microdomains on the endosome, distinct from those that mediate bulk recycling.[52] These sequence-dependent recycling microdomains are marked by a specialized actin cytoskeleton and specific membrane-modifying proteins such as sorting nexins and retromer.[53,54] This was exciting, since it challenged the traditional view that all recycling tubule populations at the early endosome have the same sorting kinetics and trafficking machinery.[52] A variety of proteins have been localized to this specific actin/sorting nexin/retromer tubular (ASRT) microdomain on the endosome.[52,54,55] A global analysis of plasma membrane proteins after depletion of SNX27 or retromer components showed a reduction in surface levels of over a 100 proteins, suggesting that this is a general recycling pathway used by many proteins.[56]

Evidence suggests that the primary role of the PDZ interactions is to link the receptor to these ASRT domains. Disrupting these interactions or depleting components of the endosomal actin cytoskeleton, retromer complex, or SNX27 inhibited B2AR recycling, similar to mutating the PDZ ligand or depleting the PDZ proteins.[52–54] Conversely, replacing the PDZ domain with an actin-binding domain from ezrin was sufficient to confer recycling to DOR and B2AR.[57] The facts that actin binding is required and sufficient for recycling, and that several GPCRs have the ability to bind to PDZ-domain proteins, suggest a conserved role of the PDZ-linked actin cytoskeleton in endosomal sorting of GPCRs.

However, why and how GPCRs are excluded from bulk recycling is still an open question. Current data suggest a kinetic basis for this exclusion (Fig. 2). Estimation of diffusion rates showed that the mobility of the B2AR on endosomes was highly restricted.[52] This might make it difficult for B2AR to diffuse into and populate the bulk recycling tubules, which undergo fast fission. The actin cytoskeleton on the ASRT domains stabilizes these domains and delays fission of these tubules. This might allow enough time for the slow-diffusing B2AR to diffuse into these tubules, where they will be concentrated by interactions with the ASRT proteins. Such a "kinetic sorting" mechanism in fact might provide a common mechanism for sorting in many different membrane compartments. Validating this model at a mechanistic level is an important future direction in our understanding of GPCR biology.

In contrast to beta adrenergic receptors, the opioid receptor family exhibits very diverse trafficking characteristics. The KOR requires a PDZ

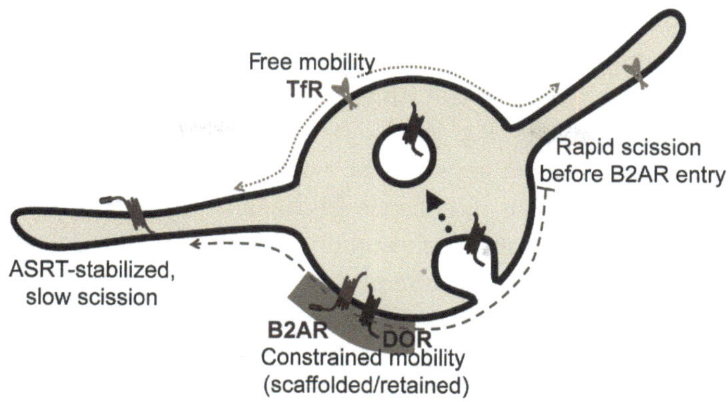

Figure 2 A kinetic model for sorting in the endosome. The mobility of sequence-dependent recycling proteins or degraded proteins on the endosome is constrained by an unknown mechanism. This slow mobility reduces their entry into bulk recycling tubules which form and undergo scission rapidly. Bulk recycling proteins have unconstrained mobility on the endosome, which allows them to enter the short-lived bulk recycling tubules. ASRT domains are stabilized by the actin cytoskeleton, which provides sequence-dependent recycling proteins enough time to diffuse into these tubules. Degraded proteins are captured by the ESCRT machinery and packaged into intraluminal vesicles. (See the color plate.)

domain for postendocytic recycling, similar to the adrenergic receptors.[22] DOR, as mentioned above, does not recycle and is degraded in lysosomes following agonist-induced endocytosis.[12,58] Interestingly, the mu-opioid receptor (MOR) recycles following agonist-induced internalization, but does not require a PDZ-ligand sequence, like KOR and the adrenergic receptors. MOR contains a unique, seven amino-acid recycling sequence in its C-terminal tail, LENLEAE. Mutation of this sequence reroutes MOR to the lysosome following endocytosis, and fusion of this sequence to the C-terminal tail of DOR is sufficient to promote its rapid recycling and prevent lysosomal degradation of DOR.[25]

Although MOR's recycling sequence was discovered a decade ago, the exact mechanism of how this sequence promotes MOR sorting and recycling remains unknown. However, a number of MOR binding partners that regulate MOR trafficking have been found. The actin-binding protein, filamin A, has been shown to interact with the C-terminal tail of MOR, and this interaction is thought to reduce MOR agonist-induced internalization.[59] Additionally, the dendritic spine protein, spinophilin, interacts with MOR in the striatum, and interestingly, knockout of spinophilin reduces sensitivity to morphine-induced analgesia.[60] Further, agonist-induced internalization of MOR is significantly reduced in spinophilin knockout cells.[60] Interestingly, spinophilin interacts with DOR, as well as MOR, and this interaction requires the third intracellular loop, the G protein-coupling domain of GPCRs, as well as the first 18 amino acids of the C-terminal tail, conserved between the two opioid receptors.[61] Additionally, this interaction seems to enhance ERK signaling through DOR, but not MOR, suggesting that the interaction with spinophilin may modulate sensitivity of MOR and DOR differentially.[61] Moreover, single nucleotide polymorphisms within the G protein-coupling domain of the third extracellular loop of MOR are also associated with calmodulin binding and an increase in basal MOR activity.[62] Interestingly, spinophilin also interacts with alpha-2 adrenergic receptors and D2 dopamine receptors, also through the third intracellular loop of these receptors,[63,64] suggesting that spinophilin may regulate several GPCRs, potentially through G protein coupling. The additional requirement of a conserved region of MOR and DOR C-terminal tails for spinophilin binding suggests a potential role in membrane trafficking, while the exact mechanism remains unknown.

Even though GPCR recycling uses diverse sequences and proteins, it is likely that there is a common mechanism that mediates the recycling of most GPCRs, which can be accessed by specific "recycling adaptors" that link receptors to this machinery. In support of this, disruption of Hrs inhibits the recycling of most GPCRs that recycle in a sequence-dependent manner. This role of Hrs seems to be independent of its role in ESCRT-mediated degradation of proteins, as the other components of the ESCRT machinery such as Tsg101 do not produce this phenotype.[65] Further, Hrs mediates its effect on B2AR recycling via its Vps27–Hrs–STAM domain, which does not play an active role in lysosomal sorting.[65] Additional support for this idea comes from the identification of mutations on the B2AR that converts the receptor into a bulk recycling protein. A sequence on the proximal part of the C-terminal tail on B2AR (EKENKLL) that resembles, but is distinct from, an acidic dileucine sequence, was required for Hrs dependence and sequence dependence of B2AR recycling.[66] Similarly, mutating a phosphorylation site on B2AR, adjacent to the EKENKLL sequence, that is phosphorylated by protein kinase A (PKA), converts B2AR recycling to be independent of actin and the PDZ-ligand sequence.[67,68] Identification of these specific sequences indicates that there is a specific machinery that retains B2AR on the endosome, which might constitute step 1 in the hierarchical sorting model, actively excluding it from bulk recycling. As of now, there are no known interacting partners to either of these sites. Once future studies identify candidates, we will be able to dissect out the mechanisms and function of this hierarchical sorting of GPCRs.

Another argument for common machineries mediating GPCR recycling is the involvement of general vesicle trafficking machineries in this process. For example, Rab GTPases regulate many steps in vesicle trafficking of GPCRs, such as vesicle budding, tethering, and docking.[69] Rab5 is a marker for the early endosome to which many GPCRs, such as the B2AR, endothelin A and B receptors, and the thyrotropin-releasing hormone receptor, localize following agonist-induced endocytosis.[70–73] Overexpression of dominant negative Rab5 mutants interferes with endocytic trafficking of B2AR, the D2 dopamine receptor, neurokinin-1 receptor (NK1R), CXC chemokine receptor 2, lysophosphatidic acid-coupled/EDG-2 receptor, and the cannabinoid receptor 2.[72,74–78] Rab5 may be involved in resensitization of B2AR and NK1R.[72,75] Further, Rab4 is thought to control rapid recycling from Rab5/Rab4 early endosomes to the plasma membrane, while Rab11 mediates a slower recycling pathway.[79] Both these have been implicated in the recycling of GPCRs as well as a variety of non-GPCR cargo, including the TfR. For example, a dominant negative Rab4 or depletion of Rab4 by RNA interference inhibits rapid B2AR recycling following agonist-induced endocytosis.[67,72] MOR is also thought to recycle in a Rab4-dependent manner, but also recycles through a slower Rab11-mediated pathway.[80] Other GPCRs are thought to traffic through the slow, Rab11-dependent pathway, such as the cannabinoid receptor 2, angiotensin II type I receptor, and the M4 muscarinic acetylcholine receptor.[78,81,82] Together, this suggests that, in addition to regulation of GPCR recycling by C-terminal recycling sequences and their respective binding partners, GPCR postendocytic trafficking is also subject to regulation by Rab GTPase activity at different endosomal compartments.

3. REGULATION OF SEQUENCE-DEPENDENT RECYCLING BY MODIFYING STEPS IN HIERARCHICAL SORTING

Recent evidence suggests that intracellular signaling cascades can control endosomal sorting of GPCRs. This provides new explanations for how cells might coordinate the diverse cellular responses mediated by different GPCRs at physiological timescales. B2AR recycling is regulated by PKA, a signaling kinase downstream of B2AR activation.[67,68] Further, B2AR signaling can homologously regulate its recycling through PKA phosphorylation of the C-terminal tail of B2AR. These two PKA sites on the C-terminal tail of B2AR regulate a switch in the type of recycling tubules that B2AR is sorted into at the early endosome. Increased PKA phosphorylation of B2AR, following sustained adrenergic signaling, restricts B2AR to the ASRT domain on the endosome. Conversely, non-phosphorylated B2AR can enter the non-ASRT, or bulk recycling tubules, traversed by nutrient receptors, like TfR.[68] This suggests a hierarchical sorting mechanism that allows a cell to fine-tune its responses to extracellular signals (Fig. 3A). For example, in the case of sustained adrenergic signaling, restriction of B2AR to the sequence-dependent pathway by PKA phosphorylation allows a cell to quickly slow down B2AR resensitization by decreasing recycling from the endosome. It is possible that a similar mechanism exists for B1AR, considering the described role of PKA phosphorylation in B1AR recycling and resensitization,[29,51] but the exact mechanisms have not been addressed in detail.

Considering that the ASRT domains contain a variety of protein complexes whose functions are highly regulated, these potentially provide additional control points for regulating GPCR recycling even beyond a simple switch between bulk and ASRT-dependent recycling. For example, cortactin, one of the key components of the actin cytoskeleton in ASRT domains, is regulated by c-Src phosphorylation.[83] This phosphorylation increases the rate of vesicle scission from these domains.[84] Because modifying vesicle scission affects the surface delivery of all cargo proteins that use

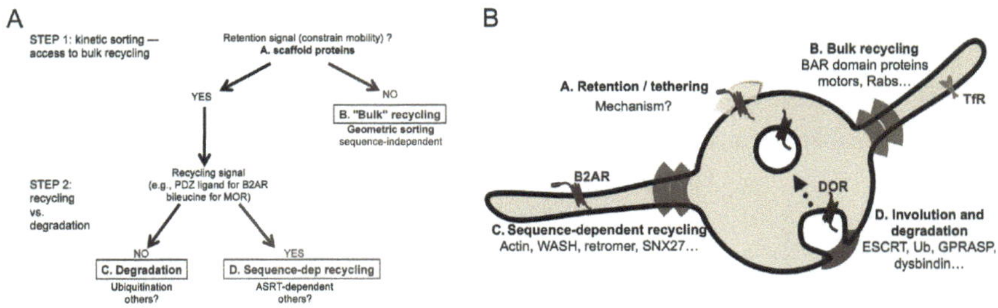

Figure 3 A hierarchical model for GPCR sorting in the endosome. (A) The first step in sorting is an unknown mechanism that tethers GPCRs on the body of the endosome, excluding them from entering bulk recycling tubules. This could be a consequence of decreased mobility, as in the kinetic sorting model described in Fig. 2. Bulk recycling proteins are not tethered by this mechanism, allowing them to recycle by geometric sorting. Tethered proteins are either recycled, if they have a recycling sequence, or degraded, if they contain one of multiple signals including ubiquitination. (B) The proposed protein complexes on the endosome that mediate these hierarchical steps in sorting are shown. (See the color plate.)

this pathway,[56] it is possible that this provides a general mechanism for controlling recycling of all these proteins. Such hierarchical steps of recycling provide a number of checkpoints for signaling receptors. They can first be sorted at the early endosome according to the recycling sequence on their C-terminal, and then phosphorylation of GPCRs themselves at their C-terminal tails, or components of the ASRT domains, by kinases allows the cell to alter recycling kinetics in response to diverse physiological situations (Fig. 3B).

Recent work suggests that MOR recycling can also be regulated by phosphorylation by kinases downstream of GPCR activation. Much like B2AR, MOR recycling in striatal neurons is decreased by forskolin, which activated cyclic AMP (cAMP), although PKA was not directly tested in these neurons.[85] MOR recycling is also regulated through a heterologous pain signaling pathway in sensory neurons. Pain signaling, through activation of the NK1R by substance P, increases MOR recycling in sensory neurons through PKC phosphorylation of two residues on the C-terminal tail of MOR, serine 363 (S363) and threonine 370 (T370).[86] This is consistent with physiological and pharmacological data that PKC inhibition reduces MOR resensitization, and that PKC modulation can change behavioral correlates of opioid tolerance.[87,88] Interestingly, the phosphorylation state at these sites can be differentially regulated by diverse opioid agonists. S363 is constitutively phosphorylated, whether or not the receptor is bound to an opioid drug. T370, on the other hand, is phosphorylated when the receptor is activated by high-efficacy opioids, like [D-Ala2, *N*-Me-Phe4, Gly5-ol]-Enkephalin (DAMGO) and fentanyl, but not morphine.[89] Further, substance P has also been shown to induce phosphorylation at MOR T370.[90] This heterologous regulation of MOR recycling further suggests the possibility that hierarchical sorting of GPCRs, through signaling regulation, allows the cell to control GPCR resensitization by regulating recycling kinetics in response to different physiological stimuli.

4. RELEVANCE OF SEQUENCE-DEPENDENT GPCR RECYCLING IN THE ENDOSOME

The exact role of endocytic trafficking in GPCR function has been often highly debated in the past. After activation, receptors are desensitized by phosphorylation and need to be dephosphorylated, to be resensitized, and made competent for ligand binding and signaling. Evidence suggests that both phosphorylation and dephosphorylation can occur at the cell surface.

Endocytic trafficking might play roles in modifying the kinetics of this process, although the direction in which endocytosis drives resensitization might vary depending on the receptor and conditions.[6,30] Nevertheless, the long-standing model for GPCR function was that they signal primarily at the cell membrane through their heterotrimeric G proteins, after which they are desensitized. Based on this, a straightforward role for endosomal GPCR sorting, that sorting of receptors into the sequence-dependent pathway caused rapid recycling and delivery of receptors to the cell surface and led to recovery of cellular responsiveness to the same signal, has been appreciated for a while.[91]

Recent evidence suggests, however, that sequence-dependent recycling plays more complex roles in tuning both spatial and temporal characteristics of GPCR signaling. GPCRs can signal through several non-G protein signaling pathways—for example, through the GPCR adaptor, beta arrestin.[92] Some GPCRs recruit and signal through arrestin at the endosome. Sustained signaling through G proteins at the endosome following GPCR endocytosis has also been demonstrated for some GPCRs.[93–95] The parathyroid hormone receptor (PTHR) continues to signal after receptor endocytosis, and different agonists differ in their ability to induce this type of signaling. Interestingly, PTHRs signaling at the endosome were also shown to associate with Gα stimulatory protein (Gαs), challenging the traditional view that G protein coupling and signaling occurs primarily at the cell surface.[93,96] Further, internalized thyroid-stimulating hormone receptors exhibit coupling to Gs and cAMP production following internalization.[94]

For "conventional" GPCRs like the B2AR, however, G protein signaling at endosomes has remained controversial, in part because the signaling profiles of B2AR are much faster compared to receptors like the PTHR and because traditional signaling readouts could not discriminate between the cell surface and subcellular compartments as signaling sources. Recent breakthroughs have generated a GFP-tagged nanobody biosensor that specifically recognizes the activated form of the Gαs. This sensor, which provides spatial resolution, showed that B2AR can activate Gαs at early endosomes.[97] This presents the first clear demonstration that even conventional GPCRs, where the focus has largely been on signaling from the cell surface, can induce G protein signaling cascades from endosomal compartments. Following up on this work, the von Zastrow group has also shown that cell surface and endosomal G protein signaling activated distinct transcriptional profiles,[98] suggesting that spatial encoding of where the cAMP signaling was generated could produce diverse cellular responses. While still early, this emerging body of work provides a new perspective on the role of endocytic trafficking in GPCR function. Direct visualization of active conformations of B2AR and Gs proteins on the endosome suggests a transformative idea that endocytosis and subsequent endosomal sorting of GPCRs, in addition to deciding whether GPCRs are degraded at the lysosome or recycled to the cell surface, could also regulate diverse intracellular signaling cascades at the early endosome following agonist activation.

5. SUMMARY AND PERSPECTIVES

The postendocytic sorting of internalized adrenergic and opioid receptors between the recycling and degradative pathways determines whether receptors are delivered back to the cell surface or degraded in the lysosome. GPCR recycling is an active process that requires specific sequences on the receptor tails. Several proteins that interact with these sequences and drive receptors into the sequence-dependent recycling pathway have been identified. Considering that many membrane proteins can recycle as part of bulk membrane flow, how and why GPCRs are excluded from bulk recycling is a clear area that needs further investigation.

A well-appreciated role for postendocytic sorting is that it controls receptor signaling by determining the number of signaling receptors that are recycled to the cell surface. Accumulating data in the recent years, however, suggest that postendocytic sorting plays more complex roles in regulating receptor function. Recycling of GPCRs is mediated by specific microdomains on the endosome that are physically separate from domains that mediate bulk membrane recycling. GPCR recycling domains also serve as organizing centers for specific signaling complexes. Importantly, signals originating from the surface and endosomes induce distinct downstream responses, suggesting that GPCR signaling is spatially encoded. The cytoplasmic interactions of GPCRs, by determining receptor localization in these microdomains, might directly determine this spatial encoding and therefore might provide control points for the cell to precisely modulate both the spatial and temporal characteristics of signaling. We have only recently begun to identify signaling pathways that modify these interactions and regulate receptor signaling. As future studies identify more examples of signaling-mediated regulation of these machineries, especially in physiologically relevant systems, we will be able to build a better picture of how cross talk between multiple signaling pathways allow cells to generate an integrated response in complex signaling environments.

REFERENCES

1. Yoburn BC, Purohit V, Patel K, Zhang Q. Opioid agonist and antagonist treatment differentially regulates immunoreactive mu-opioid receptors and dynamin-2 in vivo. *Eur J Pharmacol.* 2004;498(1–3):87–96.
2. Von Zastrow M, Kobilka BK. Ligand-regulated internalization and recycling of human beta 2-adrenergic receptors between the plasma membrane and endosomes containing transferrin receptors. *J Biol Chem.* 1992;267(5):3530–3538.
3. Lamb ME, De Weerd WF, Leeb-Lundberg LM. Agonist-promoted trafficking of human bradykinin receptors: arrestin- and dynamin-independent sequestration of the B2 receptor and bradykinin in HEK293 cells. *Biochem J.* 2001;355(Pt 3):741–750.
4. Pierce KL, Premont RT, Lefkowitz RJ. Seven-transmembrane receptors. *Nat Rev Mol Cell Biol.* 2002;3(9):639–650.
5. Von Zastrow M. Mechanisms regulating membrane trafficking of G protein-coupled receptors in the endocytic pathway. *Life Sci.* 2003;74(2–3):217–224.
6. Marchese A, Paing MM, Temple BRS, Trejo J. G protein-coupled receptor sorting to endosomes and lysosomes. *Annu Rev Pharmacol Toxicol.* 2008;48:601–629.
7. Sorkin A, Von Zastrow M. Signal transduction and endocytosis: close encounters of many kinds. *Nat Rev Mol Cell Biol.* 2002;3(8):600–614.
8. Alvarez VA, Arttamangkul S, Dang V, et al. μ-Opioid receptors: ligand-dependent activation of potassium conductance, desensitization, and internalization. *J Neurosci.* 2002; 22(13):5769–5776.
9. Dunn KW, McGraw TE, Maxfield FR. Iterative fractionation of recycling receptors from lysosomally destined ligands in an early sorting endosome. *J Cell Biol.* 1989; 109(6 Pt 2):3303–3314.
10. Mayor S, Presley JF, Maxfield FR. Sorting of membrane components from endosomes and subsequent recycling to the cell surface occurs by a bulk flow process. *J Cell Biol.* 1993;121(6):1257–1269.
11. Hao M, Maxfield FR. Characterization of rapid membrane internalization and recycling. *J Biol Chem.* 2000;275(20):15279–15286.
12. Tanowitz M, Von Zastrow M. Ubiquitination-independent trafficking of G protein-coupled receptors to lysosomes. *J Biol Chem.* 2002;277(52):50219–50222.
13. Huang F, Zeng X, Kim W, et al. Lysine 63-linked polyubiquitination is required for EGF receptor degradation. *Proc Natl Acad Sci USA.* 2013;110(39):15722–15727.
14. Babst M. A protein's final ESCRT. *Traffic.* 2005;6(1):2–9.
15. Piper RC, Katzmann DJ. Biogenesis and function of multivesicular bodies. *Annu Rev Cell Dev Biol.* 2007;23:519–547.
16. Piper RC, Lehner PJ. Endosomal transport via ubiquitination. *Trends Cell Biol.* 2011; 21(11):647–655.
17. Cao TT, Deacon HW, Reczek D, Bretscher A, von Zastrow M. A kinase-regulated PDZ-domain interaction controls endocytic sorting of the beta2-adrenergic receptor. *Nature.* 1999;401(6750):286–290.
18. Cong M. Binding of the beta 2 adrenergic receptor to N-ethylmaleimide-sensitive factor regulates receptor recycling. *J Biol Chem.* 2001;276(48):45145–45152.
19. Hu LA. β_1-Adrenergic Receptor Association with PSD-95. Inhibition of receptor internalization and facilitation of β_1-adrenergic receptor interaction with n-methyl-D-aspartate receptors. *J Biol Chem.* 2000;275(49):38659–38666.
20. He J, Bellini M, Inuzuka H, et al. Proteomic analysis of beta1-adrenergic receptor interactions with PDZ scaffold proteins. *J Biol Chem.* 2006;281(5):2820–2827.
21. Gage RM, Kim KA, Cao TT, von Zastrow M. A transplantable sorting signal that is sufficient to mediate rapid recycling of G protein-coupled receptors. *J Biol Chem.* 2001;276(48):44712–44720.
22. Li J-G, Chen C, Liu-Chen L-Y. Ezrin-radixin-moesin-binding phosphoprotein-50/Na+/H+ exchanger regulatory factor (EBP50/NHERF) blocks U50,488H-induced

down-regulation of the human kappa opioid receptor by enhancing its recycling rate. *J Biol Chem.* 2002;277(30):27545–27552.
23. Hirakawa T, Galet C, Kishi M, Ascoli M. GIPC binds to the human lutropin receptor (hLHR) through an unusual PDZ domain binding motif, and it regulates the sorting of the internalized human choriogonadotropin and the density of cell surface hLHR. *J Biol Chem.* 2003;278(49):49348–49357.
24. Hu LA, Chen W, Martin NP, Whalen EJ, Premont RT, Lefkowitz RJ. GIPC interacts with the 1-adrenergic receptor and regulates 1-adrenergic receptor-mediated ERK activation. *J Biol Chem.* 2003;278(28):26295–26301.
25. Tanowitz M, von Zastrow M. A novel endocytic recycling signal that distinguishes the membrane trafficking of naturally occurring opioid receptors. *J Biol Chem.* 2003; 278(46):45978–45986.
26. Galet C, Hirakawa T, Ascoli M. The postendocytotic trafficking of the human lutropin receptor is mediated by a transferable motif consisting of the C-terminal cysteine and an upstream leucine. *Mol Endocrinol.* 2004;18(2):434–446.
27. Huang P, Steplock D, Weinman EJ, et al. Opioid receptor interacts with Na+/H+ exchanger regulatory factor-1/ezrin-radixin-moesin-binding phosphoprotein-50 (NHERF-1/EBP50) to stimulate Na+/H+ exchange independent of Gi/Go proteins. *J Biol Chem.* 2004;279(24):25002–25009.
28. Vargas GA, Von Zastrow M. Identification of a novel endocytic recycling signal in the D1 dopamine receptor. *J Biol Chem.* 2004;279(36):37461–37469.
29. Gardner LA, Naren AP, Bahouth SW. Assembly of an SAP97-AKAP79-cAMP-dependent protein kinase scaffold at the type 1 PSD-95/DLG/ZO1 motif of the human beta(1)-adrenergic receptor generates a receptosome involved in receptor recycling and networking. *J Biol Chem.* 2007;282(7):5085–5099.
30. Hanyaloglu AC, von Zastrow M. Regulation of GPCRs by endocytic membrane trafficking and its potential implications. *Annu Rev Pharmacol Toxicol.* 2008;48: 537–568.
31. Hurley JH. The ESCRT complexes. *Crit Rev Biochem Mol Biol.* 2010;45(6):463–487.
32. Hurley JH, Hanson PI. Membrane budding and scission by the ESCRT machinery: it's all in the neck. *Nat Rev Mol Cell Biol.* 2010;11(8):556–566.
33. Shields SB, Piper RC. How ubiquitin functions with ESCRTs. *Traffic.* 2011; 12(10):1306–1317.
34. Henne WM, Stenmark H, Emr SD. Molecular mechanisms of the membrane sculpting ESCRT pathway. *Cold Spring Harb Perspect Biol.* 2013;5(9):a016766. http://dx.doi.org/10.1101/cshperspect.a016766.
35. Dores MR, Trejo J. Ubiquitination of G protein-coupled receptors: functional implications and drug discovery. *Mol Pharmacol.* 2012;82(4):563–570.
36. Verma R, Marchese A. The endosomal sorting complex required for transport pathway mediates chemokine receptor CXCR4 promoted lysosomal degradation of the mTOR antagonist DEPTOR. *J Biol Chem.* 2015;290(11):6810–6824. http://dx.doi.org/10.1074/jbc.M114.606699.
37. Henry AG, White IJ, Marsh M, von Zastrow M, Hislop JN. The role of ubiquitination in lysosomal trafficking of δ-opioid receptors. *Traffic.* 2011;12(2):170–184.
38. Dores MR, Chen B, Lin H, et al. ALIX binds a YPX(3)L motif of the GPCR PAR1 and mediates ubiquitin-independent ESCRT-III/MVB sorting. *J Cell Biol.* 2012;197(3): 407–419.
39. Gullapalli A, Wolfe BL, Griffin CT, Magnuson T, Trejo J. An essential role for SNX1 in lysosomal sorting of protease-activated receptor-1: evidence for retromer-, Hrs-, and Tsg101-independent functions of sorting nexins. *Mol Biol Cell.* 2006;17(3): 1228–1238.
40. Whistler JL, Enquist J, Marley A, et al. Modulation of postendocytic sorting of G protein-coupled receptors. *Science.* 2002;297:615–620.

41. He C, Wei Y, Sun K, et al. Beclin 2 functions in autophagy, degradation of G protein-coupled receptors, and metabolism. *Cell.* 2013;154(5):1085–1099.
42. Heydorn A, Søndergaard BP, Ersbøll B, et al. A library of 7TM receptor C-terminal tails. Interactions with the proposed post-endocytic sorting proteins ERM-binding phosphoprotein 50 (EBP50), N-ethylmaleimide-sensitive factor (NSF), sorting nexin 1 (SNX1), and G protein-coupled receptor-associated sorting protein (GASP). *J Biol Chem.* 2004;279(52):54291–54303.
43. Staehelin M, Hertel C. [3H]CGP-12177, a beta-adrenergic ligand suitable for measuring cell surface receptors. *J Recept Res.* 1983;3(1-2):35–43.
44. Hertel C, Staehelin M. Reappearance of beta-adrenergic receptors after isoproterenol treatment in intact C6-cells. *J Cell Biol.* 1983;97(5 Pt 1):1538–1543.
45. Gage RM, Matveeva EA, Whiteheart SW, von Zastrow M. Type I PDZ ligands are sufficient to promote rapid recycling of G protein-coupled receptors independent of binding to N-ethylmaleimide-sensitive factor. *J Biol Chem.* 2005;280(5):3305–3313.
46. Chen Z, Hague C, Hall RA, Minneman KP. Syntrophins regulate alpha1D-adrenergic receptors through a PDZ domain-mediated interaction. *J Biol Chem.* 2006;281(18):12414–12420.
47. Karthikeyan S, Leung T, Birrane G, Webster G, Ladias JA. Crystal structure of the PDZ1 domain of human Na(+)/H(+) exchanger regulatory factor provides insights into the mechanism of carboxyl-terminal leucine recognition by class I PDZ domains. *J Mol Biol.* 2001;308(5):963–973.
48. Paasche JD, Attramadal T, Kristiansen K, et al. Subtype-specific sorting of the ETA endothelin receptor by a novel endocytic recycling signal for G protein-coupled receptors. *Mol Pharmacol.* 2005;67(5):1581–1590.
49. Romero G, von Zastrow M, Friedman PA. Role of PDZ proteins in regulating trafficking, signaling, and function of GPCRs: means, motif, and opportunity. *Adv Pharmacol.* 2011;62:279–314.
50. Hall RA, Premont RT, Chow CW, et al. The beta2-adrenergic receptor interacts with the Na+/H+-exchanger regulatory factor to control Na+/H+ exchange. *Nature.* 1998;392(6676):626–630.
51. Tao J, Wang H-Y, Malbon CC. Src docks to A-kinase anchoring protein gravin, regulating beta2-adrenergic receptor resensitization and recycling. *J Biol Chem.* 2007;282(9):6597–6608.
52. Puthenveedu MA, Lauffer B, Temkin P, et al. Sequence-dependent sorting of recycling proteins by actin-stabilized endosomal microdomains. *Cell.* 2010;143(5):761–773.
53. Lauffer BEL, Melero C, Temkin P, et al. SNX27 mediates PDZ-directed sorting from endosomes to the plasma membrane. *J Cell Biol.* 2010;190(4):565–574.
54. Temkin P, Lauffer B, Jäger S, Cimermancic P, Krogan NJ, von Zastrow M. SNX27 mediates retromer tubule entry and endosome-to-plasma membrane trafficking of signalling receptors. *Nat Cell Biol.* 2011;13(6):715–721.
55. Gallon M, Cullen PJ. Retromer and sorting nexins in endosomal sorting. *Biochem Soc Trans.* 2015;43(1):33–47.
56. Steinberg F, Gallon M, Winfield M, et al. A global analysis of SNX27-retromer assembly and cargo specificity reveals a function in glucose and metal ion transport. *Nat Cell Biol.* 2013;15(5):461–471.
57. Lauffer BEL, Chen S, Melero C, Kortemme T, von Zastrow M, Vargas GA. Engineered protein connectivity to actin mimics PDZ-dependent recycling of G protein-coupled receptors but not its regulation by Hrs. *J Biol Chem.* 2009;284(4):2448–2458.
58. Tsao PI, von Zastrow M. Type-specific sorting of G protein-coupled receptors after endocytosis. *J Biol Chem.* 2000;275(15):11130–11140.
59. Onoprishvili I, Andria ML, Kramer HK, Ancevska-Taneva N, Hiller JM, Simon EJ. Interaction between the mu opioid receptor and filamin A is involved in receptor regulation and trafficking. *Mol Pharmacol.* 2003;64(5):1092–1100.

60. Charlton JJ, Allen PB, Psifogeorgou K, et al. Multiple actions of spinophilin regulate mu opioid receptor function. *Neuron*. 2008;58(2):238–247.
61. Fourla D-D, Papakonstantinou M-P, Vrana S-M, Georgoussi Z. Selective interactions of spinophilin with the C-terminal domains of the δ- and μ-opioid receptors and G proteins differentially modulate opioid receptor signaling. *Cell Signal*. 2012;24(12): 2315–2328.
62. Wang D, Quillan JM, Winans K, Lucas JL, Sadée W. Single nucleotide polymorphisms in the human mu opioid receptor gene alter basal G protein coupling and calmodulin binding. *J Biol Chem*. 2001;276(37):34624–34630.
63. Richman JG, Brady AE, Wang Q, Hensel JL, Colbran RJ, Limbird LE. Agonist-regulated interaction between 2-adrenergic receptors and spinophilin. *J Biol Chem*. 2001;276(18):15003–15008.
64. Smith FD, Oxford GS, Milgram SL. Association of the D2 dopamine receptor third cytoplasmic loop with spinophilin, a protein phosphatase-1-interacting protein. *J Biol Chem*. 1999;274(28):19894–19900.
65. Hanyaloglu AC, McCullagh E, von Zastrow M. Essential role of Hrs in a recycling mechanism mediating functional resensitization of cell signaling. *EMBO J*. 2005; 24(13):2265–2283.
66. Hanyaloglu AC, von Zastrow M. A novel sorting sequence in the beta2-adrenergic receptor switches recycling from default to the Hrs-dependent mechanism. *J Biol Chem*. 2007;282(5):3095–3104.
67. Yudowski GA, Puthenveedu MA, Henry AG, von Zastrow M. Cargo-mediated regulation of a rapid Rab4-dependent recycling pathway. *Mol Biol Cell*. 2009;20(11): 2774–2784.
68. Vistein R, Puthenveedu MA. Reprogramming of G protein-coupled receptor recycling and signaling by a kinase switch. *Proc Natl Acad Sci USA*. 2013;110(38): 15289–15294.
69. Tuvim MJ, Adachi R, Hoffenberg S, Dickey BF. Traffic control: Rab GTPases and the regulation of interorganellar transport. *News Physiol Sci*. 2001;16:56–61.
70. Moore RH, Tuffaha A, Millman EE, et al. Agonist-induced sorting of human beta2-adrenergic receptors to lysosomes during downregulation. *J Cell Sci*. 1999;112(Pt 3): 329–338.
71. Bremnes T, Paasche JD, Mehlum A, Sandberg C, Bremnes B, Attramadal H. Regulation and intracellular trafficking pathways of the endothelin receptors. *J Biol Chem*. 2000; 275(23):17596–17604.
72. Seachrist JL, Anborgh PH, Ferguson SS. β_2-Adrenergic receptor internalization, endosomal sorting, and plasma membrane recycling are regulated by rab GTPases. *J Biol Chem*. 2000;275(35):27221–27228.
73. Hinkle PM, Gehret AU, Jones BW. Desensitization, trafficking, and resensitization of the pituitary thyrotropin-releasing hormone receptor. *Front Neurosci*. 2012;6:180.
74. Iwata K, Ito K, Fukuzaki A, Inaki K, Haga T. Dynamin and rab5 regulate GRK2-dependent internalization of dopamine D2 receptors. *Eur J Biochem*. 1999;263(2): 596–602.
75. Schmidlin F, Dery O, DeFea KO, et al. Dynamin and Rab5a-dependent trafficking and signaling of the neurokinin 1 receptor. *J Biol Chem*. 2001;276(27):25427–25437.
76. Fan G-H, Lapierre LA, Goldenring JR, Richmond A. Differential regulation of CXCR2 trafficking by Rab GTPases. *Blood*. 2003;101(6):2115–2124.
77. Murph MM, Scaccia LA, Volpicelli LA, Radhakrishna H. Agonist-induced endocytosis of lysophosphatidic acid-coupled LPA1/EDG-2 receptors via a dynamin2- and Rab5-dependent pathway. *J Cell Sci*. 2003;116(Pt 10):1969–1980.
78. Grimsey NL, Goodfellow CE, Dragunow M, Glass M. Cannabinoid receptor 2 undergoes Rab5-mediated internalization and recycles via a Rab11-dependent pathway. *Biochim Biophys Acta*. 2011;1813(8):1554–1560.

79. Sönnichsen B, De Renzis S, Nielsen E, Rietdorf J, Zerial M. Distinct membrane domains on endosomes in the recycling pathway visualized by multicolor imaging of Rab4, Rab5, and Rab11. *J Cell Biol.* 2000;149(4):901–914.
80. Wang F, Chen X, Zhang X, Ma L. Phosphorylation state of μ-opioid receptor determines the alternative recycling of receptor via Rab4 or Rab11 pathway. *Mol Endocrinol.* 2008;22(8):1881–1892.
81. Li H, Li H-F, Felder RA, Periasamy A, Jose PA. Rab4 and Rab11 coordinately regulate the recycling of angiotensin II type I receptor as demonstrated by fluorescence resonance energy transfer microscopy. *J Biomed Opt.* 2008;13(3):031206.
82. Volpicelli LA, Lah JJ, Fang G, Goldenring JR, Levey AI. Rab11a and myosin Vb regulate recycling of the M4 muscarinic acetylcholine receptor. *J Neurosci.* 2002; 22(22):9776–9784.
83. Martinez-Quiles N, Ho H-YH, Kirschner MW, Ramesh N, Geha RS. Erk/Src phosphorylation of cortactin acts as a switch on-switch off mechanism that controls its ability to activate N-WASP. *Mol Cell Biol.* 2004;24(12):5269–5280.
84. Vistein R, Puthenveedu MA. Src regulates sequence-dependent beta-2 adrenergic receptor recycling via cortactin phosphorylation. *Traffic.* 2014;15(11):1195–1205.
85. Roman-Vendrell C, Yu YJ, Yudowski GA. Fast modulation of μ-opioid receptor (MOR) recycling is mediated by receptor agonists. *J Biol Chem.* 2012;287(18): 14782–14791.
86. Bowman SL, Soohoo AL, Shiwarski DJ, Schulz S, Pradhan AA, Puthenveedu MA. Cell autonomous regulation of mu-opioid receptor recycling by substance P. *Cell Rep.* 2015;10:1925–1936.
87. Inoue M, Ueda H. Protein kinase C-mediated acute tolerance to peripheral mu-opioid analgesia in the bradykinin-nociception test in mice. *J Pharmacol Exp Ther.* 2000; 293(2):662–669.
88. Dang VC, Williams JT. Chronic morphine treatment reduces recovery from opioid desensitization. *J Neurosci.* 2004;24(35):7699–7706.
89. Doll C, Konietzko J, Pöll F, Koch T, Höllt V, Schulz S. Agonist-selective patterns of μ-opioid receptor phosphorylation revealed by phosphosite-specific antibodies. *Br J Pharmacol.* 2011;164(2):298–307.
90. Illing S, Mann A, Schulz S. Heterologous regulation of agonist-independent μ-opioid receptor phosphorylation by protein kinase C. *Br J Pharmacol.* 2014;171(5):1330–1340.
91. Lefkowitz RJ, Pitcher J, Krueger K, Daaka Y. Mechanisms of beta-adrenergic receptor desensitization and resensitization. *Adv Pharmacol.* 1998;42:416–420.
92. Shenoy SK, Lefkowitz RJ. Seven-transmembrane receptor signaling through beta-arrestin. *Sci STKE.* 2005;2005(308):cm10.
93. Ferrandon S, Feinstein TN, Castro M, et al. Sustained cyclic AMP production by parathyroid hormone receptor endocytosis. *Nat Chem Biol.* 2009;5(10):734–742.
94. Calebiro D, Nikolaev VO, Gagliani MC, et al. Persistent cAMP-signals triggered by internalized G-protein–coupled receptors. *PLoS Biol.* 2009;7(8):e1000172.
95. Calebiro D, Nikolaev VO, Persani L, Lohse MJ. Signaling by internalized G-protein-coupled receptors. *Trends Pharmacol Sci.* 2010;31(5):221–228.
96. Okazaki M, Ferrandon S, Vilardaga J-P, Bouxsein ML, Potts Jr JT, Gardella TJ. Prolonged signaling at the parathyroid hormone receptor by peptide ligands targeted to a specific receptor conformation. *Proc Natl Acad Sci USA.* 2008;105(43):16525–16530.
97. Irannejad R, Tomshine JC, Tomshine JR, et al. Conformational biosensors reveal GPCR signalling from endosomes. *Nature.* 2013;495(7442):534–538.
98. Tsvetanova NG, von Zastrow M. Spatial encoding of cyclic AMP signaling specificity by GPCR endocytosis. *Nat Chem Biol.* 2014;10(12):1061–1065.

CHAPTER NINE

α_2 Adrenergic Receptor Trafficking as a Therapeutic Target in Antidepressant Drug Action

Christopher Cottingham*[,1], Craig J. Ferryman*, Qin Wang[†,1]
*Department of Biology and Chemistry, Morehead State University, Morehead, Kentucky, USA
[†]Department of Cell, Developmental and Integrative Biology, University of Alabama at Birmingham, Birmingham, Alabama, USA
[1]Corresponding authors: e-mail address: c.cottingham@moreheadstate.edu; qinwang@uab.edu

Contents

Abstract

Antidepressant drugs remain poorly understood, especially with respect to pharmacological mechanisms of action. This lack of knowledge results from the extreme complexity inherent to psychopharmacology, as well as to a corresponding lack of knowledge regarding depressive disorder pathophysiology. While the final analysis is likely to be multifactorial and heterogeneous, compelling evidence exists for upregulation of brain α_2 adrenergic receptors (ARs) in depressed patients. This evidence has sparked a line of research into actions of a particular antidepressant drug class, the tricyclic antidepressants (TCAs), as direct ligands at α_{2A}ARs. Our findings, as outlined herein, demonstrate that TCAs function as arrestin-biased ligands at α_{2A}ARs. Importantly, TCA-induced α_{2A}AR/arrestin recruitment leads to receptor endocytosis and downregulation of α_{2A}AR expression with prolonged exposure. These findings represent a novel mechanism linking α_2AR trafficking with antidepressant pharmacology.

1. INTRODUCTION

Psychopharmacology is an exceedingly intricate discipline, aiming as it does to modulate human emotion, behavior, and other complex cognitive processes. Although our understanding of brain–behavior relationships has

Progress in Molecular Biology and Translational Science, Volume 132
ISSN 1877-1173
http://dx.doi.org/10.1016/bs.pmbts.2015.03.007

improved dramatically in recent decades, modern neuroscience still struggles to provide coherent pathophysiological mechanisms for psychiatric disorders. Further complicating matters, at a molecular level, the psychopharmacologist must contend with an almost bewildering array of neurotransmitters, ionotropic and metabotropic receptors, transporters, and enzymes both catabolic and anabolic. These myriad considerations help to explain why our understanding of the therapeutic actions underlying psychoactive agents remains frustratingly limited decades after the psychopharmacology revolution of the 1950s.

Antidepressant drugs (ADs) are among the most widely used psychopharmacological agents. These therapeutics have application not just in managing the depressive disorders but also in treating conditions such as migraine headaches[1–3] and chronic neuropathic pain.[1,3–5] All currently available ADs function by affecting, in some fashion, monoamine neurotransmitter systems in the brain.[6] The central monoamine neurotransmitters, which include serotonin (5-HT), dopamine (DA), and norepinephrine (NE), come complete with a corresponding set of enzymes, transporters, and primarily metabotropic receptors. All of these proteins can and have served as molecular antidepressant targets. Indeed, the earliest ADs were the monoamine oxidase inhibitors, which inhibit the activity of the monoamine oxidase enzyme responsible for catabolic inactivation of released monoamines. While these drugs still enjoy some limited therapeutic use, the vast majority of contemporary ADs function primarily as transporter blockers, a therapeutic mechanism known as reuptake inhibition. Transporter blockers include the selective serotonin reuptake inhibitors (SSRIs), the serotonin-norepinephrine reuptake inhibitors (SNRIs), and the tricyclic antidepressants (TCAs).

Regardless of what enzyme or transporter is inhibited, one can expect a resulting increase in synaptic availability of these monoamine neurotransmitters. Beyond this statement of fact, the mechanisms of action remain largely mysterious, as one must then consider which of the fourteen 5-HT receptors,[7] five DA receptors,[7] or nine adrenergic receptors[7] are being affected by the increased neurotransmitter level. In addition, some of these drugs, particularly the TCAs, are classic examples of pharmacologically "dirty" drugs, having multiple direct ligand interactions with these receptors alongside their reuptake inhibition.[6]

Given the extreme complexity inherent to antidepressant pharmacology outlined above, it seems prudent to apply a measured, systematic approach to understanding how these therapeutic agents affect their various molecular targets. To that end, we have expended considerable energy in recent years to characterize how TCAs function as direct ligands at the α_{2A} adrenergic receptor (AR), a G protein-coupled receptor (GPCR) with numerous physiological roles throughout the central nervous system (CNS) and which has itself been implicated in the neurobiology of depressive disorders. Our research efforts have uncovered extensive novel information, especially regarding the induction of arrestin recruitment to and subsequent arrestin-mediated trafficking of α_{2A}ARs by TCAs. This chapter will therefore be aimed at summarizing our findings on the arrestin-biased behavior of TCAs at the α_{2A}AR, and placing these findings in their proper physiological and pharmacological context.

2. PHYSIOLOGICAL BASIS FOR STUDYING α_{2A}AR TRAFFICKING IN DEPRESSION

A careful review of the basic and clinical literature reveals broad support for the study of α_{2A}AR trafficking in the context of antidepressant psychopharmacology. While it is important to note that the neurobiology of depressive disorders is likely to be multifactorial and heterogeneous, currently available evidence demonstrates that dysregulation of α_2AR expression and/or function level is involved in at least some cases of depression. More specifically, the evidence supports a link between upregulation of α_2AR expression and/or function and major depressive disorder (MDD). For a more detailed explanation and discussion of this evidence, the reader is referred to our recent review article on this subject.[8] Here, we will focus on a concise overview of the findings.

Many studies in this area have been unable to look specifically at the α_{2A}AR, meaning that some clarification on α_2ARs is required before proceeding. The α_{2A}AR is one of three subtypes (α_{2A}AR, α_{2B}AR, and α_{2C}AR) comprising the α_2AR subfamily,[9,10] itself one of three AR subfamilies (α_1ARs, α_2ARs, and βARs).[7] Our ability to investigate with subtype specificity is hindered by a lack of subtype-selective ligands for the different α_2ARs. This is significant given that most published studies have relied on radioligand binding to determine α_2AR levels. However, it is well established that the α_{2A}AR is the most predominantly expressed α_2AR subtype within the CNS,[11–13] and is largely responsible for mediating classic α_2AR functions such as the regulation of neurotransmitter release from presynaptic terminals[14,15] and the centrally mediated sedative, hypotensive, and analgesic effects of α_{2A}AR agonists.[16–20]

Clinical studies on α_2AR levels in depression have largely taken one of two major approaches, assaying receptor density through radioligand binding in platelet samples obtained from living patients[21–30] or in postmortem brain tissue obtained mainly from depressed suicide completers. Regardless of approach, these studies as a whole support the existence of α_2AR upregulation in cases of MDD, with the upregulation being observed in brain regions including the frontal and prefrontal cortex,[31–37] hippocampus,[31] and locus coeruleus.[38–40] One study specifically found an upregulation of α_{2A}AR mRNA in the prefrontal cortex of depressed suicide completers.[41]

The neurobiological consequences of α_2AR upregulation are likely to be far-reaching and complex, given the broad reach of noradrenergic projections from the locus coeruleus throughout the brain. Furthermore, while the presynaptic role of α_{2A}AR autoreceptors in regulating NE release from and firing activity of locus coeruleus neurons has long been appreciated, more recent evidence has revealed significant heteroreceptor and postsynaptic roles for α_{2A}ARs in the CNS.[20,42–45] As a general rule, α_2ARs have an inhibitory effect on CNS function, as they classically couple to $G\alpha_{i/o}$-containing heterotrimeric G proteins, thereby linking to inhibition of adenylyl cyclase.[46,47] More specifically, at the cellular level, α_2AR activation links to inhibition of neurotransmitter release via inhibition of voltage-gated Ca^{2+} channels and neuronal hyperpolarization via activation of G protein-coupled inwardly rectifying K^+ channels.[44,46,47] On a larger physiological scale, α_2ARs mediate sedative, hypotensive, and anti-epileptogenic[48] effects. It can therefore be reasonably assumed that α_{2A}AR upregulation will have a generally inhibitory effect on central neurotransmission.

If α_{2A}AR upregulation represents at least a component of MDD pathophysiology, it stands to reason that receptor downregulation represents a valid AD therapeutic strategy. Indeed, additional evidence indicates that the chronic exposure to ADs necessary for symptom relief in MDD is associated with decreases in α_2AR expression and/or activity level. This evidence comes from both clinical studies[21,22,26,29,33,38,49–52] and preclinical modeling[53–60] of AD pharmacological activity and is especially associated with chronic exposure to TCAs.

It is here that we find the potential link between MDD and GPCR trafficking, as it has long been appreciated that downregulation of receptor expression can result from prolonged exposure to ligands and ligand-induced receptor endocytosis. Although it would be logical to assume that downregulation of neurotransmitter receptor expression could result from the elevated neurotransmitter levels caused by AD-mediated reuptake inhibition, there is a dearth of evidence to support that assumption. In fact, with respect to ARs, chronic exposure to NE reuptake inhibitors has been shown to sustain only a mild increase in brain NE levels to 10 nM from a typical baseline near 1 nM,[57,61,62] suggesting that an additional mechanism is necessary to explain TCA-induced α_2AR downregulation. As mentioned previously, TCAs exhibit a large degree of molecular promiscuity, with nontransporter targets that include α_2ARs,[6,63–65] providing an attractive potential additional mechanism.

3. ARRESTIN-BIASED REGULATION OF THE α_{2A}AR BY THE TCA DRUG CLASS

The need for a mechanistic explanation regarding AD-induced α_2AR downregulation, coupled with the knowledge that TCAs can bind directly to these highly relevant GPCRs, has sparked a productive line of research in our laboratories. Our initial work sought to characterize the immediate molecular consequences of TCA binding to α_{2A}ARs. Although we began with a focus on a specific TCA, desipramine (DMI),[60] we have since expanded our work to include two additional representative members of the TCA drug class, imipramine (IMI) and amitriptyline (AMI).[66] TCAs share a common basic chemical structure which has been slightly modified to generate the various specific drugs (Fig. 1), and each of these TCAs exhibits a slightly different molecular profile in terms of their relative affinities for the α_{2A}AR (both murine, as assessed experimentally, and human, as reported in the literature), the NE transporter (NET), and the 5-HT transporter (SERT), data which are summarized in Table 1. DMI and IMI share a similar α_{2A}AR affinity in the low micromolar range, similar to α_{2A}AR affinity for the endogenous ligand NE, while AMI has an approximately 14-fold stronger α_{2A}AR affinity, in the nanomolar range. It is interesting to note that using transporter affinities results in different groupings, as DMI exhibits a strong selectivity for NET over SERT, while IMI and AMI exhibit a strong selectivity for SERT over NET. This finding suggests that different chemical determinants are relevant for receptor interactions versus transporter interactions.

Beyond their molecular profile, our work has sought to investigate what, if any, actions these TCAs might have when acting as direct α_{2A}AR ligands. With respect to signaling activity, the TCAs appear to behave similarly to classic neutral antagonists, as they drive no appreciable coupling of

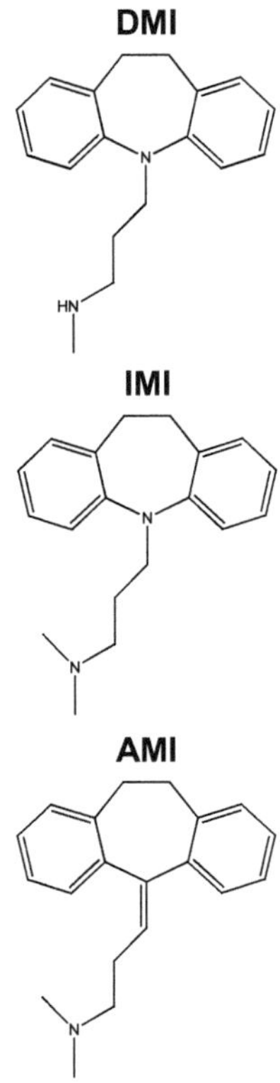

Figure 1 Chemical structures of the three TCAs.

heterotrimeric G proteins to the α_{2A}AR (Table 1). Furthermore, we have been unable to observe any activation of typical α_{2A}AR downstream signaling targets such as MAP kinases[69,70] or Akt[60] in response to TCA stimulation of α_{2A}ARs,[60] although we certainly cannot and should not rule out the possibility of novel signal transduction at this time. As well, it is important to note that TCAs do not appear to function simply as classic neutral antagonists. While attempting to determine whether DMI could block NE-induced α_{2A}AR signaling, as would be expected for an antagonist, we found that DMI actually potentiates NE-induced MAP kinase activation by α_{2A}ARs,[71] indicating a more complex pharmacological interaction.

Further complicating the attempt to pharmacologically classify TCAs as direct α_{2A}AR ligands, we discovered that these agents exhibit agonist-like properties with respect to receptor-arrestin recruitment. Through a combination of coimmunoprecipitation and fluorescence lifetime imaging (FLIM)-based FRET approaches, we have shown that TCAs have a variable

Table 1 Summary of TCA Pharmacological Properties as Direct α_{2A}AR Ligands

Parameter	DMI[60]	IMI[66]	AMI[66]
K_i (murine α_{2A}AR)	4.62 μM	5.07 μM	330 nM
K_i (human α_2AR)	3.44 μM[67,68]	3.10 μM[67]	402 nM[67,68]
Transporter binding[6]			
NET K_i (nM)	0.8	37	34.5
SERT K_i (nM)	17.5	1.4	4.3
G protein coupling	No	No	No
Endocytosis			
Peak internalization (%)	35	40	40
Arrestin dependence	Yes	Yes	Yes
Arrestin recruitment (by FLIM-FRET)			
Arrestin2	+	ND	++
Arrestin3	+++	+++	+

Data for murine K_i, G protein coupling, endocytosis, and arrestin recruitment come from Cottingham *et al.*[60] or Cottingham *et al.*,[66] as indicated. Human K_i values are as reported in Cusack *et al.*[67] and/or Owens *et al.*[68]; values for DMI and AMI are an average of values reported in both publications. Transporter binding values are as reported in Baldessarini.[6]

capacity to drive recruitment of arrestins to the α_{2A}AR.[60,66] The nonvisual arrestins, arrestin2 and arrestin3, also sometimes referred to as β-arrestin1 and β-arrestin-2, are well appreciated as key regulators of GPCR function. These adaptor or scaffolding-type proteins are often essential to classical GPCR endocytosis via clathrin-coated pits,[72–75] a process which leads to receptor desensitization,[76,77] and have more recently been appreciated as multi-functional proteins capable of transducing intracellular signaling independent of heterotrimeric G proteins.[78–81] Additionally, expression of the nonvisual arrestins has been reported to be dysregulated in cases of MDD[82,83] and to be affected by chronic exposure to ADs.[82–84] Coupled with the evidence on α_2AR dysregulation described earlier, such findings highlight the potential importance of α_{2A}AR arrestin-mediated trafficking to MDD and its psychopharmacological management.

As mentioned above, and summarized in Table 1, our research has revealed that TCAs are able to drive α_{2A}AR-arrestin recruitment at concentrations near or below their α_{2A}AR affinity values. Both DMI (at concentrations of 1 and 10 μM) and IMI (10 μM) drive a similar degree of robust arrestin3 recruitment to the receptor, when the process is observed via FLIM-FRET in live cells. AMI (1 μM) drives arrestin3 recruitment less strongly than the others, though still significantly. Furthermore, we have found that certain TCAs preferentially recruit one arrestin over the other. Specifically, we have uncovered clear preferential arrestin3 recruitment to the α_{2A}AR by DMI, with much weaker arrestin2 recruitment by this TCA.[60] Meanwhile, AMI drives a more robust recruitment of arrestin2 to the α_{2A}AR compared with arrestin3.[66] This finding provides yet another example of the subtle biomolecular differences between these chemically similar pharmacological agents and has additional significance in light of evidence supporting differential expression patterns for arrestin2 versus arrestin3 in the brain.[85]

Importantly, we have been able to establish that TCA-dependent arrestin recruitment to the α_{2A}AR has a functional consequence, namely to induce receptor endocytosis. By stimulating both heterologous cells stably expressing α_{2A}ARs and primarily cultured cortical neurons endogenously expressing α_{2A}ARs, we demonstrated that TCA stimulation drives α_{2A}AR endocytosis in an arrestin-dependent fashion.[60,66] Each of the TCAs assayed was able to drive a similar degree of receptor internalization, peaking at approximately 35–40% loss of cell surface receptor density. Further, we have demonstrated that TCA-induced α_{2A}AR endocytosis occurs in a clathrin-dependent fashion,[60] confirming the involvement of the classical arrestin- and clathrin-mediated pathway for GPCR endocytosis.

Collectively, the data summarized in the preceding paragraphs establish the TCAs as arrestin-biased ligands at the α_{2A}AR, meaning that they are able to selectively drive arrestin recruitment to the receptor while stimulating no detectable heterotrimeric G protein coupling activity. Arrestin bias is a more recent phenomenon in the GPCR field,[80,81] and one that had not previously been observed at the α_{2A}AR specifically. As well, this particular form of arrestin bias selects for a trafficking response but not for any readily apparent arrestin-mediated signal transduction. Most previously reported arrestin-biased ligands acted as agonists, simply initiating signal transduction via arrestin rather than heterotrimeric G proteins. Nonetheless, as stated above, we cannot rule out the possibility of novel TCA-induced arrestin-mediated signaling by α_{2A}ARs.

Given the evidence for α_2AR upregulation in MDD outlined earlier, the potential clinical significance of our findings to this point should be readily apparent. It has long been appreciated that prolonged or chronic exposure, on the order of several hours to days, to agonists can drive a downregulation of receptor expression,[72,74] an effect which has clear therapeutic relevance in the context of MDD. Indeed, we have shown that prolonged exposure, on the order of 4–24 h, to a TCA can drive downregulation of overall α_{2A}AR expression in cultured heterologous cells and primary cortical neurons.[60] This downregulation effect reaches a peak of approximately 50% reduction in α_{2A}AR expression at the 24-h time point, a reduction which is sufficient to attenuate NE-stimulated α_{2A}AR-mediated MAP kinase activation in the treated cells. To be clear, this downregulation effect occurs in response to TCA treatment alone in non-noradrenergic cells, which would not be releasing NE into the culture environment. Additionally, prolonged treatment of cultured cells with NE at 10 nM, the concentration reported to result from chronic NE reuptake inhibition,[57,61] is not sufficient to drive α_{2A}AR downregulation.

A final set of *in vivo* experiments underscores the physiological relevance of TCA-induced α_{2A}AR downregulation. Through the use of subcutaneous osmotic mini-pumps, we exposed mice to a 2-week course of chronic TCA treatment. After the 2-week period, mice were sacrificed and their cerebral cortices were used to generate a crude synaptosomal membrane preparation. In these preparations, we found a significant decrease in α_{2A}AR expression via radioligand binding, indicating receptor downregulation.[60] Most importantly, the effect on receptor expression was completely lost in arrestin3-null mice, suggesting the *in vivo* occurrence of arrestin-mediated α_{2A}AR trafficking and downregulation. This last piece of evidence also supports the *in vivo* relevance of the preferential recruitment of arrestin3 over arrestin2 to α_{2A}ARs by DMI observed *in vitro*.

Although additional study is necessary to directly observe this phenomenon in the brains of depressed human patients, our data strongly implicate arrestin-biased regulation of α_{2A}ARs by TCAs leading to receptor downregulation in the therapeutic mechanism of action of these ADs. It is important to note that our experiments have largely utilized physiologically relevant drug concentrations, a fact which is apparent when reported clinical therapeutic levels for these drugs (summarized in Table 2) are considered.

Table 2 Typical Therapeutic Ranges for TCAs

Clinical Therapeutic Range	DMI	IMI	AMI
ng/mL	125–600	175–300	100–250
μM	0.5–2.3	0.4–1.1	0.4–0.9

Values are clinically determined plasma concentrations as reported in Baldessarini.[6]

It is likely, then, that α_{2A}AR upregulation associated with MDD can be normalized, at least in part, due to the arrestin-biased interaction of a TCA with the receptor directly, leading to receptor trafficking and, ultimately, stable downregulation of receptor expression. Nevertheless, given the extreme complexity inherent to psychopharmacology, this putative mechanism is almost certainly just one piece of a massive puzzle that has yet to be fully solved.

As an addendum to the preceding evidence directly regarding receptor trafficking and downregulation, we have also demonstrated *in vivo* relevance for an α_{2A}AR- and arrestin3-mediated TCA mechanism through the use of behavioral pharmacology.[86] Porsolt's forced swim test[87,88] is a gold-standard preclinical model for assaying antidepressant effects on rodent behavior in which effective ADs show activity by reducing immobility, a form of learned helplessness behavior. Our work has shown that DMI induces an antidepressant effect on mouse behavior (i.e., reduces immobility) in an α_{2A}AR-dependent fashion, with the antidepressant effect lost in α_{2A}AR-null mice. Furthermore, we have shown that antidepressant responsiveness to DMI is significantly attenuated in arrestin3-null mice. While it is difficult to draw a clear mechanistic link to the trafficking line of evidence, these behavioral pharmacology findings nevertheless provide *in vivo* support for the importance of α_{2A}ARs and arrestin to TCA drug actions.

4. THERAPEUTIC IMPLICATIONS

The accumulated evidence presented thus far, including both clinical data and preclinical data from the literature and extensive work from our own laboratories, makes a compelling case for an α_{2A}AR-dependent AD therapeutic mechanism of action relying on physiologically normalizing arrestin-mediated downregulation of receptor expression. This putative mechanism is summarized in Fig. 2. In short, the arrestin-biased nature of TCAs at the α_{2A}AR means that these drugs do not stimulate heterotrimeric G protein coupling and would therefore be unable to contribute to the generally inhibitory neuronal effects classically driven by α_{2A}AR signaling. These TCAs are, however, capable of driving arrestin recruitment to the receptor, an effect which leads to receptor endocytosis and ultimately to α_{2A}AR downregulation following prolonged exposure. It is important to remember that this downregulation would likely represent a neuroadaptive effect, normalizing or correcting a pathophysiological α_{2A}AR upregulation underlying the occurrence of MDD. This is a particularly attractive

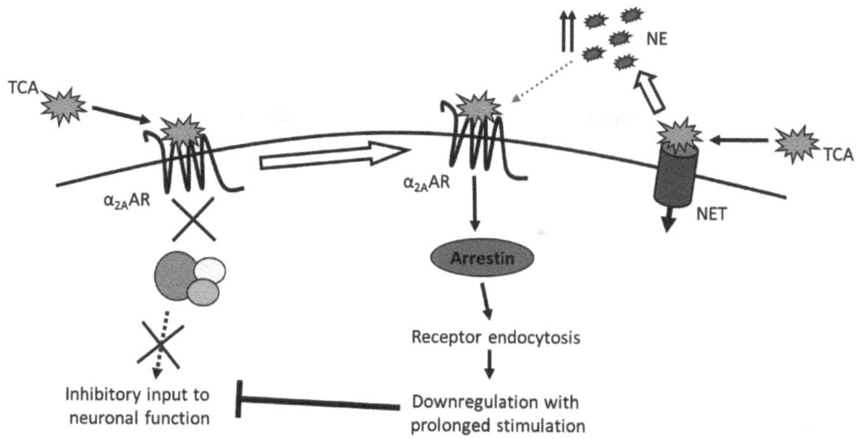

Figure 2 Working model for a TCA-induced therapeutic mechanism relying on α_{2A}AR downregulation. Exposure to a TCA leads to acute α_{2A}AR endocytosis in an arrestin-dependent fashion. With chronic exposure, this endocytic response transitions into a stable downregulation of receptor expression. This direct TCA-induced mechanism is necessary given that NE levels achieved by reuptake inhibition are insufficient to drive receptor downregulation. Downregulation of α_{2A}ARs would be expected to reduce overall signaling activity by this receptor, activity which represents generally inhibitory input to neuronal function. (See the color plate.)

mechanism given the well-appreciated clinical lag time of 3–6 weeks between the start of antidepressant therapy and the onset of symptom relief.[6] Our findings are essential to progress in the field of AD psychopharmacology, given the lack of clear mechanistic models for AD therapeutics and the significant limitations of contemporary ADs.

All presently available ADs represent massive and broad biochemical alterations to the brain, making the typical AD a very blunt pharmacological instrument. This characteristic is clear for the TCAs and their molecular promiscuity, interacting with numerous receptor and transporter targets. However, even a clean drug such as an SSRI is not really clean at all, as the resulting increase in 5-HT levels will be global, affecting not only the serotonergic raphe nuclei but also any brain region receiving serotonergic inputs, and has the potential to increase activity of fourteen different 5-HT receptors. We believe that it is essential for the next generation of AD therapeutics to be more pharmacologically precise and finely tuned, thereby affecting a drastically narrower range of molecular targets.

Pharmacological imprecision is only one limitation of currently available AD therapeutics. While these drugs can certainly be quite effective at managing MDD symptoms with a minimum of undesirable side effects, far too many patients are still unable to achieve symptom relief. Available evidence suggests that approximately half of all clinically depressed patients either have incomplete symptom relief or fail to respond at all to AD therapy.[89–91] Furthermore, the decades since the psychopharmacology revolution have seen precious little improvement in overall AD efficacy. Recent findings suggest that newer generation drugs such as the SSRIs and SNRIs are not significantly better at achieving symptom relief than older therapeutics such as the TCAs.[92–95] To be sure, the SSRIs and SNRIs do exhibit better tolerability and smaller side effect profiles when compared with the TCAs. Nevertheless, our apparent inability to improve antidepressant efficacy underscores the extreme difficulty inherent to understanding this complex pharmacological problem.

All of these limitations have understandably led to doubts regarding the value of monoamine-based therapeutic strategies. Much attention has been given of late to the application of the glutamatergic compound ketamine, which antagonizes NMDA-type ionotropic glutamate receptors, as a rapid-acting AD.[96–98] As well, the mixed MT_1/MT_2 melatonin receptor agonist-5-HT_{2C} receptor antagonist agomelatine has shown promise as an AD, with an efficacy that approximates traditional ADs,[99] and has been approved for use in Europe, although not yet in the United States. While there is significant evidence to support these therapeutic applications, it is important to caution that there is just as little information supporting an etiological role for glutamatergic or melatonergic systems in MDD as there is for monoaminergic systems. Therefore, the basis for ketamine and agomelatine as ADs comes primarily from the symptomatic side, as was true for the first monoaminergic drugs. Furthermore, it seems unwise to completely discount monoaminergic mechanisms, which have been the foundation of AD psychopharmacology for decades. Continuing to improve our knowledge base on existing therapeutics, as we have done in our line of research, should allow for both a better understanding of the neurobiology of MDD and improved design of future therapeutics.

Our research has demonstrated that slight chemical variations on the common base tricyclic structure leads to differing molecular pharmacological profiles, both in terms of relative α_{2A}AR versus transporter affinity and ability to drive arrestin recruitment. This evidence raises the possibility that, with further chemical tweaking, a more selective arrestin-biased α_{2A}AR ligand can be developed. Such a hypothetical drug would meet our desired characteristic of greater pharmacological precision and would be able to more cleanly target and normalize upregulated α_2AR expression without the broad-spectrum biomolecular effects of the original TCAs. Given the current and ever-expanding GPCR structural revolution,[100,101] which is now capable of probing the receptor/arrestin interaction itself,[102] the time is ripe for just such a drug design effort.

REFERENCES

1. Verdu B, Decosterd I, Buclin T, Stiefel F, Berney A. Antidepressants for the treatment of chronic pain. *Drugs*. 2008;68(18):2611–2632.
2. Jackson JL, Shimeall W, Sessums L, et al. Tricyclic antidepressants and headaches: systematic review and meta-analysis. *BMJ*. 2010;341(1):c5222.
3. Dharmshaktu P, Tayal V, Kalra BS. Efficacy of antidepressants as analgesics: a review. *J Clin Pharmacol*. 2012;52(1):6–17. http://dx.doi.org/10.1177/0091270010394852.
4. Moore RA, Derry S, Aldington D, Cole P, Wiffen PJ. Amitriptyline for neuropathic pain and fibromyalgia in adults. *Cochrane Database Syst Rev*. 2012;12http://dx.doi.org/10.1002/14651858.CD008242.pub2, CD008242.
5. Hearn L, Moore RA, Derry S, Wiffen PJ, Phillips T. Desipramine for neuropathic pain in adults. *Cochrane Database Syst Rev*. 2014;9http://dx.doi.org/10.1002/14651858.CD011003.pub2, CD011003.
6. Baldessarini RJ. Drug therapy of depression and anxiety disorders. In: Brunton L, Lazo JS, Parker KL, eds. *Goodman & Gilman's The Pharmacological Basis of Therapeutics*. 11th ed. New York: McGraw-Hill, Inc.; 2006:429–460.
7. Alexander S, Benson H, Faccenda E, et al. The concise guide to pharmacology 2013/14: G protein-coupled receptors. *Br J Pharmacol*. 2013;170:1459–1581.
8. Cottingham C, Wang Q. alpha(2) adrenergic receptor dysregulation in depressive disorders: implications for the neurobiology of depression and antidepressant therapy. *Neurosci Biobehav Rev*. 2012;36:2214–2225. http://dx.doi.org/10.1016/j.neubiorev.2012.07.011.
9. Cottingham C, Chen H, Chen Y, Peng Y, Wang Q. Genetic variations of alpha2-adrenergic receptors illuminate the diversity of receptor functions. In: Wang Q, ed. Current Topics in Membranes. vol 67. Amsterdam: Elsevier Inc.; 2011.
10. Wang Q. Alpha2-adrenergic receptors. In: Robertson D, Biaggioni I, Burnstock G, Low P, Paton J, eds. *Primer on the Autonomic Nervous System*. San Diego, CA, USA: Elsevier Inc.; 2011.
11. De Vos H, Vauquelin G, De KJ, De Backer JP, Van LI. Regional distribution of alpha 2A- and alpha 2B-adrenoceptor subtypes in postmortem human brain. *J Neurochem*. 1992;58(4):1555–1560.
12. Sastre M, García-Sevilla JA. Alpha 2-adrenoceptor subtypes identified by [3H] RX821002 binding in the human brain: the agonist guanoxabenz does not discriminate different forms of the predominant alpha 2A subtype. *J Neurochem*. 1994;63(3):1077–1085.
13. Wang R, Macmillan LB, Fremeau Jr RT, Magnuson MA, Lindner J, Limbird LE. Expression of alpha 2-adrenergic receptor subtypes in the mouse brain: evaluation of spatial and temporal information imparted by 3 kb of 5' regulatory sequence for the alpha 2A AR-receptor gene in transgenic animals. *Neuroscience*. 1996;74(1):199–218.
14. Hein L, Altman JD, Kobilka BK. Two functionally distinct alpha2-adrenergic receptors regulate sympathetic neurotransmission. *Nature*. 1999;402(6758):181–184.
15. Knaus AE, Muthig V, Schickinger S, et al. Alpha2-adrenoceptor subtypes—unexpected functions for receptors and ligands derived from gene-targeted mouse models. *Neurochem Int*. 2007;51(5):277–281. http://dx.doi.org/10.1016/j.neuint.2007.06.036.

16. MacMillan LB, Hein L, Smith MS, Piascik MT, Limbird LE. Central hypotensive effects of the alpha2a-adrenergic receptor subtype. *Science*. 1996;273(5276):801–803.
17. Stone LS, MacMillan LB, Kitto KF, Limbird LE, Wilcox GL. The alpha2a adrenergic receptor subtype mediates spinal analgesia evoked by alpha2 agonists and is necessary for spinal adrenergic-opioid synergy. *J Neurosci Off J Soc Neurosci*. 1997;17(18):7157–7165.
18. Altman JD, Trendelenburg AU, MacMillan L, et al. Abnormal regulation of the sympathetic nervous system in alpha2A-adrenergic receptor knockout mice. *Mol Pharmacol*. 1999;56(1):154–161.
19. Franowicz JS, Kessler LE, Borja CMD, Kobilka BK, Limbird LE, Arnsten AFT. Mutation of the alpha2A-adrenoceptor impairs working memory performance and annuls cognitive enhancement by guanfacine. *J Neurosci Off J Soc Neurosci*. 2002;22(19):8771–8777.
20. Gilsbach R, Roser C, Beetz N, et al. Genetic dissection of alpha2-adrenoceptor functions in adrenergic versus nonadrenergic cells. *Mol Pharmacol*. 2009;75(5):1160–1170.
21. Healy D, Carney PA, O'Halloran A, Leonard BE. Peripheral adrenoceptors and serotonin receptors in depression. Changes associated with response to treatment with trazodone or amitriptyline. *J Affect Disord*. 1985;9(3):285–296.
22. Garcia-Sevilla JA, Udina C, Fuster MJ, Alvarez E, Casas M. Enhanced binding of [3H] (-) adrenaline to platelets of depressed patients with melancholia: effect of long-term clomipramine treatment. *Acta Psychiatr Scand*. 1987;75(2):150–157.
23. Pandey GN, Janicak PG, Javaid JI, Davis JM. Increased 3H-clonidine binding in the platelets of patients with depressive and schizophrenic disorders. *Psychiatry Res*. 1989;28(1):73–88.
24. Takeda T, Harada T, Otsuki S. Platelet 3H-clonidine and 3H-imipramine binding and plasma cortisol level in depression. *Biol Psychiatry*. 1989;26(1):52–60.
25. Piletz JE, Halaris A, Saran A, Marler M. Elevated 3H-para-aminoclonidine binding to platelet purified plasma membranes from depressed patients. *Neuropsychopharmacol Off Publ Am Coll Neuropsychopharmacol*. 1990;3(3):201–210.
26. Piletz JE, Halaris A, Saran A, Marler MR. Desipramine lowers tritiated para-aminoclonidine binding in platelets of depressed patients. *Arch Gen Psychiatry*. 1991;48(9):813–820.
27. Kaneko M, Kanno T, Honda K, et al. Platelet alpha-2 adrenergic receptor binding and plasma free 3-methoxy-4-hydroxyphenylethylene glycol in depressed patients before and after treatment with mianserin. *Neuropsychobiology*. 1992;25(1):14–19. http://dx.doi.org/10.1159/000118803.
28. Werstiuk ES, Coote M, Griffith L, Shannon H, Steiner M. Effects of electroconvulsive therapy on peripheral adrenoceptors, plasma, noradrenaline, MHPG and cortisol in depressed patients. *Br J Psychiatry J Ment Sci*. 1996;169(6):758–765.
29. Gurguis GN, Vo SP, Griffith JM, Rush AJ. Platelet alpha2A-adrenoceptor function in major depression: Gi coupling, effects of imipramine and relationship to treatment outcome. *Psychiatry Res*. 1999;89(2):73–95.
30. García-Sevilla JA, Ventayol P, Pérez V, et al. Regulation of platelet alpha 2A-adrenoceptors, Gi proteins and receptor kinases in major depression: effects of mirtazapine treatment. *Neuropsychopharmacol Off Publ Am Coll Neuropsychopharmacol*. 2004;29(3):580–588. http://dx.doi.org/10.1038/sj.npp.1300356.
31. González AM, Pascual J, Meana JJ, et al. Autoradiographic demonstration of increased alpha 2-adrenoceptor agonist binding sites in the hippocampus and frontal cortex of depressed suicide victims. *J Neurochem*. 1994;63(1):256–265.
32. Callado LF, Meana JJ, Grijalba B, Pazos A, Sastre M, García-Sevilla JA. Selective increase of alpha2A-adrenoceptor agonist binding sites in brains of depressed suicide victims. *J Neurochem*. 1998;70(3):1114–1123.

33. García-Sevilla JA, Escribá PV, Ozaita A, et al. Up-regulation of immunolabeled alpha2A-adrenoceptors, Gi coupling proteins, and regulatory receptor kinases in the prefrontal cortex of depressed suicides. *J Neurochem.* 1999;72(1):282–291.
34. Meana JJ, García-Sevilla JA. Increased alpha 2-adrenoceptor density in the frontal cortex of depressed suicide victims. *J Neural Transm.* 1987;70(3–4):377–381.
35. Meana JJ, Barturen F, García-Sevilla JA. Alpha 2-adrenoceptors in the brain of suicide victims: increased receptor density associated with major depression. *Biol Psychiatry.* 1992;31(5):471–490.
36. González-Maeso J, Rodríguez-Puertas R, Meana JJ, García-Sevilla JA, Guimón J. Neurotransmitter receptor-mediated activation of G-proteins in brains of suicide victims with mood disorders: selective supersensitivity of alpha(2A)-adrenoceptors. *Mol Psychiatry.* 2002;7(7):755–767. http://dx.doi.org/10.1038/sj.mp.4001067.
37. Valdizan EM, Diez-Alarcia R, Gonzalez-Maeso J, et al. α_2 Adrenoceptor functionality in postmortem frontal cortex of depressed suicide victims. *Biol Psychiatry.* 2010;68(9):869–872. http://dx.doi.org/10.1016/j.biopsych.2010.07.023.
38. De Paermentier F, Mauger JM, Lowther S, Crompton MR, Katona CL, Horton RW. Brain alpha-adrenoceptors in depressed suicides. *Brain Res.* 1997;757(1):60–68.
39. Ordway GA, Widdowson PS, Smith KS, Halaris A. Agonist binding to alpha 2-adrenoceptors is elevated in the locus coeruleus from victims of suicide. *J Neurochem.* 1994;63(2):617–624.
40. Ordway GA, Schenk J, Stockmeier CA, May W, Klimek V. Elevated agonist binding to alpha2-adrenoceptors in the locus coeruleus in major depression. *Biol Psychiatry.* 2003;53(4):315–323.
41. Escribá PV, Ozaita A, García-Sevilla JA. Increased mRNA expression of alpha2A-adrenoceptors, serotonin receptors and mu-opioid receptors in the brains of suicide victims. *Neuropsychopharmacol Off Publ Am Coll Neuropsychopharmacol.* 2004; 29(8):1512–1521. http://dx.doi.org/10.1038/sj.npp.1300459.
42. Wang M, Ramos BP, Paspalas CD, et al. Alpha2A-adrenoceptors strengthen working memory networks by inhibiting cAMP-HCN channel signaling in prefrontal cortex. *Cell.* 2007;129(2):397–410. http://dx.doi.org/10.1016/j.cell.2007.03.015.
43. Shields AD, Wang Q, Winder DG. alpha2A-adrenergic receptors heterosynaptically regulate glutamatergic transmission in the bed nucleus of the stria terminalis. *Neuroscience.* 2009;163(1):339–351.
44. Gilsbach R, Albarrán-Juárez J, Hein L. Pre- versus postsynaptic signaling by α(2)-adrenoceptors. *Curr Top Membr.* 2011;67:139–160. http://dx.doi.org/10.1016/B978-0-12-384921-2.00007-0.
45. Gilsbach R, Hein L. Are the pharmacology and physiology of α_2 adrenoceptors determined by α_2-heteroreceptors and autoreceptors respectively? *Br J Pharmacol.* 2012;165(1):90–102. http://dx.doi.org/10.1111/j.1476-5381.2011.01533.x.
46. Limbird LE. Receptors linked to inhibition of adenylate cyclase: additional signaling mechanisms. *FASEB J Off Publ Fed Am Soc Exp Biol.* 1988;2(11):2686–2695.
47. Kobilka B. Adrenergic receptors as models for G protein-coupled receptors. *Annu Rev Neurosci.* 1992;15:87–114. http://dx.doi.org/10.1146/annurev.ne.15.030192.000511.
48. Wilson MH, Puranam RS, Ottman R, et al. Evaluation of the alpha(2A)-adrenergic receptor gene in a heritable form of temporal lobe epilepsy. *Neurology.* 1998;51(6):1730–1731.
49. García-Sevilla JA, Zis AP, Hollingsworth PJ, Greden JF, Smith CB. Platelet alpha 2-adrenergic receptors in major depressive disorder. Binding of tritiated clonidine before and after tricyclic antidepressant drug treatment. *Arch Gen Psychiatry.* 1981;38(12):1327–1333.

50. Smith CB, Hollingsworth PJ, Garcia-Sevilla JA, Zis AP. Platelet alpha 2 adrenoreceptors are decreased in number after antidepressant therapy. *Prog Neuro-psychopharmacol Biol Psychiatry*. 1983;7(2–3):241–247.
51. García-Sevilla JA, Padró D, Giralt MT, Guimón J, Areso P. Alpha 2-adrenoceptor-mediated inhibition of platelet adenylate cyclase and induction of aggregation in major depression. Effect of long-term cyclic antidepressant drug treatment. *Arch Gen Psychiatry*. 1990;47(2):125–132.
52. Karege F, Bovier P, Widmer J, Gaillard JM, Tissot R. Platelet membrane alpha 2-adrenergic receptors in depression. *Psychiatry Res*. 1992;43(3):243–252.
53. Nomura S, Duman RS, Enna SJ. In vivo or in vitro exposure to imipramine reduces alpha 2-adrenoceptor-mediated inhibition of cyclic AMP production in rat brain cerebral cortical slices. *Brain Res*. 1987;410(1):195–198.
54. Giralt MT, García-Sevilla JA. Acute and long-term regulation of brain alpha 2-adrenoceptors after manipulation of noradrenergic transmission in the rat. *Eur J Pharmacol*. 1989;164(3):455–466.
55. Barturen F, Garcia-Sevilla JA. Long term treatment with desipramine increases the turnover of alpha 2-adrenoceptors in the rat brain. *Mol Pharmacol*. 1992;42(5):846–855.
56. Esteban S, Lladó J, Sastre-Coll A, García-Sevilla JA. Activation and desensitization by cyclic antidepressant drugs of alpha2-autoreceptors, alpha2-heteroreceptors and 5-HT1A-autoreceptors regulating monamine synthesis in the rat brain in vivo. *Naunyn Schmiedebergs Arch Pharmacol*. 1999;360(2):135–143.
57. Mateo Y, Fernández-Pastor B, Meana JJ. Acute and chronic effects of desipramine and clorgyline on alpha(2)-adrenoceptors regulating noradrenergic transmission in the rat brain: a dual-probe microdialysis study. *Br J Pharmacol*. 2001;133(8):1362–1370. http://dx.doi.org/10.1038/sj.bjp.0704196.
58. Subhash MN, Nagaraja MR, Sharada S, Vinod KY. Cortical alpha-adrenoceptor downregulation by tricyclic antidepressants in the rat brain. *Neurochem Int*. 2003;43(7):603–609.
59. Giaroni C, Canciani L, Zanetti E, et al. Effects of chronic desipramine treatment on alpha2-adrenoceptors and mu-opioid receptors in the guinea pig cortex and hippocampus. *Eur J Pharmacol*. 2008;579(1–3):116–125. http://dx.doi.org/10.1016/j.ejphar.2007.10.007.
60. Cottingham C, Chen Y, Jiao K, Wang Q. The antidepressant desipramine is an arrestin-biased ligand at the alpha(2A)-adrenergic receptor driving receptor down-regulation in vitro and in vivo. *J Biol Chem*. 2011;286(41):36063–36075. http://dx.doi.org/10.1074/jbc.M111.261578.
61. Mateo Y, Pineda J, Meana JJ. Somatodendritic alpha2-adrenoceptors in the locus coeruleus are involved in the in vivo modulation of cortical noradrenaline release by the antidepressant desipramine. *J Neurochem*. 1998;71(2):790–798.
62. Parini S, Renoldi G, Battaglia A, Invernizzi RW. Chronic reboxetine desensitizes terminal but not somatodendritic alpha2-adrenoceptors controlling noradrenaline release in the rat dorsal hippocampus. *Neuropsychopharmacol Off Publ Am Coll Neuropsychopharmacol*. 2005;30(6):1048–1055. http://dx.doi.org/10.1038/sj.npp.1300661.
63. Wong DT, Threlkeld PG, Best KL, Bymaster FP. A new inhibitor of norepinephrine uptake devoid of affinity for receptors in rat brain. *J Pharmacol Exp Ther*. 1982;222(1):61–65.
64. Richelson E, Nelson A. Antagonism by antidepressants of neurotransmitter receptors of normal human brain in vitro. *J Pharmacol Exp Ther*. 1984;230(1):94–102.
65. Deupree JD, Montgomery MD, Bylund DB. Pharmacological properties of the active metabolites of the antidepressants desipramine and citalopram. *Eur J Pharmacol*. 2007;576(1–3):55–60. http://dx.doi.org/10.1016/j.ejphar.2007.08.017.

66. Cottingham C, Percival S, Birky T, Wang Q. Tricyclic antidepressants exhibit variable pharmacological profiles at the α(2A) adrenergic receptor. *Biochem Biophys Res Commun*. 2014;451(3):461–466. http://dx.doi.org/10.1016/j.bbrc.2014.08.024.
67. Cusack B, Nelson A, Richelson E. Binding of antidepressants to human brain receptors: focus on newer generation compounds. *Psychopharmacology (Berl)*. 1994;114(4):559–565.
68. Owens MJ, Morgan WN, Plott SJ, Nemeroff CB. Neurotransmitter receptor and transporter binding profile of antidepressants and their metabolites. *J Pharmacol Exp Ther* 1997;283(3):1305–1322.
69. Wang Q, Zhao J, Brady AE, et al. Spinophilin blocks arrestin actions in vitro and in vivo at G protein-coupled receptors. *Science*. 2004;304(5679):1940–1944. http://dx.doi.org/10.1126/science.1098274.
70. Wang Q, Lu R, Zhao J, Limbird LE. Arrestin serves as a molecular switch, linking endogenous alpha2-adrenergic receptor to SRC-dependent, but not SRC-independent, ERK activation. *J Biol Chem*. 2006;281(36):25948–25955. http://dx.doi.org/10.1074/jbc.M605415200.
71. Cottingham C, Jones A, Wang Q. Desipramine selectively potentiates norepinephrine-elicited ERK1/2 activation through the alpha2A adrenergic receptor. *Biochem Biophys Res Commun*. 2012;420(1):161–165. http://dx.doi.org/10.1016/j.bbrc.2012.02.135.
72. Hanyaloglu AC, von Zastrow M. Regulation of GPCRs by endocytic membrane trafficking and its potential implications. *Annu Rev Pharmacol Toxicol*. 2008;48:537–568. http://dx.doi.org/10.1146/annurev.pharmtox.48.113006.094830.
73. Shenoy SK, Lefkowitz RJ. Multifaceted roles of beta-arrestins in the regulation of seven-membrane-spanning receptor trafficking and signalling. *Biochem J*. 2003; 375(pt 3):503–515. http://dx.doi.org/10.1042/BJ20031076.
74. Tan CM, Brady AE, Nickols HH, Wang Q, Limbird LE. Membrane trafficking of G protein-coupled receptors. *Annu Rev Pharmacol Toxicol*. 2004;44:559–609. http://dx.doi.org/10.1146/annurev.pharmtox.44.101802.121558.
75. Kang DS, Tian X, Benovic JL. Role of β-arrestins and arrestin domain-containing proteins in G protein-coupled receptor trafficking. *Curr Opin Cell Biol*. 2014;27:63–71. http://dx.doi.org/10.1016/j.ceb.2013.11.005.
76. Premont RT, Gainetdinov RR. Physiological roles of G protein-coupled receptor kinases and arrestins. *Annu Rev Physiol*. 2007;69:511–534. http://dx.doi.org/10.1146/annurev.physiol.69.022405.154731.
77. Shenoy SK, Lefkowitz RJ. β-Arrestin-mediated receptor trafficking and signal transduction. *Trends Pharmacol Sci*. 2011;32(9):521–533. http://dx.doi.org/10.1016/j.tips.2011.05.002.
78. Shukla AK, Xiao K, Lefkowitz RJ. Emerging paradigms of β-arrestin-dependent seven transmembrane receptor signaling. *Trends Biochem Sci*. 2011;36(9):457–469. http://dx.doi.org/10.1016/j.tibs.2011.06.003.
79. DeWire SM, Ahn S, Lefkowitz RJ, Shenoy SK. Beta-arrestins and cell signaling. *Annu Rev Physiol*. 2007;69:483–510. http://dx.doi.org/10.1146/annurev.ph.69.013107.100021.
80. Violin JD, Lefkowitz RJ. Beta-arrestin-biased ligands at seven-transmembrane receptors. *Trends Pharmacol Sci*. 2007;28(8):416–422. http://dx.doi.org/10.1016/j.tips.2007.06.006.
81. Rajagopal S, Rajagopal K, Lefkowitz RJ. Teaching old receptors new tricks: biasing seven-transmembrane receptors. *Nat Rev Drug Discov*. 2010;9(5):373–386. http://dx.doi.org/10.1038/nrd3024.
82. Avissar S, Matuzany-Ruban A, Tzukert K, Schreiber G. Beta-arrestin-1 levels: reduced in leukocytes of patients with depression and elevated by antidepressants in rat brain.

Am J Psychiatry. 2004;161(11):2066–2072. http://dx.doi.org/10.1176/appi.ajp.161.11.2066.
83. Matuzany-Ruban A, Avissar S, Schreiber G. Dynamics of beta-arrestin1 protein and mRNA levels elevation by antidepressants in mononuclear leukocytes of patients with depression. *J Affect Disord*. 2005;88(3):307–312. http://dx.doi.org/10.1016/j.jad.2005.08.007.
84. Golan M, Schreiber G, Avissar S. Antidepressants elevate GDNF expression and release from C6 glioma cells in a β-arrestin1-dependent, CREB interactive pathway. *Int J Neuropsychopharmacol Off Sci J Coll Int Neuropsychopharmacol CINP*. 2011;14(10):1289–1300. http://dx.doi.org/10.1017/S1461145710001550.
85. Gurevich EV, Benovic JL, Gurevich VV. Arrestin2 and arrestin3 are differentially expressed in the rat brain during postnatal development. *Neuroscience*. 2002;109(3):421–436.
86. Cottingham C, Li X, Wang Q. Noradrenergic antidepressant responses to desipramine in vivo are reciprocally regulated by arrestin3 and spinophilin. *Neuropharmacology*. 2012;62(7):2354–2362. http://dx.doi.org/10.1016/j.neuropharm.2012.02.011.
87. Porsolt RD, Le PM, Jalfre M. Depression: a new animal model sensitive to antidepressant treatments. *Nature*. 1977;266(5604):730–732.
88. Porsolt RD, Bertin A, Jalfre M. Behavioral despair in mice: a primary screening test for antidepressants. *Arch Int Pharmacodyn Ther*. 1977;229(2):327–336.
89. Trivedi MH, Rush AJ, Wisniewski SR, et al. Evaluation of outcomes with citalopram for depression using measurement-based care in STAR*D: implications for clinical practice. *Am J Psychiatry*. 2006;163(1):28–40. http://dx.doi.org/10.1176/appi.ajp.163.1.28.
90. Turner EH, Matthews AM, Linardatos E, Tell RA, Rosenthal R. Selective publication of antidepressant trials and its influence on apparent efficacy. *N Engl J Med*. 2008;358(3):252–260. http://dx.doi.org/10.1056/NEJMsa065779.
91. Fournier JC, DeRubeis RJ, Hollon SD, et al. Antidepressant drug effects and depression severity: a patient-level meta-analysis. *JAMA*. 2010;303(1):47–53. http://dx.doi.org/10.1001/jama.2009.1943.
92. Anderson IM. Selective serotonin reuptake inhibitors versus tricyclic antidepressants: a meta-analysis of efficacy and tolerability. *J Affect Disord*. 2000;58(1):19–36.
93. Arroll B, Macgillivray S, Ogston S, et al. Efficacy and tolerability of tricyclic antidepressants and SSRIs compared with placebo for treatment of depression in primary care: a meta-analysis. *Ann Fam Med*. 2005;3(5):449–456. http://dx.doi.org/10.1370/afm.349.
94. Fava M, Rush AJ, Wisniewski SR, et al. A comparison of mirtazapine and nortriptyline following two consecutive failed medication treatments for depressed outpatients: a STAR*D report. *Am J Psychiatry*. 2006;163(7):1161–1172. http://dx.doi.org/10.1176/appi.ajp.163.7.1161.
95. McGrath PJ, Stewart JW, Fava M, et al. Tranylcypromine versus venlafaxine plus mirtazapine following three failed antidepressant medication trials for depression: a STAR*D report. *Am J Psychiatry*. 2006;163(9):1531–1541. http://dx.doi.org/10.1176/appi.ajp.163.9.1531, quiz 1666.
96. Browne CA, Lucki I. Antidepressant effects of ketamine: mechanisms underlying fast-acting novel antidepressants. *Front Pharmacol*. 2013;4:161. http://dx.doi.org/10.3389/fphar.2013.00161.
97. Salvadore G, Singh JB. Ketamine as a fast acting antidepressant: current knowledge and open questions. *CNS Neurosci Ther*. 2013;19(6):428–436. http://dx.doi.org/10.1111/cns.12103.
98. Naughton M, Clarke G, O'Leary OF, Cryan JF, Dinan TG. A review of ketamine in affective disorders: current evidence of clinical efficacy, limitations of use and

pre-clinical evidence on proposed mechanisms of action. *J Affect Disord.* 2014;156:24–35. http://dx.doi.org/10.1016/j.jad.2013.11.014.

99. Taylor D, Sparshatt A, Varma S, Olofinjana O. Antidepressant efficacy of agomelatine: meta-analysis of published and unpublished studies. *BMJ.* 2014;348(5):g1888. http://dx.doi.org/10.1136/bmj.g1888.
100. Kobilka BK. Structural insights into adrenergic receptor function and pharmacology. *Trends Pharmacol Sci.* 2011;32(4):213–218. http://dx.doi.org/10.1016/j.tips.2011.02.005.
101. Granier S, Kobilka B. A new era of GPCR structural and chemical biology. *Nat Chem Biol.* 2012;8(8):670–673. http://dx.doi.org/10.1038/nchembio.1025.
102. Shukla AK, Westfield GH, Xiao K, et al. Visualization of arrestin recruitment by a G-protein-coupled receptor. *Nature.* 2014;512(7513):218–222. http://dx.doi.org/10.1038/nature13430.

CHAPTER TEN

Regulation of α_{2B}-Adrenerigc Receptor Export Trafficking by Specific Motifs

Guangyu Wu[1], Jason E. Davis, Maoxiang Zhang
Department of Pharmacology and Toxicology, Medical College of Georgia, Georgia Regents University, Augusta, Georgia, USA
[1]Corresponding author: e-mail address: guwu@gru.edu

Contents

Abstract

Intracellular trafficking and precise targeting to specific locations of G protein-coupled receptors (GPCRs) control the physiological functions of the receptors. Compared to the extensive efforts dedicated to understanding the events involved in the endocytic and recycling pathways, the molecular mechanisms underlying the transport of the GPCR superfamily from the endoplasmic reticulum (ER) through the Golgi to the plasma membrane are relatively less well defined. Over the past years, we have used α_{2B}-adrenergic receptor (α_{2B}-AR) as a model to define the factors that control GPCR export trafficking. In this chapter, we will review specific motifs identified to mediate the export of nascent α_{2B}-AR from the ER and the Golgi and discuss the possible underlying mechanisms. As these motifs are highly conserved among GPCRs, they may provide common mechanisms for export trafficking of these receptors.

Progress in Molecular Biology and Translational Science, Volume 132
ISSN 1877-1173
http://dx.doi.org/10.1016/bs.pmbts.2015.03.004

1. EXPORT TRAFFICKING OF GPCRs

The transport of cell surface proteins (such as receptors, channels, and transporters) has been considered as a constitutive process. However, over the past decade, many studies have shown that their transport is regulatable, in a cell type- and cargo-specific manner, and can be mediated through non-classical routes.[1–5] As the largest superfamily of cell surface receptors involved in signal regulation in cells, the precise function of G protein-coupled receptors (GPCRs) is under control by spatial–temporal regulation of their intracellular trafficking which dictates the amount of receptor expression at their functional destinations and the magnitude of the cellular response to a given signal. Over the past decades, most studies on GPCR trafficking have focused on their internalization, recycling, and degradation pathways.[6–9] However, the molecular mechanisms underlying the anterograde transport of newly synthesized GPCRs from the endoplasmic reticulum (ER) to the cell surface have just begun to be revealed.

It has become increasingly apparent that similar to the endocytic pathway, the anterograde trafficking of GPCRs is a complicated and highly regulated cellular process which is orchestrated by the structural features of the receptors and by many regulatory proteins. Specifically, (1) GPCR transport to the cell surface is regulated by extracellular stimuli[10–12] and mediated through multiple pathways[13–15]; (2) GPCR export from the ER and the Golgi is dictated by highly conserved motifs[16–29]; (3) dimerization and posttranslational modifications (such as N-linked glycosylation) also play important roles in GPCR export from the ER to the cell surface[30]; (4) GPCR export is modulated by a multitude of regulatory proteins such as Rab GTPases, ER chaperones, and receptor activity-modifying proteins, which may facilitate receptor maturation, stabilize receptor conformation, and promote receptor delivery to the plasma membrane.[13–15,31–42]

α_2-Adrenergic receptors (α_2-ARs) are prototypic GPCRs that have three subtypes, α_{2A}-AR, α_{2B}-AR, and α_{2C}-AR, all of which have an important role in regulating sympathetic nervous system, both peripherally and centrally. All three α_2-ARs have similar structural features: the third intracellular loop (ICL3) is quite large with more than 170 amino-acid residues, whereas other loops and the termini are relatively short with less than 25 residues (Fig. 1). Over the past several years, we have mainly used α_{2B}-AR as a representative to investigate the export trafficking of GPCRs.

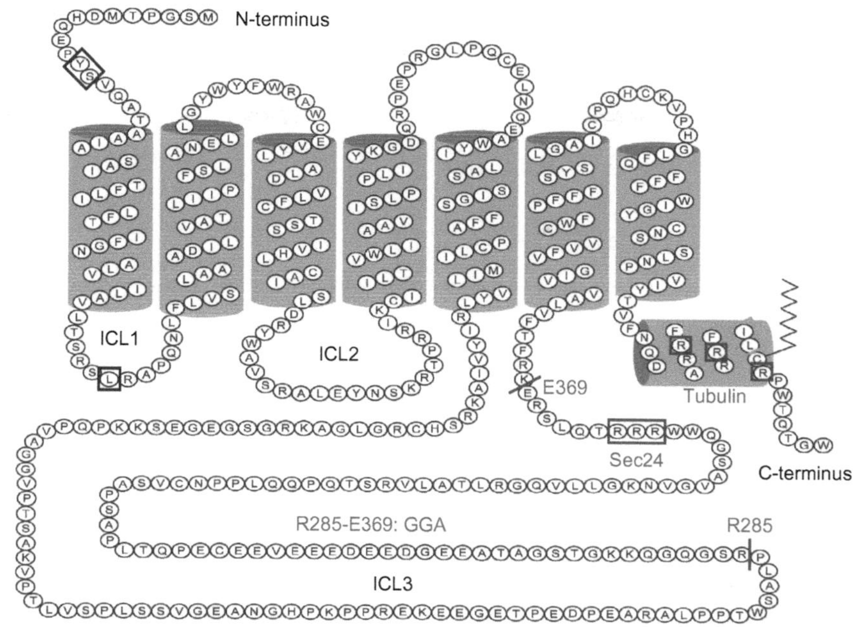

Figure 1 Specific motifs required for α_{2B}-AR export trafficking (see text for detail). (See the color plate.)

We have identified several highly conserved motifs, in both termini and intracellular loops, which are essential for the receptors to exit from the ER and the Golgi apparatus and subsequent transport to the cell surface. The locations of these motifs in α_{2B}-AR are summarized in Fig. 1. In this chapter, we will review the roles of these motifs in export trafficking of α_{2B}-AR, as well as other GPCRs, and discuss the possible underlying mechanisms.

2. REGULATION OF α_{2B}-AR EXPORT FROM THE ER

Similar to many other plasma membrane proteins, the life of GPCRs begins in the ER where they are synthesized. Once correctly folded and properly assembled, nascent receptors are able to pass the ER quality-control system and exit from the ER, beginning their long journey of intracellular trafficking. The receptors then move through several successive intracellular compartments, which include the ER–Golgi intermediate compartment, the *cis*/medial/*trans*-Golgi apparatus, and the *trans*-Golgi network (TGN), *en route* to the cell surface. As the first step in intracellular trafficking of GPCRs, export from the ER is the rate-limiting step in receptor targeting to the cell surface and markedly affects the kinetics of receptor maturation.[43] Our studies have identified a single Leu residue in the first intracellular loop (ICL1), a triple Arg motif (3R) in the ICL3, and a hydrophobic motif $F(x)_6IL$ as well as a positively charged motif $R(x)_3R(x)_4R$ in the C-terminus, all of which are required for α_{2B}-AR export from the ER.

2.1 Single Leu Residue in the ICL1

The ICL1 of α_{2B}-AR is very short (Fig. 1) containing Leu48 in the center. This single Leu residue is remarkably conserved among the GPCR superfamily (Fig. 2): about 85% of the family A GPCRs in humans and 83% in all species have this single Leu residue.[44] We found that deleting the ICL1 or mutating Leu48 residue in the ICL1 caused extensive ER accumulation of α_{2B}-AR, indicating that Leu48 plays a crucial role in α_{2B}-AR exit from the ER. Because mutation of Leu48 to Ile, Val, Tyr, and Trp similarly abolished

GPCRs	ICL1	GPCRs	ICL1	GPCRs	ICL1
5HT1A	ACNRH**L**RTPT	DRD2	SREKA**L**QTTT	MTR1A	YRNKK**L**RNAG
5HT1B	YRTRK**L**HTPA	DRD3	LKERA**L**QTTT	MTR1B	LRNRK**L**RNAG
5HT1D	LLTRK**L**HTPA	DRD4	ATERA**L**QTPT	OPSG	MKFKK**L**RHPL
5HT1E	GTTKK**L**HQPA	DRD5	VRSRH**L**RANM	OPSB	LRYKK**L**RQPL
5HT1F	IVTRK**L**HHPA	HRH1	RSERK**L**HTVG	OPSR	MKFKK**L**RHPL
5HT2A	SLEKK**L**QNAT	HRH2	GLNRR**L**RNLT	OPSD	VQHKK**L**RTPL
5HT2B	SLEKK**L**QYAT	HRH3	VADSS**L**RTQN	AT1R	YFYMK**L**KTVA
5HT2C	SMEKK**L**HNAT	SSR3	CFVKK**L**RQPS	CXCR4	GYQKK**L**RSMT
5HT4R	CWDRQ**L**RKIK	TAAR5	SYFKA**L**HTPT	CCR1	VQYKR**L**KNMT
5HT6R	CTQPA**L**RNTS	TAAR8	LHFKQ**L**HSPT	CCR2	INCKK**L**KCLT
5HT7	CFVKK**L**RQPS	TAAR9	LHFKQ**L**HTPT	CCR3	IKYRR**L**RIMT
α_{1A}-AR	ACHRH**L**HSVT	ACM1	KVNTE**L**KTVN	CCR5	INCKR**L**KSMT
α_{1B}-AR	ACNRH**L**RTPT	ACM2	KVNRH**L**QTVN	CCR4	FKYKR**L**RSMT
α_{1D}-AR	ACNRH**L**QTVT	ACR4	KVNRQ**L**QTVN	CCR7	IYFKR**L**KTMT
α_{2A}-AR	FTSRA**L**KAPQ	MC3R	VRNGN**L**HSPM	CCR8	VVCKK**L**RSIT
α_{2B}-AR	LTSRS**L**RAPQ	MC4R	AKNKN**L**HSPM	XCR1	VKYES**L**ESLT
α_{2C}-AR	LTSRA**L**RAPQ	MC5R	VKNKN**L**HSPM	GASR	GLSRR**L**RTVT
β_1-AR	AKTPR**L**QTLT	MSHR	AKNRN**L**HSPM	GHSR	SRFRE**L**RTTT
β_2-AR	AKFER**L**QTVT	AA1R	KVNQA**L**RDAT	PRLHR	ARVRR**L**HNVT
β_3-AR	AWTPR**L**QTMT	AA2A	WLNSN**L**QNVT	OPN3	YKFQR**L**RTPT
DRD1	IRFRH**L**RSKV	AA2B	GTANT**L**QTPT	OPN4	CRSRS**L**RTPA

Figure 2 The conserved Leu residues in the ICL1 of GPCRs.[44] 5HT, 5-hydroxytryptamine receptor; AR, adrenergic receptor; DR, dopamine receptor; HR, histamine receptor; SSR3, somatostatin receptor type 3; TAAR, trace amine-associated receptor; ACM, muscarinic acetylcholine receptor; MCR, melanocortin receptor; MSHR, melanocyte-stimulating hormone receptor; AR, adenosine receptor; MTR, melatonin receptor; OPSG, green-sensitive opsin; OPSB, blue-sensitive opsin; OPSR, red-sensitive opsin; OPSD, rhodopsin; AT1R, angiotensin II type 1A receptor; CXCR4, CXC chemokine receptor type 4; CCR, C–C chemokine receptor; XCR1, chemokine XC receptor 1; GASR, gastrin/cholecystokinin type B receptor; GHSR, growth hormone secretagogue receptor type 1; PRLHR, prolactin-releasing peptide receptor; OPN, opsin.

α_{2B}-AR export from the ER while substitution of Leu48 for Phe preserved α_{2B}-AR transport, the function of the Leu48 residue in modulating α_{2B}-AR export is mediated by its specific physiochemical and structural features including overall size of the side chain, spacing between the bulky portion of the side chains and the α-carbon, and polarity. In addition to α_{2B}-AR export, mutation of the Leu residue in the ICL1 also significantly attenuated the cell surface expression of β_2-AR, α_{1B}-AR, and angiotensin II type 1 receptor (AT1R).[44] Based on the crystal structures of GPCRs, the side chain of this conserved Leu residue inserts into the transmembrane bundle, suggesting that this Leu residue is most likely involved in receptor folding to a specific conformation which is crucial for receptor export from the ER.

2.2 The Triple Arg Motif (3R) in the ICL3

Protein export from the ER is exclusively mediated through the COPII-coated vesicles. In order to be efficiently exported in COPII vesicles, cargo proteins, particularly transmembrane proteins, may use ER export motifs in their cytoplasmic C-termini to bind to the components of COPII vesicles, particularly Sec24 subunits. There are four Sec24 isoforms (Sec24A, Sec24B, Sec24C, and Sec24D) identified in human cells, and these can be further divided into Sec24A/B and Sec24C/D subclasses based on the sequence homology. Interaction of ER export motifs with Sec24 enhances the recruitment of the cargo on ER exit sites and facilitates cargo export from the ER. Although the diacidic, dihydrophobic, and dibasic motifs have all been identified to function as ER export motifs in non-GPCR membrane proteins,[45–57] the diacidic motifs are the most well characterized to control ER export of several proteins, including vesicular stomatitis viral glycoprotein (VSVG), cystic fibrosis transmembrane conductance regulator, and potassium channels. It has also been described that different ER export motifs may preferentially interact with certain Sec24 isoforms. For example, the FF motif may bind to all four Sec24 isoforms, whereas the IxM, LxxLE, and DxE export signals can interact with different Sec24 isoforms.[55]

We have identified three basic Arg residues (3R) located in the ICL3 as a novel ER export code to direct α_{2B}-AR exit from the ER.[28] Similar to the single Leu residue in the ICL1, triple basic residues are also highly conserved in the ICL3 of GPCRs. The 3R motif has several important properties that are very similar to well-characterized ER export motifs in non-GPCR membrane proteins. First, the 3R motif mediates α_{2B}-AR interaction with Sec24 isoforms as determined in GST fusion protein pull-down and co-immunoprecipitation assays using exogenously expressed, endogenous and purified Sec24. Thus, the 3R motif represents the first motif in the GPCR superfamily which is able to physically associate with components of the COPII vesicles. More interestingly, the interaction of the 3R motif with Sec24C/D isoforms was much stronger than Sec24A/B isoforms (Fig. 3A), supporting the notion that the ER sorting signals may have selectivity toward distinct Sec24 isoforms. Based on the studies showing that mutation of the 3R motif to Ala, Gln, or Glu markedly attenuated or abolished the interaction, while substitution of the 3R motif to Lys preserved the interaction (Fig. 3B), we can conclude that the interaction is likely ionic.

Second, the export function of the 3R motif is independent of its position within α_{2B}-AR. The localization of the 3R motif in the ICL3 of α_{2B}-AR is in marked contrast to other Sec24-interacting ER export motifs identified thus far which are exclusively localized in the C-terminal regions of membrane proteins. We found that the addition of the 3R motif to the C-terminus not only promoted the cell surface export of α_{2B}-AR but also completely rescued the defective cell surface transport of mutated α_{2B}-AR, in which the 3R motif of the ICL3 was mutated. These data strongly

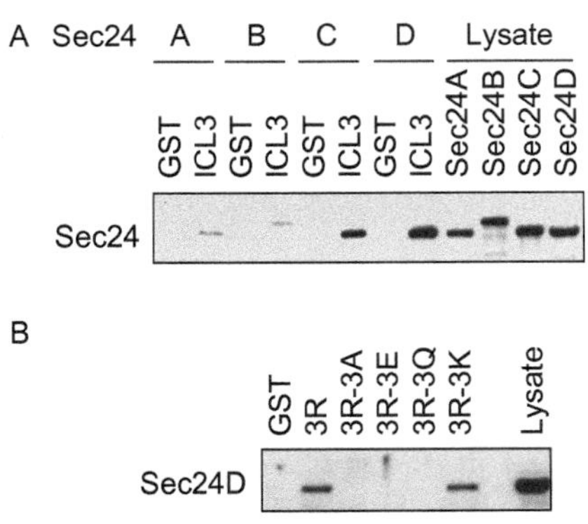

Figure 3 Interaction of the ICL3 of α_{2B}-AR with different Sec24 isoforms. (A) Interaction of the ICL3 with Sec24A, Sec24B, Sec24C, and Sec24D isoforms as determined in GST fusion protein pull-down assays. Each Sec24 isoform tagged with green fluorescent protein was expressed in HEK293 cells, and total cell lysates were incubated with GST or GST-ICL3 fusion proteins. (B) Effect of the combinational mutation of three Arg residues at positions 361, 362, and 363 (3R) to three Ala (3A), three Glu (3E), three Gln (3Q), and three Lys (3K) on the ICL3 interaction with Sec24D. The ICL3 fragment G349-E369 and its mutants were generated as GST fusion proteins and their interaction with Sec24D were determined. *The data are adapted from Ref. 28.*

indicate that the 3R motif, either in the ICL3 or in the C-terminus, can function as an export signal to direct receptor exit from the ER.

Third, export function of the 3R motif is transferable to other proteins. It has been well defined that linear ER export motifs are capable of conferring their transport function to other proteins which are normally expressed in the ER[46,47,53] as long as they are accessible to the COPII transport machinery. For example, the export function of the diacidic motifs can be conferred to CD8 glycoprotein. We demonstrated that transplantation of an ICL3 fragment containing the 3R motif was able to translocate CD8 glycoprotein to the cell surface and mutation of the 3R motif abolished this function. These data further support that the 3R motif is an independent and linear ER export signal.

It is interesting to note that dibasic motifs such as KKxx and RxR motifs have been demonstrated to function as ER retention or retrieval signals for ER-resident proteins which is likely mediated through interacting with members of the COPI vesicles.[58] The RxR motif has also been found to be responsible for ER retention of γ-aminobutyric acid type B receptor.[59] These data suggest that basic residues may have multiple or even opposing effects on GPCR trafficking.

2.3 The Hydrophobic Motif $F(x)_6IL$ in the C-Terminus

Similar to many other GPCRs, the C-terminal tail of α_{2B}-AR consists of a putative amphipathic 8th α-helix in the membrane-proximal region and a nonstructural membrane-distal region (Fig. 1). Our studies have demonstrated that dileucine residues I443/L444 together with the F436 residue in the membrane-proximal portion are essential for α_{2B}-AR export from the ER and transport to the cell surface[60] (Fig. 1); and the function of F436 and I443/L444 in mediating α_{2B}-AR export cannot be fully substituted by any other hydrophobic residues.[26] Furthermore, the $F(x)_6LL$ motif (where x can be any residue and L leucine or isoleucine) is highly conserved in the membrane-proximal C-termini of many family A GPCRs.[60] Indeed, further studies have shown that this motif is also required for ER export of α_{1B}-AR, β_2-AR, AT1R, and M1-muscarinc receptor.[19,26]

The function of the C-terminus, particularly the membrane-proximal 8th α-helical portion, in regulating cell surface transport of the receptors has been described for a number of GPCRs. Mutagenesis studies of the membrane-proximal C-termini have led to the identification of LL motif and several other hydrophobic motifs such as $E(x)_3LL$, $FN(x)_2LL(x)_3L$, $F(x)_3F(x)_3F$, $L(x)_3F(x)_3F$, and $Y(x)_3F(x)_3F$. These motifs are required for the ER export and cell surface transport of α_{1B}-AR, serotonin 5HT1A and 5-HT1B receptors, dopamine D1 receptor, vasopressin V2 and V3 receptors, and neuropeptide Y2 receptor.[18–24,60–62]

Although $F(x)_6LL$ and other hydrophobic motifs are well demonstrated to mediate GPCR export from the ER, the underlying molecular mechanisms remain elusive. Because these motifs are located in the membrane-proximal C-terminal α-helical region, it is reasonable to assume that mutation of these motifs will most likely disrupt proper receptor folding in the ER; therefore, mutated receptors are unable to export from the ER. This possibility is supported by the fact that mutating the motifs caused the receptors to lose their ligand-binding abilities, and the defective transport of the mutated receptors can be rescued by pharmacological and molecular chaperones.[24,26,59,61] However, it remains to be determined if wild-type receptors, once arrested in the ER by mutating these hydrophobic motifs, are still able to bind to their ligands. Furthermore, we have demonstrated that the LL motif mediates β_2-AR interaction with Rab8,[36] a small GTPase involved in the post-Golgi transport, suggesting that the hydrophobic motifs in the C-termini of GPCRs may have multiple functions (see Section 3.3).

2.4 The Positively Charged Motif $R(x)_3R(x)_4R$ in the C-Terminus

To elucidate the molecular mechanisms responsible for the function of the C-termini in GPCR transport from the ER to the cell surface, we searched for proteins interacting with the C-terminus of α_{2B}-AR by using peptide-conjugated affinity matrix combined with proteomics. This strategy identified tubulin directly interacting with α_{2B}-AR.[63] Subsequent studies revealed that tubulin also bound to the C-termini of α_{2A}-AR and AT1R, but not β_2-AR[64] (Fig. 4A and B). Mutagenesis analysis of the C-terminus identified R437, R441, and R446, which form the motif $R(x)_3R(x)_4R$ in the membrane-proximal region, as responsible for tubulin interaction. Importantly, mutation of these three Arg residues abolished receptor transport to the cell surface, and receptor mutants were extensively arrested in the ER. These data provide evidence indicating that the cargo GPCRs may directly contact with the microtubule network to coordinate their own ER-to-cell surface traffic.

Based on the structural homology modeling using high-resolution crystal structure of β_2-AR,[65] the positively charged residues R437/R441/R446 in

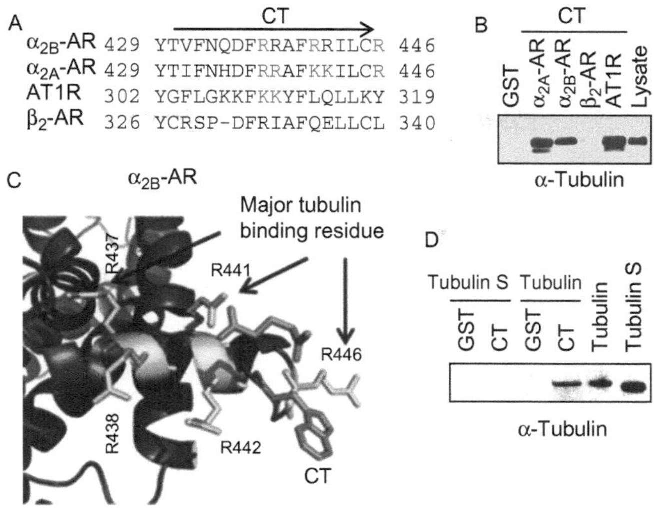

Figure 4 Interaction of the C-termini of GPCRs with tubulin. (A) Alignment of the C-termini (CT) of four GPCRs. (B) Interaction of the C-termini with purified tubulin measured in GST fusion protein pull-down assays. (C) Homology modeling of α_{2B}-AR based on the crystal structure of β_2-AR. (D) Tubulin S lacking the C-terminus does not bind to the α_{2B}-AR C-terminus. *The data are adapted from Refs. 63,64.* (See the color plate.)

the α_{2B}-AR C-terminus likely project from the same side on the cytosolic face (Fig. 4C). This structural feature of positively charged residues is shared by many family A GPCRs. These basic residues appear to cluster on helix 8 in an amphipathic pattern, spaced apart by one or two residues, to ensure their presentation on the same face of the helical structure. It is possible that these highly conserved positively charged residues in the C-termini of GPCRs have multiple functions, including regulating proper receptor folding in the ER and subsequent export from the ER through interaction with tubulin.

It is well known that the microtubule network modulates many intracellular trafficking processes including ER export and ER-to-Golgi transport.[66–70] Microtubules are characterized by hollow tubes of polymerized α- and β-tubulin dimers. Both α- and β-tubulin contain the highly acidic EExEEY/F motif in the flexible CT. This motif does not play a major role in maintaining the microtubule structure. Instead, this motif coats the outer surface of microtubules to mediate microtubule interaction with many proteins, including motor proteins, and to regulate microtubule dynamics. For example, microtubule plus-end tracking proteins, such as cytoplasmic linker protein 170, use highly basic grooves in the CAP-Gly domain to target the tubulin C-terminal acidic motif.[71,72] The α_{2B}-AR-binding domain was also mapped to the tubulin C-terminus, as the tubulin mutant lacking the C-terminus prepared by limited proteolysis did not interact with α_{2B}-AR (Fig. 4D).[63] These data indicate that the interaction of α_{2B}-AR with tubulin is ionic in nature. These data also suggest that the microtubule association with GPCRs and many other proteins likely share similar structural features.

3. REGULATION OF α_{2B}-AR POST-GOLGI TRANSPORT

The Golgi/TGN compartment is often referred to as the "sorting center" where newly synthesized proteins are sorted to be delivered to their final cellular destinations such as endosomes, lysosomes, and the plasma membrane. As discussed above, protein exit from the ER is mediated through COPII-coated vesicles and directed by ER export signals. Similarly, post-Golgi transport can be mediated through clathrin-coated transport vesicles, and specific motifs can enhance cargo recruitment onto the vesicles. Clathrin-coated vesicles are composed of clathrin and various adaptor proteins, including heterotetrameric adaptor protein (AP) complexes, GGAs (Golgi-associated, γ-adaptin homologous, ARF-interacting proteins), and hepatocyte growth factor receptor substrate (Hrs). The tyrosine-based motifs, NPxY and YxxØ (where x can be any residue and Ø is a hydrophobic residue), and the dileucine-based motifs, [D/E]xxxL[L/I] and DxxLL, are well-defined endosomal sorting signals which mediate cargo protein interaction with the AP complex of TGN-derived transport vesicles and sort the cargo into the TGN-to-endosomes pathway.[73–75] Whereas YxxØ and [D/E]xxxL[L/I] motifs are recognized by the AP complexes, DxxLL is recognized by GGAs.[76] Although clathrin/AP-coated vesicles have been demonstrated to mediate endocytosis of agonist-occupied GPCRs from the plasma membrane to the endosomes, the vesicles involved in the transport of newly synthesized GPCRs from the Golgi/TGN to the plasma membrane remain unknown.

3.1 The GGA-Biding Motif in the ICL3

There are three GGA isoforms in humans (GGA1, GGA2, and GGA3) with identical domain organizations. The GGA protein is composed of the N-terminal VHS (the Vps27, Hrs, Stam) domain followed by the GAT (GGAs and TOM1) domain, the hinge region, and the C-terminal GAE (γ-adaptin ear) domain. Each domain of GGAs has been shown to interact with specific proteins to coordinate their trafficking functions. The VHS domain has been well defined to interact with the DxxLL-type sorting motifs of cargo proteins such as cation-dependent and cation-independent mannose 6-phosphate receptors,[76–80] sortilin,[81] sorting-protein-related receptor,[82] low-density lipoprotein receptor-related proteins,[83,84] and β-secretase.[85] These highly coordinated VHS-DxxLL signal interactions specifically sort cargo proteins into the TGN-to-endosome pathway.[86] The GAT domain has been identified to interact with GTP-bound ARF1, and this interaction, together with PIP4, provides molecular anchors for the recruitment of GGAs onto the TGN.[87–91] The hinge region interacts with clathrin which is responsible for the recruitment of clathrin onto the TGN, leading to the formation of clathrin-coated vesicles. The C-terminal GAE domain of GGAs associates with a number of accessory proteins.[92–94]

Our recent studies have shown that shRNA-mediated depletion of individual GGAs strongly arrested α_{2B}-AR in the TGN and significantly reduced receptor cell surface expression (unpublished data). We further demonstrated that GGAs physically associated with α_{2B}-AR through specific domains. The GGA-binding domain was mapped to the ICL3 of the receptor, and progressive deletion of the ICL3 identified that the fragment R285-E369 mainly mediated α_{2B}-AR interaction with GGAs (Fig. 1). Surprisingly, the α_{2B}-AR-binding domains were identified to different regions of all three GGAs, specifically the GGA1 hinge, the GGA2 GAE, and the GGA3 VHS domains (unpublished data). These studies demonstrate novel functions of the GGA family proteins in the TGN-to-plasma membrane transport of GPCRs which is likely mediated through specific interactions.

3.2 The YS Motif in the N-Terminus

Similar to the C-termini, the N-termini are also important in the export trafficking of GPCRs. For example, the deletion of the N-termini facilitates the cell surface expression of α_{1D}-AR and α_{2C}-AR, suggesting that the N-termini may contain signals retaining the receptors in the ER.[95] A hydrophobic sequence in the N-terminus of α_{2C}-AR was then proven to function as an ER retention motif,[96] which provides a mechanism responsible for the intracellular accumulation of α_{2C}-AR in some cell types.

We found that the YS motif in the membrane-proximal N-terminal region is required for the transport of α_{2B}-AR to the cell surface.

Substitution of Y12 and S13 residues significantly reduced the cell surface expression of α_{2B}-AR and the mutated receptors were retained in the Golgi apparatus,[27] suggesting that the YS motif mediates α_{2B}-AR export at the level of the Golgi. The YS motif only exists in the membrane-proximal N-termini of all three α_2-AR family members, and indeed, it exerts a similar function on α_{2A}-AR trafficking.[27] Therefore, the YS motif may function as an export signal specifically modulating the Golgi export of the members of α_2-AR subfamily.

As the N-terminus is positioned toward the lumen of ER and Golgi during the export process, the YS motif is not able to directly interact with components of transport machinery in the cytoplasm. Furthermore, the fact that YS mutant receptors are able to exit from the ER to reach the Golgi compartment suggests that they are properly folded. Although it is clear that the defective transport induced by mutation of the YS motif is unlikely caused by misfolding, the exact underlying mechanism remains unknown.

3.3 The Rab8-Binding Motif in the C-Terminus

Rab8 has been extensively investigated in protein transport from the TGN under polarized conditions. For GPCRs, Rab8 modulates the post-Golgi transport of rhodopsin in *Xenopus*[97] and the internalization of metabotropic glutamate receptor subtype 1.[98] We have demonstrated that Rab8 GTPase regulates the post-Golgi transport of α_{2B}-AR and β_2-AR in several cell lines and primary neurons which is mediated through physical interactions.[36] More interestingly, distinct motifs in the C-termini of α_{2B}-AR and β_2-AR were identified to interact with Rab8, and these motifs probably dictate differential regulation of these two receptors by Rab8.[36] The dileucine LL motif was shown to be required for β_2-AR interaction with Rab8, whereas multiple residues in the fragments TVFN and PWTQTGW of the C-terminus were identified to modulate α_{2B}-AR interaction with Rab8. The fact that mutation of the LL motif selectively influences Rab8 interaction with β_2-AR, but not α_{2B}-AR,[36] suggests that it may have different roles in the regulation of post-Golgi transport of distinct GPCRs. It is possible that, similar to the function of the diacidic ExD motif in VSVG transport,[99] a single LL motif may modulate export trafficking of some, but not all, GPCRs (e.g., β_2-AR) at multiple intracellular compartments. In addition to regulating ER export, it may also coordinate GPCR exit from the Golgi/TGN.

4. CONCLUDING REMARKS

It is apparent that export trafficking of α_{2B}-AR is a selective process which is under control at multiple levels by a number of specific motifs embedded within the receptors. Receptor export from a specific intracellular compartment or transport along a specific pathway may be coordinated by several different motifs, whereas one motif may have multiple functions that influence receptor export trafficking at distinct transport steps. Although the precise mechanisms underlying the function of these motifs in regulating receptor export remains largely to be determined, these motifs may be involved in proper receptor folding/assembly to ensure that the receptors are able to pass through the ER quality-control system, mediate receptor interaction with specific components of the transport machinery which will enhance receptor recruitment onto specific vesicles and sort the receptors to specific transport pathways, and/or modulate receptor interaction with some regulatory proteins involved in the export processing. Further investigation, including search for proteins interacting with these motifs in the cytoplasm, particularly components of the transport machinery or other trafficking-related regulatory proteins, will help elucidate the molecular mechanisms of α_{2B}-AR export trafficking. Although above-mentioned motifs are mainly identified in α_{2B}-AR, the export function of these motifs has been determined in several other GPCRs. Importantly, these motifs are highly conserved among a group or family of GPCRs, and thus they may provide common mechanisms for export trafficking of these receptors. As it has become increasingly appreciated that defective export trafficking of GPCRs is linked directly to the pathogenesis of several human diseases, understanding the mechanisms of GPCR cell surface transport may provide an important foundation for the development of new therapeutic strategies in treating diseases involving abnormal GPCR trafficking.

ACKNOWLEDGMENT

This work was supported by National Institutes of Health Grant GM076167 (G.W).

REFERENCES

1. Sun Y, Chiu TT, Foley KP, Bilan PJ, Klip A. Myosin Va mediates Rab8A-regulated GLUT4 vesicle exocytosis in insulin-stimulated muscle cells. *Mol Biol Cell.* 2014;25(7):1159–1170.

2. Lu W, Man H, Ju W, Trimble WS, MacDonald JF, Wang YT. Activation of synaptic NMDA receptors induces membrane insertion of new AMPA receptors and LTP in cultured hippocampal neurons. *Neuron*. 2001;29(1):243–254.
3. Ward DT, Hammond TG, Harris HW. Modulation of vasopressin-elicited water transport by trafficking of aquaporin2-containing vesicles. *Annu Rev Physiol*. 1999;61:683–697.
4. Patterson MA, Szatmari EM, Yasuda R. AMPA receptors are exocytosed in stimulated spines and adjacent dendrites in a Ras-ERK-dependent manner during long-term potentiation. *Proc Natl Acad Sci USA*. 2010;107(36):15951–15956.
5. Brewer PD, Habtemichael EN, Romenskaia I, Mastick CC, Coster AC. Insulin-regulated Glut4 translocation: membrane protein trafficking with six distinctive steps. *J Biol Chem*. 2014;289(25):17280–17298.
6. von Zastrow M. Mechanisms regulating membrane trafficking of G protein-coupled receptors in the endocytic pathway. *Life Sci*. 2003;74(2–3):217–224.
7. Marchese A, Paing MM, Temple BR, Trejo J. G protein-coupled receptor sorting to endosomes and lysosomes. *Annu Rev Pharmacol Toxicol*. 2008;48:601–629.
8. Wolfe BL, Trejo J. Clathrin-dependent mechanisms of G protein-coupled receptor endocytosis. *Traffic*. 2007;8(5):462–470.
9. Reiter E, Ahn S, Shukla AK, Lefkowitz RJ. Molecular mechanism of beta-arrestin-biased agonism at seven-transmembrane receptors. *Annu Rev Pharmacol Toxicol*. 2012;52:179–197.
10. Yin H, Li Q, Qian G, et al. Rab1 GTPase regulates phenotypic modulation of pulmonary artery smooth muscle cells by mediating the transport of angiotensin II type 1 receptor under hypoxia. *Int J Biochem Cell Biol*. 2011;43(3):401–408.
11. Li Y, Wang G, Lin K, et al. Rab1 GTPase promotes expression of beta-adrenergic receptors in rat pulmonary microvascular endothelial cells. *Int J Biochem Cell Biol*. 2010;42(7):1201–1209.
12. Bie B, Zhang Z, Cai YQ, et al. Nerve growth factor-regulated emergence of functional delta-opioid receptors. *J Neurosci*. 2010;30(16):5617–5628.
13. Wu G, Zhao G, He Y. Distinct pathways for the trafficking of angiotensin II and adrenergic receptors from the endoplasmic reticulum to the cell surface: Rab1-independent transport of a G protein-coupled receptor. *J Biol Chem*. 2003;278(47):47062–47069.
14. Filipeanu CM, Zhou F, Fugetta EK, Wu G. Differential regulation of the cell-surface targeting and function of beta- and alpha1-adrenergic receptors by Rab1 GTPase in cardiac myocytes. *Mol Pharmacol*. 2006;69(5):1571–1578.
15. Wang G, Wu G. Small GTPase regulation of GPCR anterograde trafficking. *Trends Pharmacol Sci*. 2012;33(1):28–34.
16. Dong C, Filipeanu CM, Duvernay MT, Wu G. Regulation of G protein-coupled receptor export trafficking. *Biochim Biophys Acta*. 2007;1768(4):853–870.
17. Zhang X, Dong C, Wu QJ, Balch WE, Wu G. Di-acidic motifs in the membrane-distal C termini modulate the transport of angiotensin II receptors from the endoplasmic reticulum to the cell surface. *J Biol Chem*. 2011;286(23):20525–20535.
18. Carrel D, Hamon M, Darmon M. Role of the C-terminal di-leucine motif of 5-HT1A and 5-HT1B serotonin receptors in plasma membrane targeting. *J Cell Sci*. 2006;119 (Pt 20):4276–4284.
19. Sawyer GW, Ehlert FJ, Shults CA. A conserved motif in the membrane proximal C-terminal tail of human muscarinic m1 acetylcholine receptors affects plasma membrane expression. *J Pharmacol Exp Ther*. 2010;332(1):76–86.
20. Guo Y, Jose PA. C-terminal di-leucine motif of dopamine D(1) receptor plays an important role in its plasma membrane trafficking. *PLoS One*. 2011;6(12):e29204.

21. Donnellan PD, Kimbembe CC, Reid HM, Kinsella BT. Identification of a novel endoplasmic reticulum export motif within the eighth alpha-helical domain (alpha-H8) of the human prostacyclin receptor. *Biochim Biophys Acta.* 2011;1808(4):1202–1218.
22. Bermak JC, Li M, Bullock C, Zhou QY. Regulation of transport of the dopamine D1 receptor by a new membrane-associated ER protein. *Nat Cell Biol.* 2001;3(5):492–498.
23. Robert J, Auzan C, Ventura MA, Clauser E. Mechanisms of cell-surface rerouting of an endoplasmic reticulum-retained mutant of the vasopressin V1b/V3 receptor by a pharmacological chaperone. *J Biol Chem.* 2005;280(51):42198–42206.
24. Schulein R, Hermosilla R, Oksche A, et al. A dileucine sequence and an upstream glutamate residue in the intracellular carboxyl terminus of the vasopressin V2 receptor are essential for cell surface transport in COS.M6 cells. *Mol Pharmacol.* 1998;54(3):525–535.
25. Dong C, Nichols CD, Guo J, Huang W, Lambert NA, Wu G. A triple Arg motif mediates alpha2B-adrenergic receptor interaction with Sec24C/D and export. *Traffic.* 2012;13(6):857–868.
26. Duvernay MT, Dong C, Zhang X, Zhou F, Nichols CD, Wu G. Anterograde trafficking of G protein-coupled receptors: function of the C-terminal F(X)6LL motif in export from the endoplasmic reticulum. *Mol Pharmacol.* 2009;75(4):751–761.
27. Dong C, Wu G. Regulation of anterograde transport of alpha2-adrenergic receptors by the N termini at multiple intracellular compartments. *J Biol Chem.* 2006;281(50):38543–38554.
28. Dong C, Nichols CD, Guo J, Huang W, Lambert NA, Wu G. A triple arg motif mediates alpha(2B)-adrenergic receptor interaction with Sec24C/D and export. *Traffic.* 2012;13(6):857–868.
29. Hirota N, Yasuda D, Hashidate T, et al. Amino acid residues critical for endoplasmic reticulum export and trafficking of platelet-activating factor receptor. *J Biol Chem.* 2010;285(8):5931–5940.
30. Terrillon S, Bouvier M. Roles of G-protein-coupled receptor dimerization. *EMBO Rep.* 2004;5(1):30–34.
31. McLatchie LM, Fraser NJ, Main MJ, et al. RAMPs regulate the transport and ligand specificity of the calcitonin-receptor-like receptor. *Nature.* 1998;393(6683):333–339.
32. Colley NJ, Baker EK, Stamnes MA, Zuker CS. The cyclophilin homolog ninaA is required in the secretory pathway. *Cell.* 1991;67(2):255–263.
33. Dwyer ND, Troemel ER, Sengupta P, Bargmann CI. Odorant receptor localization to olfactory cilia is mediated by ODR-4, a novel membrane-associated protein. *Cell.* 1998;93(3):455–466.
34. Ferreira PA, Nakayama TA, Pak WL, Travis GH. Cyclophilin-related protein RanBP2 acts as chaperone for red/green opsin. *Nature.* 1996;383(6601):637–640.
35. Achour L, Scott MG, Shirvani H, et al. CD4-CCR5 interaction in intracellular compartments contributes to receptor expression at the cell surface. *Blood.* 2009;113(9):1938–1947.
36. Dong C, Yang L, Zhang X, et al. Rab8 interacts with distinct motifs in alpha2B- and beta2-adrenergic receptors and differentially modulates their transport. *J Biol Chem.* 2010;285(26):20369–20380.
37. Filipeanu CM, Zhou F, Claycomb WC, Wu G. Regulation of the cell surface expression and function of angiotensin II type 1 receptor by Rab1-mediated endoplasmic reticulum-to-Golgi transport in cardiac myocytes. *J Biol Chem.* 2004;279(39):41077–41084.
38. Dong C, Zhang X, Zhou F, et al. ADP-ribosylation factors modulate the cell surface transport of G protein-coupled receptors. *J Pharmacol Exp Ther.* 2010;333(1):174–183.
39. Dong C, Zhou F, Fugetta EK, Filipeanu CM, Wu G. Endoplasmic reticulum export of adrenergic and angiotensin II receptors is differentially regulated by Sar1 GTPase. *Cell Signal.* 2008;20(6):1035–1043.

40. Dong C, Wu G. Regulation of anterograde transport of adrenergic and angiotensin II receptors by Rab2 and Rab6 GTPases. *Cell Signal.* 2007;19(11):2388–2399.
41. Xia Z, Gray JA, Compton-Toth BA, Roth BL. A direct interaction of PSD-95 with 5-HT2A serotonin receptors regulates receptor trafficking and signal transduction. *J Biol Chem.* 2003;278(24):21901–21908.
42. Zhang X, Wang G, Dupre DJ, et al. Rab1 GTPase and dimerization in the cell surface expression of angiotensin II type 2 receptor. *J Pharmacol Exp Ther.* 2009;330(1):109–117.
43. Petaja-Repo UE, Hogue M, Laperriere A, Bhalla S, Walker P, Bouvier M. Newly synthesized human delta opioid receptors retained in the endoplasmic reticulum are retrotranslocated to the cytosol, deglycosylated, ubiquitinated, and degraded by the proteasome. *J Biol Chem.* 2001;276(6):4416–4423.
44. Duvernay MT, Dong C, Zhang X, Robitaille M, Hebert TE, Wu G. A single conserved leucine residue on the first intracellular loop regulates ER export of G protein-coupled receptors. *Traffic.* 2009;10(5):552–566.
45. Wang X, Matteson J, An Y, et al. COPII-dependent export of cystic fibrosis transmembrane conductance regulator from the ER uses a di-acidic exit code. *J Cell Biol.* 2004;167(1):65–74.
46. Ma D, Zerangue N, Lin YF, et al. Role of ER export signals in controlling surface potassium channel numbers. *Science.* 2001;291(5502):316–319.
47. Votsmeier C, Gallwitz D. An acidic sequence of a putative yeast Golgi membrane protein binds COPII and facilitates ER export. *EMBO J.* 2001;20(23):6742–6750.
48. Zuzarte M, Rinne S, Schlichthorl G, Schubert A, Daut J, Preisig-Muller R. A di-acidic sequence motif enhances the surface expression of the potassium channel TASK-3. *Traffic.* 2007;8(8):1093–1100.
49. Mikosch M, Kaberich K, Homann U. ER export of KAT1 is correlated to the number of acidic residues within a triacidic motif. *Traffic.* 2009;10(10):1481–1487.
50. Giraudo CG, Maccioni HJ. Endoplasmic reticulum export of glycosyltransferases depends on interaction of a cytoplasmic dibasic motif with Sar1. *Mol Biol Cell.* 2003;14(9):3753–3766.
51. Farhan H, Reiterer V, Korkhov VM, Schmid JA, Freissmuth M, Sitte HH. Concentrative export from the endoplasmic reticulum of the gamma-aminobutyric acid transporter 1 requires binding to SEC24D. *J Biol Chem.* 2007;282(10):7679–7689.
52. Miller EA, Beilharz TH, Malkus PN, et al. Multiple cargo binding sites on the COPII subunit Sec24p ensure capture of diverse membrane proteins into transport vesicles. *Cell.* 2003;114(4):497–509.
53. Nishimura N, Balch WE. A di-acidic signal required for selective export from the endoplasmic reticulum. *Science.* 1997;277(5325):556–558.
54. Nishimura N, Bannykh S, Slabough S, et al. A di-acidic (DXE) code directs concentration of cargo during export from the endoplasmic reticulum. *J Biol Chem.* 1999;274(22):15937–15946.
55. Mancias JD, Goldberg J. Structural basis of cargo membrane protein discrimination by the human COPII coat machinery. *EMBO J.* 2008;27(21):2918–2928.
56. Mossessova E, Bickford LC, Goldberg J. SNARE selectivity of the COPII coat. *Cell.* 2003;114(4):483–495.
57. Wendeler MW, Paccaud JP, Hauri HP. Role of Sec24 isoforms in selective export of membrane proteins from the endoplasmic reticulum. *EMBO Rep.* 2007;8(3):258–264.
58. Jackson MR, Nilsson T, Peterson PA. Identification of a consensus motif for retention of transmembrane proteins in the endoplasmic reticulum. *EMBO J.* 1990;9(10):3153–3162.
59. Margeta-Mitrovic M, Jan YN, Jan LY. A trafficking checkpoint controls GABA(B) receptor heterodimerization. *Neuron.* 2000;27(1):97–106.
60. Duvernay MT, Zhou F, Wu G. A conserved motif for the transport of G protein-coupled receptors from the endoplasmic reticulum to the cell surface. *J Biol Chem.* 2004;279(29):30741–30750.

61. Robert J, Clauser E, Petit PX, Ventura MA. A novel C-terminal motif is necessary for the export of the vasopressin V1b/V3 receptor to the plasma membrane. *J Biol Chem.* 2005;280(3):2300–2308.
62. Walther C, Lotze J, Beck-Sickinger AG, Morl K. The anterograde transport of the human neuropeptide Y2 receptor is regulated by a subtype specific mechanism mediated by the C-terminus. *Neuropeptides.* 2012;46(6):335–343.
63. Duvernay MT, Wang H, Dong C, Guidry JJ, Sackett DL, Wu G. Alpha2B-adrenergic receptor interaction with tubulin controls its transport from the endoplasmic reticulum to the cell surface. *J Biol Chem.* 2011;286(16):14080–14089.
64. Zhang X, Wang H, Duvernay MT, Zhu S, Wu G. The angiotensin II type 1 receptor C-terminal Lys residues interact with tubulin and modulate receptor export trafficking. *PLoS One.* 2013;8(2):e57805.
65. Rosenbaum DM, Cherezov V, Hanson MA, et al. GPCR engineering yields high-resolution structural insights into beta2-adrenergic receptor function. *Science.* 2007;318(5854):1266–1273.
66. Driskell OJ, Mironov A, Allan VJ, Woodman PG. Dynein is required for receptor sorting and the morphogenesis of early endosomes. *Nat Cell Biol.* 2007;9(1):113–120.
67. Mizuno M, Singer SJ. A possible role for stable microtubules in intracellular transport from the endoplasmic reticulum to the Golgi apparatus. *J Cell Sci.* 1994;107(Pt 5):1321–1331.
68. Presley JF, Cole NB, Schroer TA, Hirschberg K, Zaal KJ, Lippincott-Schwartz J. ER-to-Golgi transport visualized in living cells. *Nature.* 1997;389(6646):81–85.
69. Ralston E, Ploug T, Kalhovde J, Lomo T. Golgi complex, endoplasmic reticulum exit sites, and microtubules in skeletal muscle fibers are organized by patterned activity. *J Neurosci.* 2001;21(3):875–883.
70. Saraste J, Svensson K. Distribution of the intermediate elements operating in ER to Golgi transport. *J Cell Sci.* 1991;100(Pt 3):415–430.
71. Slep KC, Vale RD. Structural basis of microtubule plus end tracking by XMAP215, CLIP-170, and EB1. *Mol Cell.* 2007;27(6):976–991.
72. Mishima M, Maesaki R, Kasa M, et al. Structural basis for tubulin recognition by cytoplasmic linker protein 170 and its autoinhibition. *Proc Natl Acad Sci USA.* 2007;104(25):10346–10351.
73. Johnson KF, Kornfeld S. The cytoplasmic tail of the mannose 6-phosphate/insulin-like growth factor-II receptor has two signals for lysosomal enzyme sorting in the Golgi. *J Cell Biol.* 1992;119(2):249–257.
74. Boucher R, Larkin H, Brodeur J, Gagnon H, Theriault C, Lavoie C. Intracellular trafficking of LRP9 is dependent on two acidic cluster/dileucine motifs. *Histochem Cell Biol.* 2008;130(2):315–327.
75. Hou JC, Suzuki N, Pessin JE, Watson RT. A specific dileucine motif is required for the GGA-dependent entry of newly synthesized insulin-responsive aminopeptidase into the insulin-responsive compartment. *J Biol Chem.* 2006;281(44):33457–33466.
76. Puertollano R, Aguilar RC, Gorshkova I, Crouch RJ, Bonifacino JS. Sorting of mannose 6-phosphate receptors mediated by the GGAs. *Science.* 2001;292(5522):1712–1716.
77. Doray B, Bruns K, Ghosh P, Kornfeld S. Interaction of the cation-dependent mannose 6-phosphate receptor with GGA proteins. *J Biol Chem.* 2002;277(21):18477–18482.
78. Zhu Y, Doray B, Poussu A, Lehto VP, Kornfeld S. Binding of GGA2 to the lysosomal enzyme sorting motif of the mannose 6-phosphate receptor. *Science.* 2001; 292(5522):1716–1718.
79. Shiba T, Takatsu H, Nogi T, et al. Structural basis for recognition of acidic-cluster dileucine sequence by GGA1. *Nature.* 2002;415(6874):937–941.
80. Misra S, Puertollano R, Kato Y, Bonifacino JS, Hurley JH. Structural basis for acidic-cluster-dileucine sorting-signal recognition by VHS domains. *Nature.* 2002; 415(6874):933–937.

81. Nielsen MS, Madsen P, Christensen EI, et al. The sortilin cytoplasmic tail conveys Golgi-endosome transport and binds the VHS domain of the GGA2 sorting protein. *EMBO J.* 2001;20(9):2180–2190.
82. Nielsen MS, Gustafsen C, Madsen P, et al. Sorting by the cytoplasmic domain of the amyloid precursor protein binding receptor SorLA. *Mol Cell Biol.* 2007; 27(19):6842–6851.
83. Takatsu H, Katoh Y, Shiba Y, Nakayama K. Golgi-localizing, gamma-adaptin ear homology domain, ADP-ribosylation factor-binding (GGA) proteins interact with acidic dileucine sequences within the cytoplasmic domains of sorting receptors through their Vps27p/Hrs/STAM (VHS) domains. *J Biol Chem.* 2001;276(30):28541–28545.
84. Doray B, Knisely JM, Wartman L, Bu G, Kornfeld S. Identification of acidic dileucine signals in LRP9 that interact with both GGAs and AP-1/AP-2. *Traffic.* 2008;9(9):1551–1562.
85. He X, Chang WP, Koelsch G, Tang J. Memapsin 2 (beta-secretase) cytosolic domain binds to the VHS domains of GGA1 and GGA2: implications on the endocytosis mechanism of memapsin 2. *FEBS Lett.* 2002;524(1–3):183–187.
86. Bonifacino JS, Traub LM. Signals for sorting of transmembrane proteins to endosomes and lysosomes. *Annu Rev Biochem.* 2003;72:395–447.
87. Dell'Angelica EC, Puertollano R, Mullins C, et al. GGAs: a family of ADP ribosylation factor-binding proteins related to adaptors and associated with the Golgi complex. *J Cell Biol.* 2000;149(1):81–94.
88. Puertollano R, Randazzo PA, Presley JF, Hartnell LM, Bonifacino JS. The GGAs promote ARF-dependent recruitment of clathrin to the TGN. *Cell.* 2001;105(1):93–102.
89. Wang J, Sun HQ, Macia E, et al. PI4P promotes the recruitment of the GGA adaptor proteins to the trans-Golgi network and regulates their recognition of the ubiquitin sorting signal. *Mol Biol Cell.* 2007;18(7):2646–2655.
90. Lefrancois S, McCormick PJ. The Arf GEF GBF1 is required for GGA recruitment to Golgi membranes. *Traffic.* 2007;8(10):1440–1451.
91. Shiba T, Kawasaki M, Takatsu H, et al. Molecular mechanism of membrane recruitment of GGA by ARF in lysosomal protein transport. *Nat Struct Biol.* 2003;10(5):386–393.
92. Wasiak S, Legendre-Guillemin V, Puertollano R, et al. Enthoprotin: a novel clathrin-associated protein identified through subcellular proteomics. *J Cell Biol.* 2002;158(5):855–862.
93. Kalthoff C, Groos S, Kohl R, Mahrhold S, Ungewickell EJ. Clint: a novel clathrin-binding ENTH-domain protein at the Golgi. *Mol Biol Cell.* 2002;13(11):4060–4073.
94. Page LJ, Sowerby PJ, Lui WW, Robinson MS. Gamma-synergin: an EH domain-containing protein that interacts with gamma-adaptin. *J Cell Biol.* 1999; 146(5):993–1004.
95. Hague C, Chen Z, Pupo AS, Schulte NA, Toews ML, Minneman KP. The N terminus of the human alpha1D-adrenergic receptor prevents cell surface expression. *J Pharmacol Exp Ther.* 2004;309(1):388–397.
96. Angelotti T, Daunt D, Shcherbakova OG, Kobilka B, Hurt CM. Regulation of G-protein coupled receptor traffic by an evolutionary conserved hydrophobic signal. *Traffic.* 2010;11(4):560–578.
97. Moritz OL, Tam BM, Hurd LL, Peranen J, Deretic D, Papermaster DS. Mutant rab8 impairs docking and fusion of rhodopsin-bearing post-Golgi membranes and causes cell death of transgenic Xenopus rods. *Mol Biol Cell.* 2001;12(8):2341–2351.
98. Esseltine JL, Ribeiro FM, Ferguson SS. Rab8 modulates metabotropic glutamate receptor subtype 1 intracellular trafficking and signaling in a protein kinase C-dependent manner. *J Neurosci.* 2012;32(47):16933–16942a.
99. Nishimura N, Plutner H, Hahn K, Balch WE. The delta subunit of AP-3 is required for efficient transport of VSV-G from the trans-Golgi network to the cell surface. *Proc Natl Acad Sci USA.* 2002;99(10):6755–6760.

CHAPTER ELEVEN

Temperature-Sensitive Intracellular Traffic of α_{2C}-Adrenergic Receptor

Catalin M. Filipeanu[1]
Department of Pharmacology, College of Medicine, Howard University, Washington, District of Columbia, USA
[1]Corresponding author: e-mail address: catalin.filipeanu@howard.edu

Contents

Abstract

α_{2C}-Adrenergic receptor (α_{2C}-AR) is the least characterized adrenergic receptor subtype and still very little is known about the intracellular traffic properties and pathophysiological roles of this receptor. α_{2C}-AR has an atypical subcellular localization. At 37 °C, in the vascular smooth muscle cells and in fibroblasts, the receptor is poorly localized at the plasma membrane and accumulates inside the cell. Exposure to lower temperatures stimulates α_{2C}-AR transport to the cell surface. This particular intracellular trafficking of α_{2C}-AR is significant in the pathology of Raynaud phenomenon. In this brief review, I will present general information on the tissue distribution and cellular localization of α_{2C}-AR. Also, I will discuss the mechanisms involved in the receptor transport by focusing on the trafficking motifs and on the molecular chaperones.

Progress in Molecular Biology and Translational Science, Volume 132
ISSN 1877-1173
http://dx.doi.org/10.1016/bs.pmbts.2015.02.008

1. α_2-ADRENERGIC RECEPTORS CLASSIFICATION

α_2-Adrenergic receptors (α_2-ARs) are members of the G protein-coupled receptor (GPCR) superfamily and components of adrenergic receptor class. The existence of two types of adrenergic receptors (α and β) has been known for six decades based on the seminal work of Ahlquist and Black.[1,2] However, the subclassification of α-AR into α_1-AR and α_2-AR was proposed by Langer in 1974 based on presynaptic and postsynaptic localization of these two receptor types.[3] In the 1970s, apparition of specific ligands like prazosin for α_1-ARs[4] and rauwolscine for α_2-ARs[5] offered new tools to study the physiological and pathological roles of adrenergic receptors *in vivo* and *in vitro*. This led to the demonstration that α_2-ARs are expressed not only at the central level but throughout the entire body.[6] The further subclassification into three distinct α_2-AR subtypes was based on the molecular cloning at the end of the 1980s by Lefkowitz, Caron, and Kobilka.[7–9] Human platelets served as the first source for cloning the first α_2-AR, which was initially named α_2C10, because it was located on chromosome 10.[7] Subsequently, two other genes were identified on chromosomes 2 and 4 and named α_2C2 and α_2C4,[8,9] respectively. The current nomenclature was introduced soon after, designating as α_{2A}-AR (α_2C10), α_{2B}-AR (α_2C4), and α_{2C}-AR (α_2C2), which is currently the universally accepted terminology.[10] Still, even from these pioneering days of molecular pharmacology, it was clear that the responses to α_2-AR ligands are specie-dependent. For instance, rauwolscine binds 10 times more potently to the human α_{2A}-AR compared to the rat orthologue.[11] The lack of selective ligands discriminating between each receptor subtype is still a problem today in differentiating the roles of each receptor subtype. This issue has been circumvented mostly by using transfected cell lines with specific cDNA for each receptor subtype. All three α_2-ARs have an estimated molecular weight in the range of 48–49 kDa and are further processed by glycosylation and palmitoylation, with the noticeable exception of α_{2B}-AR, which lacks N-linked glycosylation sites.[12] Regarding α_{2C}-AR, we found that this receptor is better N-glycosylated at reduced temperature (Fig. 1A), indicating that it requires additional conditions to attain full maturation. This conclusion is supported by the fact that in HEK293T cells transfected with α_{2C}-AR, most of the receptor is localized inside the cell at physiological temperature of 37 °C (Fig. 1B). However, after exposure to low temperature (30 °C), α_{2C}-AR is localized at the plasma membrane, as expected for members of the GPCR family (Fig. 1B). These results indicate that under physiological conditions, α_{2C}-AR transport to the cell surface is impaired.

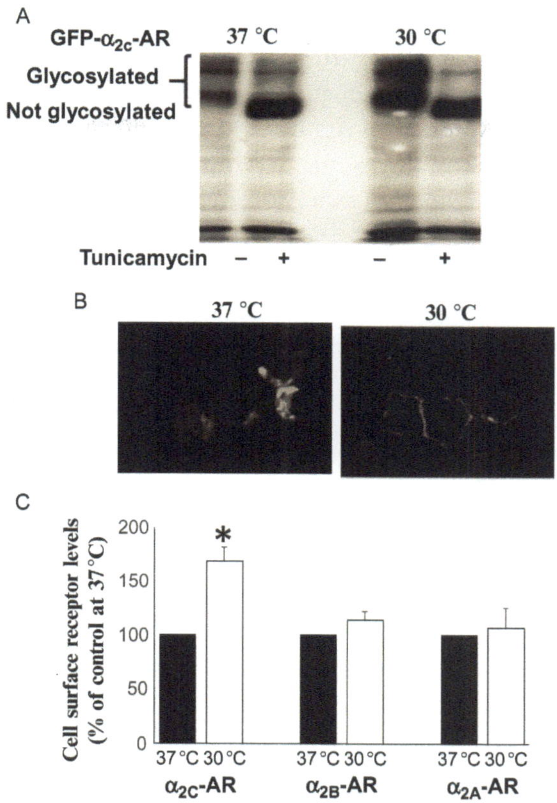

Figure 1 Subcellular localization of α_{2C}-AR. (A) The levels of N-glycosylation of α_{2C}-AR at 37 °C (left) and 30 °C (right). HEK293T cells were transfected with 5 μg of GFP-tagged α_{2C}-AR for 48 h at 37 °C. Subsequently, the cells were serum starved and incubated at 30 °C for 18 h in the presence or absence of tunicamycin, an inhibitor of N-glycosylation (10 μg/ml). α_{2C}-AR species were detected in cell lysates by Western blot using a specific GFP antibody. (B) Examination of α_{2C}-AR subcellular localization demonstrating that in HEK293 cells transfected with GFP-tagged α_{2C}-AR at 37 °C, the majority of the receptor is localized in the cytosol (left), whereas at 30 °C the receptor is clearly present at the cell surface (right). (C) The cell surface levels of the three α_2-ARs in transfected HEK293 cells at 37 °C (black columns) and at 30 °C (white columns) were measured by radioligand binding using [^{3}H]RX821002 as described.[13] Only α_{2C}-AR plasma membrane levels were significantly upregulated at reduced temperature. $n > 6$ in each case, $*p < 0.05$. (See the color plate.)

2. α_{2C}-AR TISSUE LOCALIZATION

The tissue distribution of α_{2C}-AR was first investigated using reverse transcription of total RNA. The largest amount of receptor was found in the brain and kidney. α_{2C}-AR was not found in the liver, whereas conflicting results were reported on whether or not α_{2C}-AR is expressed in the blood vessels, lung, skeletal muscle, heart, or spleen.[14–17] A more recent study profiling the relative abundance of mRNA of 354 nonodorant members of GPCR class in 41 different tissues from mice confirmed that α_{2C}-AR has the highest expression in the central nervous system.[18] Still, at lower levels, α_{2C}-AR mRNA was found in every tissue examined with the exception of liver, gall bladder, thymus, and bone marrow.[18] However, the protein levels in each tissue are still difficult to evaluate due to lack of reliable antibodies.

Surprisingly, genetic manipulation of α_{2C}-AR levels did not produce important pathological responses. For example, overexpression of α_{2C}-AR in the mouse brain demonstrated only minor behavioral changes.[19] Also, deletion of α_{2C}-AR in the mouse heart did not result in major modifications.[20] Furthermore, in α_{2C}-AR knockout mice, no major pathological changes were found, although careful examination indicated some subtle modifications like increased amphetamine-induced locomotor activity, startle reactivity, aggression, and enhanced movement in the forced swimming test[21] or involvement in GABA release in mouse striatum.[22] In other cases, like wound healing[23] or uncoupling of nitric oxide synthase,[24] simultaneous deletion of the α_{2A}-AR and α_{2C}-AR subtypes was necessary to cause modifications, indicating that each of these receptor subtypes can substitute the other to mediate the respective effect.

Interestingly, in humans α_{2C}-AR is expressed in two polymorphic variants. The second variant is designated as α_{2C}-AR Del322-325 because the sequence ^{322}GAGP^{325} in the third intracellular loop is missing.[25] Both variants displayed similar intracellular accumulation at 37 °C, and exposure to low temperature enhanced the cell surface levels of both variants to the same extent.[13] The pathophysiological importance of the α_{2C}-AR polymorphism is still under debate. Some studies reported that α_{2C}-AR Del322-325 expression was accompanied by an impaired cellular signaling,[25] an enhanced risk of congestive heart failure,[25–27] an increased sympathetic tone, an increased adrenomedullary hormonal activity,[28] and an augmented predisposition to silent brain infarction.[29] In contrast, others did not find significant differences in the activation of signaling[30] or development of cardiovascular diseases between the two polymorphic variants.[31,32] One possible explanation for these discrepancies might be due to the differential racial responses to the presence of one of the alleles.[25,31]

Also, considering that at the physiological temperature, α_{2C}-AR is localized intracellularly in vasculature and fibroblasts (Fig. 1B), where the circulating catecholamines cannot activate it, one may speculate that α_{2C}-AR is a "discrete" receptor. The exception is the involvement of α_{2C}-AR in Raynaud phenomenon, a disease associated with exposure to cold.[33,34]

3. α_{2C}-AR AND RAYNAUD PHENOMENON

Raynaud phenomenon was described for the first time in 1862 by Maurice Raynaud.[35] This disease is characterized by painful triple changes of the skin color in the affected people after exposure to cold, emotional stress, or mechanical vibrations.[33–35] The most common manifestation is observed in the fingers, but other localizations are also reported. Against the common thought, Raynaud phenomenon is not a rare disease. According to Framingham study, about 3–10% of the general population are affected and there is a larger incidence in women.[33,34] Cold is the best-characterized inducer of the disease, and this fact explains the different frequency of Raynaud phenomenon over the globe.[33] Peripheral vasoconstriction in response to cold is a highly regulated physiological process helping the preservation of body heat.[36] However, people with Raynaud phenomenon have an exaggerated response to cold leading first to powerful vasoconstriction with the affected areas turning white due to reduced local perfusion. Next, the skin color turns blue due to hypoxia, and after the episode subsides (usually in 10–15 min), the blood flow is restored and the skin color changes to red with aching painful sensations.[33–35] Raynaud phenomenon is classified as primary when it is not associated with other diseases, or secondary when it is caused by an underlying condition, most common being scleroderma, lupus, and rheumatoid arthritis.[33,34] Currently, the disease is treated with palliative measures consisting of nonspecific vasodilators like calcium channel blockers or nitrates and a specific treatment is still missing.[33,34]

Several cellular mechanisms, such as hyperactivation of renin–angiotensin system,[37] enhanced endothelin levels,[38] or augmented serotonin release,[39] were proposed to cause Raynaud phenomenon. However, cellular and clinical studies have excluded the participation of these hormonal systems in this disease.[37–39] In contrast, the importance of α_{2C}-AR in the genesis of the disease was demonstrated by multiple molecular and clinical investigations. The initial studies demonstrated that cold-induced vasoconstriction is mediated by local enhancement in the sympathetic activity and increased release of norepinephrine.[40] In the 1990s, an investigation performed in patients with Raynaud phenomenon had shown that α_2-AR blockade, but not α_1-AR inhibition, prevented the vasospastic episodes to controlled cold exposure.[41] The important role of α_{2C}-AR in cold-induced vasoconstriction was then confirmed in the rat and mouse tail arteries in which α_{2C}-AR is endogenously expressed.[42,43] Subsequently, the use of transfected cell lines with smooth muscle or fibroblast phenotypes[13,44,45] demonstrated that at 37 °C, α_{2C}-AR is localized inside the cell (Fig. 1B), particularly in the endoplasmic reticulum.[13] Interestingly, in cell lines with neuro-endocrine phenotypes like PC12 and AtT20 cells, α_{2C}-AR was shown to be efficiently transported to the plasma membrane.[46]

To mimic the exposure to low temperature in primary cells and cell lines, the cells are maintained at 28–30 °C because lower temperatures will lead to a total blockade of the cellular transport. To test if the effects of low temperature are specific for α_{2C}-AR, we determined the cell surface levels of α_{2C}-AR, α_{2B}-AR, and α_{2A}-AR at 37 and 30 °C in transfected HEK293T cells. We found that only α_{2C}-AR displayed a temperature-dependent upregulation in the plasma membrane levels (Fig. 1C), whereas the other α_2-AR subtypes did not significantly change the cell surface levels under the same conditions (Fig. 1C). These data indicate that α_{2C}-AR is a particular example within the GPCR class, having temperature-sensitive transport to the cell surface, which contributes, at least in part, to the pathology of Raynaud phenomenon.

4. MECHANISMS INVOLVED IN THE TRANSPORT OF α_{2C}-AR TO THE CELL SURFACE

4.1 Trafficking Motifs Embedded in α_{2C}-AR Structure

Research in the last 15 years recognizes that specific amino-acid sequences embedded in the GPCR structure serve as structural determinants for the receptor subcellular localization.[47,48] Whereas the mechanisms involved in the receptor internalization/degradation are quite well defined, the motifs involved in the transport from the endoplasmic reticulum to the plasma membrane are still in need of characterization. Wu's group has identified several such motifs, particularly for α_2-AR traffic[49–51] like the highly conserved motifs F(x)6L[49] and RRK[50] which are present in many GPCRs. Because these motifs are present in all three α_2-ARs, probably they are not involved in the temperature-sensitive traffic of α_{2C}-AR. Also, the same group demonstrates that all three α_2-ARs contain a conserved YS motif in the N-termini (positions 56 and 57 of α_{2C}-AR,[51] Fig. 2A), which is responsible for the traffic of the receptors from the Golgi to the plasma membrane. As this motif is also present in all three α_2-ARs, it is unlikely involved in the temperature-sensitive traffic of α_{2C}-AR.

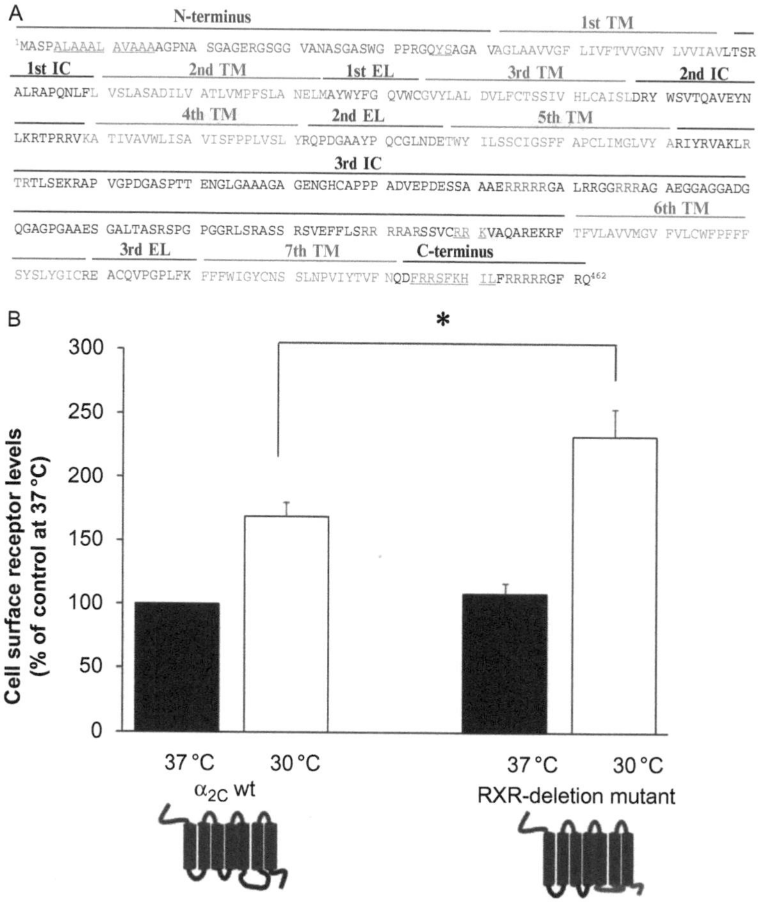

Figure 2 Trafficking motifs embedded in α_{2C}-AR. (A) Amino-acid sequences of human α_{2C}-AR. The extracellular loops are shown in blue, the transmembrane domains are shown in brown, and the intracellular loops are shown in black. The previously reported traffic motifs[45,49–51] embedded in the receptor structure are shown in green. The putative arginine-based retention motifs in the third intracellular loop and the C-terminus are shown in red. (B) The effects of exposure to low temperature (30 °C) on the plasma membrane levels of α_{2C}-AR and its deletion mutant lacking all the 11 RXR motifs by deleting 125 amino acids in the third intracellular loop and the last 9 amino acids in the C-terminus. The experiments were performed in transfected HEK293 cells in similar manner as described.[13] $n>6$ in each case, $^{*}p<0.05$. *Adapted from Ref. 52.* (See the color plate.)

In addition, Angelotti's group described a unique hydrophobic motif ALAAALAAAAA involved in the localization of α_{2C}-AR in the endoplasmic reticulum at 37 °C.[45] Deletion of this motif resulted in the partial release of the intracellular retention of the receptor.[45] However, the role of this motif in the temperature-sensitive receptor traffic was not investigated. We focused on the roles of arginine-based motifs in the temperature-sensitive transport of α_{2C}-AR because unlike other members of GPCR class, α_{2C}-AR is rich in arginine residues, mostly within the third intracellular loop and the C-terminus (Fig. 2A). Interestingly, most of these arginine residues are grouped to form RXR motifs, which have been shown to function as endoplasmic reticulum retention signals in other proteins, including members of GPCR class.[53–56] We generated a deletion α_{2C}-AR mutant missing 125 amino acids in the third intracellular loop and the last 9 amino acids in the C-terminus, excluding all putative 11 RXR motifs present in this receptor.[52] This mutant had similar cell surface levels as wild-type receptor, indicating that these RXR motifs are not involved in the receptor intracellular retention at 37 °C.[52] This finding was surprising, but it demonstrated that the deletion mutant was correctly folded. However, when the experiments were carried out at 30 °C, we found that the deletion mutant receptor had significant enhanced plasma membrane levels compared to wild type (Fig. 2B).[52] These data indicate that the RXR motifs embedded in α_{2C}-AR may be important in the prevention of the exaggerated transport of the receptor to the cell surface, which could be significant for the pathology of Raynaud phenomenon. Further, progressive alanine mutagenesis of the RXR motifs in the human α_{2C}-AR identified the clusters localized in the positions ^{294}RRRRR^{298} and ^{454}RRRRR^{458} (Fig. 2A) to be required for the upregulation of the plasma membrane receptor levels at low temperature, whereas mutation of other arginine clusters did not have any effects.[52] Interestingly, the ^{294}RRRRR^{298} and ^{454}RRRRR^{458} clusters in the mouse and rat α_{2C}-ARs are shifted with three amino acids to the left, which largely attenuates the temperature sensitivity of these receptors.[52]

The general mechanism underlying the actions of trafficking motifs within the GPCR superfamily is represented by interactions with specific proteins, broadly named molecular chaperones.[57–59]

4.2 The Concept of Molecular Chaperones

Molecular chaperones were initially defined as proteins required for correct folding of the newly synthesized proteins about 40 years ago.[57] As discussed in different chapters of this book, it is generally accepted that molecular chaperones are involved in the GPCR synthesis, anterograde traffic to the plasma membrane, internalization, and degradation.[58,59] To avoid the confusion with the initial definition, some publications use the term of GPCR interacting or accessory proteins,[59,60] for the same concept. A multitude of molecular chaperones were described for the GPCR class.[58,61] Some of these molecular chaperones are able to interact with all GPCRs, and these interactions are obligatory for the receptor maturation or degradation.[58,61] Still, specific molecular chaperones for distinct GPCR were also described.[58,61,62] Actually, targeting the expression and the activity of these specific molecular chaperones may constitute a therapeutic alternative to modulate the function of the respective receptors.[52,58,62] Here, a brief summary of the molecular chaperones involved in the expression, traffic, and function of α_{2C}-AR, including the ones shared with the entire GPCR class, will be presented.

4.3 Common Molecular Chaperones

Calnexin and calreticulin, two endoplasmic reticulum resident proteins, are best known for their calcium-binding properties.[63,64] These proteins are also fundamental molecules in the correct folding of newly formed N-glycosylated proteins, including GPCRs. Calnexin (which is endoplasmic reticulum membrane bound), calreticulin (an endoplasmic reticulum soluble protein), and ERp57 (a protein disulfide isomerase) are serially interacting with the nascent proteins until these proteins achieve the native conformation competent for exportation from the endoplasmic reticulum.[63,64] Noteworthy, N-glycosylation-deficient GPCR mutants remain associated with calnexin and are inefficiently trafficked to the plasma membrane.[65] α_{2C}-AR has two consensus N-glycosylation sites localized in the N-terminus at the positions ^{19}NAS^{21} and ^{33}NAS^{35}, and mutations of these sites severely reduced the receptor plasma membrane levels (Fig. 3A). However, the nonglycosylated receptor mutant is still transported to the cell surface in a temperature-sensitive manner, indicating that N-glycosylation is not mandatory for the effects of low temperature on the intracellular trafficking of α_{2C}-AR (Fig. 3A). We tested the interactions of α_{2C}-AR with two endoplasmic reticulum resident chaperones, calnexin and BiP (a homologue of HSP70,[66] also named GRP78 for glucose-regulated protein 78). The α_{2C}-AR interaction with calnexin was temperature-independent, whereas its interaction with BiP decreased by $43 \pm 11\%$ after exposure to 30 °C (Fig. 3B). However, BiP overexpression did not change the receptor plasma membrane levels at 37 °C or at 30 °C in HEK293T cells (Fig. 3C), indicating that the endogenous levels of the chaperones are sufficient to mediate α_{2C}-AR intracellular transport. We also determined the effects of DRiP78

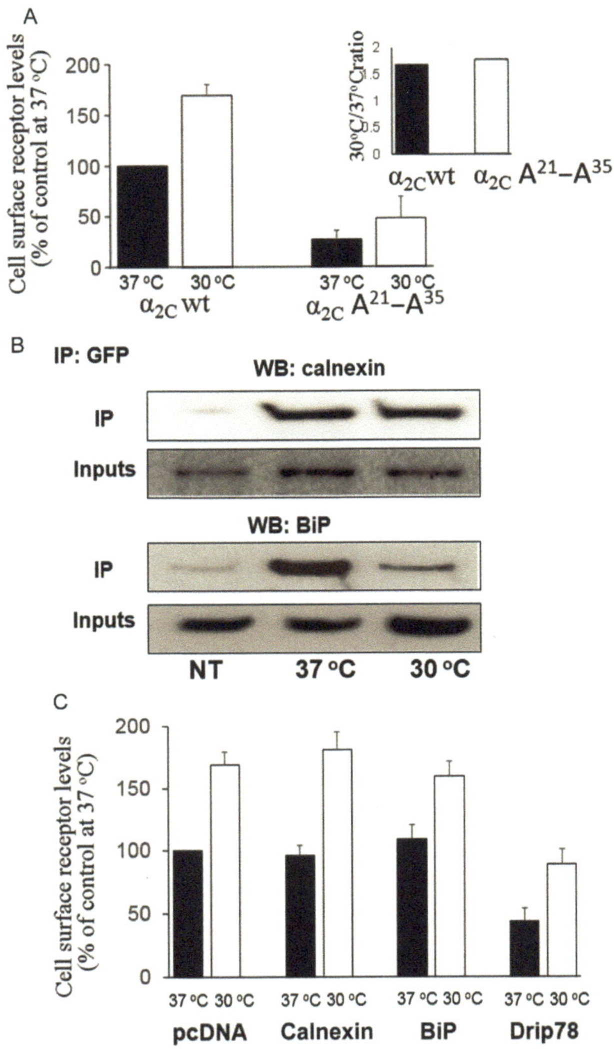

Figure 3 Common molecular chaperones in α_{2C}-AR transport. (A) Comparison between plasma membrane levels of α_{2C}-AR wild-type and α_{2C}-AR A^{21}–A^{35} mutant which lacks the N-glycosylation sites in HEK293 cells. The cell surface receptor numbers were determined by ligand binding using [^{3}H]RX821002. In the inset, the ratios of the expression levels of the receptors at 30 and 37 °C are shown. (B) Interactions between α_{2C}-AR and calnexin (top) or BiP (bottom) in nontransfected cells (NT) at 37 or 30 °C. The experiments were performed as previously described.[13,52,61] (C) The effects of overexpression of calnexin, BiP, and Drip78 on the α_{2C}-AR plasma membrane levels at 37 °C (black columns) or 30 °C (white columns) determined by radioligand binding using [^{3}H] RX821002.

(the HSP40 endoplasmic reticulum homologue) on α_{2C}-AR transport. DRiP78 is the first historically identified endoplasmic resident chaperone interacting with members of the GPCR class. This molecule has opposite effects on the transport of different receptors, inhibiting the traffic of D1 dopamine receptor and CCR5 chemokine receptor,[67,68] but stimulating the plasma membrane expression of AT1 receptor.[69] We found that in cotransfected HEK293T cells, DRiP78 strongly reduced α_{2C}-AR cell surface levels at both 37 and 30 °C, excluding its participation in the temperature-sensitive traffic of the receptor (Fig. 3C). Overall, these data indicate that the best-known endoplasmic reticulum resident chaperones modulate the total cellular levels of α_{2C}-AR, but are not involved in the receptor temperature-sensitive intracellular traffic.

4.4 Specific Chaperones of α_{2C}-AR

4.4.1 Dimerization of α_{2C}-AR with Members of GPCR Superfamily

The first proposal that homo- or hetero-interactions between members of GPCR class may have a major significance in the receptor traffic and signaling is more than 30 years old.[70] Work over two decades fully demonstrated the importance of homo- and heterodimerization in signaling and subcellular localization of many members of the GPCR superfamily, including adrenergic, opioid, muscarinic, and dopamine receptors.[71–74] Likewise, homodimerization has been shown to modulate not only the effects of β2-AR on the adenylate cyclase activity but also the internalization of the receptor.[74] The concept that heterodimerization may improve the traffic of intracellularly retained receptors was first demonstrated in the case of $GABA_BR1$.[75] In cells without coexpression of $GABA_BR2$, $GABA_BR1$ is poorly transported to the cell surface and accumulated in the endoplasmic reticulum.[75] However, coexpression of $GABA_BR2$ increased the traffic of $GABA_BR1$ to the plasma membrane by 100-fold.[75] Moreover, $GABA_BR2$ has been demonstrated to activate very weakly G proteins, and thus far, no independent pathophysiological function has been identified for $GABA_BR2$ in the absence of $GABA_BR1$, indicating that its primary role is to facilitate the cell surface transport of $GABA_BR1$.[76] The mechanisms underlying this cooperation have been shown to involve the dimerization of the two receptor subtypes. Dimerization with $GABA_BR2$ will mask the RXR motif present in the C-terminus of $GABA_BR1$, releasing its retention in endoplasmic reticulum and allowing the transport of this receptor to the cell surface. In a very elegant study, Hall's group evaluated the dimerization of rat α_{2C}-AR with other 29 members of GPCR family.[77] Only cotransfection with

β2-AR enhanced the plasma membrane levels and signaling activity of α_{2C}-AR.[77] Further, while under normal conditions α_{2C}-AR is resistant to ligand-induced internalization, dimerization with β2-AR resulted in receptor internalization after treatment with either α_2- or β2-AR ligands. In contrast, treatment with α_2-AR ligands did not induce β2-AR endocytosis, indicating that dimer internalization may depend on the activation of individual receptors in the dimer.[77] Yet, as this study did not analyzed the effects of heterodimerization on the α_{2C}-AR temperature-sensitive intracellular trafficking, future experiments are necessary to answer the roles of interactions with other members of GPCR family in the regulation of the transport of α_{2C}-AR to the cell surface.

4.4.2 Receptor Expression Enhancing Proteins

The RTP (receptor-transporting protein) and REEP (receptor expression-enhancing protein) were first reported to enhance the plasma membrane expression of olfactory receptors in HEK293T cells, but in contrast to the related protein family receptor activity-modifying protein (RAMP), without altering the pharmacological properties of the receptors.[78,79] Recently, Angelotti's group has shown that REEP1, REEP2, and REEP6 do not directly modulate GPCR plasma membrane levels, but enhance the endoplasmic reticulum cargo capacity of specific receptors, including α_{2C}-AR, but not α_{2A}-AR.[80] The mechanism underlying these effects involves the enhanced specific glycosylation of the receptors. Further, REEP1 and REEP2 appear to have a localization restricted to the neuronal tissues.[81] This exciting observation may explain the enhanced α_{2C}-AR plasma membrane localization in the cells with neuronal or endocrine phenotypes,[46,82] but definite experiments to prove this hypothesis remain to be performed.

4.4.3 Filamin-2

Filamins are members of actin-binding proteins directly interacting with actin to enhance the stability of cytoskeleton. Using the C-terminus of α_{2C}-AR as a bait in yeast two-hybrid screening, filamin-2 was identified as a binding partner of the receptor.[83] The interactions were dependent on the activation of Rap1A and phosphorylation of one specific serine residue located in the position 2113 within filamin-2.[83] Previously, the same group demonstrated that Rap1A stimulation by cAMP activation of Epac leads to an enhanced α_{2C}-AR translocation to the cell surface.[84] However, in these experiments, the true plasma membrane levels of the receptors were difficult to estimate because the measurements were performed in the plasma membrane immunoprecipitates using tag antibodies and no attempts for direct quantification were made.[83,84] In addition, previous data from the same group showed that increases in cAMP cellular levels also enhanced the total α_{2C}-AR cellular expression levels and such increases are fully expected to result in an enhancement of the plasma membrane levels of the receptor.[85] Also it is not certain if these effects are specific for α_{2C}-AR, as filamin was shown to mediate the plasma membrane delivery of Ca^{2+}-activated K^+ channels,[86] cystic fibrosis transmembrane conductance regulator,[87] and calcium sensing receptor.[88] Furthermore, microtubules are essential to maintain cellular architecture and are also involved in the anterograde trafficking of newly synthesized proteins. For instance, tubulin directly interacts with α_{2B}-AR through ionic interactions and modulates receptor cell surface levels.[89] Thus, the specificity of filamin in the temperature-dependent traffic of α_{2C}-AR needs further investigation.

4.4.4 Heat-Shock Protein 90

Heat-shock proteins (HSPs) were identified as stress-induced proteins and are a major family of molecular chaperones which based on the molecular weight can be subdivided into six classes, namely small HSP (<34 kDa), HSP40 (35–54 kDa), HSP60 (55–64 kDa), HSP70 (65–80 kDa), HSP90 (81–99 kDa), and HSP100 (>100 kDa), and each of them has distinct functions.[90,91] Although induced by various cellular stressors, members of HSP90 class are very abundant even in normal cells, accounting for about 1% of the total cellular proteins.[90,91] HSP90 class has four distinct members: two cytosolic forms including the mostly inducible Hsp90α and the constitutively expressed Hsp90β, one located in the mitochondria named Trap1, and one in the endoplasmic reticulum designated GRP94.[90–92] All Hsp90 homologues have the same domain architecture consisting of an ATP-binding domain, a middle domain, and a C-terminal domain involved in dimerization.[90–92] In contrast to the HSP70 family, the cytosolic HSP90 is not involved in the folding of newly synthesized proteins but facilitate the maturation and transport to the functional site of a limited number of proteins which they interact.[90–92] Unlike other HSPs, the chaperoning activity of HSP90 is strictly dependent on the coexpression of various cochaperones.[62,92] Over 15 such cochaperones have been identified, many with distinct roles in the regulation of the HSP90 activity and in the intracellular traffic of HSP90 targets.[62,92,93] The best-characterized "client" proteins of HSP90 are steroid hormone receptors or protein kinases. Because these proteins are involved in the pathology of various types of cancer, during the last 10 years, the search for specific HSP90 inhibitors has been a subject of intensive investigation and resulted in the development of several drugs, some of them currently in clinical trial.[91,93] However, very little is known about the roles of HSP90 in the GPCR intracellular traffic and degradation. An initial report indicated that HSP90 interaction with PAR-1 receptor is essential for thrombin-mediated morphological changes in astrocytes, but this interaction is not involved in $[Ca^{2+}]_i$ responses and does not affect PAR-1 subcellular localization.[94] Our group demonstrated that treatment with specific HSP90 inhibitors like radicicol, macbecin, and 17-DMAG enhanced the plasma membrane α_{2C}-AR levels at 37 °C, but had no effect at 30 °C[13] (Fig. 4A). These data indicate that interactions with HSP90 are a limiting step in the transport of the receptor to the cell surface at the physiological temperature. In addition, we demonstrated that HSP90 interacted with α_{2C}-AR in a temperature-sensitive manner, but such interactions were not observed in the case of α_{2B}-AR.[13] Also, treatment with HSP90 inhibitors enhanced the contraction of the rat tail artery in response to α_{2C}-AR stimulation at 37 °C, but not at 30 °C. Importantly, in vascular smooth muscle cells isolated from the rat tail artery, exposure to 30 °C decreased the cellular levels of HSP90, but did not affect HSP70.[13] Overall, our data indicate that the reduction of the cellular levels or the activity of HSP90 may constitute a biomarker to diagnose people predisposed to Raynaud phenomenon.

Recently, HSP90 has been shown to interact with two other members of GPCR class and inhibition of these interactions modulates the subcellular localization and signaling of the receptors. A large fraction of adenosine A2A receptor has been reported to accumulate in the endoplasmic reticulum in HEK293T and PC12 cells.[95] Similar to the effects on α_{2C}-AR, treatment with the HSP90 inhibitors 17-DMAG and radicicol significantly enhanced adenosine A2A receptor plasma membrane levels.[95] However, currently it is not known if adenosine A2A receptor traffic is temperature-sensitive. In contrast to α_{2C}-AR, adenosine A2A receptor traffic to the plasma membrane also depends on the interactions with HSP70.[95] On the other hand, another study demonstrated that the prostaglandin D2 receptor, DP1 receptor, forms complexes with HSP90[96] and the interactions are modulated by the expression of L-PGDS, an enzyme involved in the synthesis of the endogenous ligand of the receptor.[96] As reported in the cases of α_{2C}-AR[13] and adenosine A2A receptors,[95] interactions with HSP90 promote DP1 transport to the cell surface.[96] Together, these reports suggest that interactions with HSP90 might be an important factor controlling the subcellular localization and signaling of a subset of GPCRs.

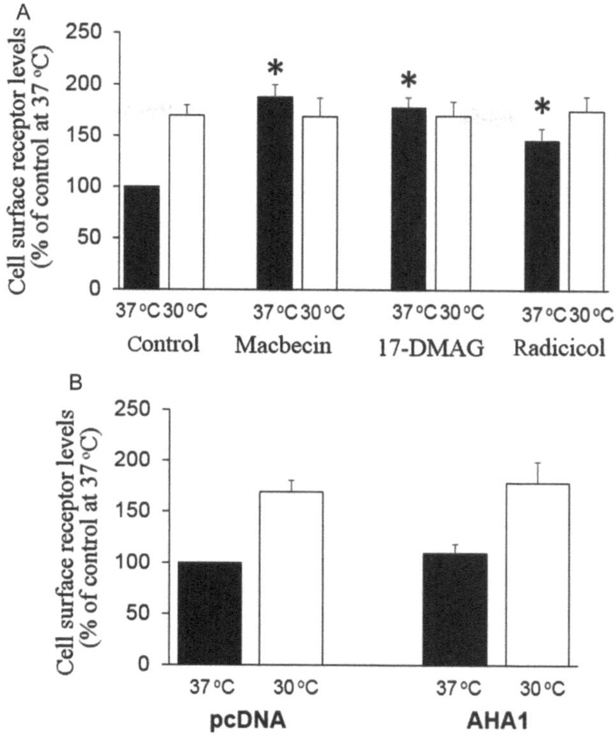

Figure 4 The effects of manipulation of HSP90 activity on the temperature-sensitive traffic of α_{2C}-AR. (A) The effects of three distinct HSP90 inhibitors, macbecin (5 μM), 17-DMAG (0.5 μM), and radicicol (10 μM), on the α_{2C}-AR plasma membrane levels at 37 °C (black columns) and 30 °C (white columns) in HEK293 cells determined by radioligand binding. The inhibitors were incubated for 18 h. *Significant differences compared to control cells. (B) Overexpression of AHA1,[62] a HSP90 cochaperone, did not change the temperature-dependent traffic of α_{2C}-AR. HEK293 cells were transfected with α_{2C}-AR and AHA1 in a ratio of 1:3 and the cell surface receptor levels determined as described.[13,52] *Adapted from Ref. 13.*

One point that remains to be addressed is which HSP90 isoform is involved in α_{2C}-AR temperature-sensitive traffic because current HSP90 inhibitors are not able to discriminate between HSP90α and HSP90β. We previously excluded the participation in this process of GRP94 and Trap1.[13] Also, of interest will be to identify the HSP90 cochaperones which are involved in the modulation of α_{2C}-AR traffic. We started addressing this question by testing the effects of AHA1, a cochaperone that we have previously shown to selectively modulate the intracellular traffic of CB1 cannabinoid receptor, but not that of CB2 cannabinoid receptor.[62] However, overexpression of AHA1 did not change the α_{2C}-AR levels at

37 °C or at 30 °C in HEK293 cells, indicating that AHA1 is not involved in the temperature-sensitive receptor traffic (Fig. 4B).

4.4.5 Pontin

We have recently demonstrated that pontin (also named RuvBL-1 or TIP49) has stronger interaction with human α_{2C}-AR compared to rat α_{2C}-AR.[52] This observation may have pathological significance in Raynaud phenomenon because rat α_{2C}-AR has a reduced export to cell surface in response to low temperature.[52] Indeed, pharmacological or genetic inhibition of pontin activity enhanced the cell surface levels of human α_{2C}-AR at 37 °C.[52] In contrast, reducing pontin activity did not change the plasma membrane receptor levels at 30 °C, indicating that this chaperone limits the human α_{2C}-AR transport to the plasma membrane only at the physiological temperature.[52] The enhanced cell surface levels of α_{2C}-AR after inhibition of pontin activity at 37 °C were in parallel with the effects on the cellular cAMP production and the ERK1/2 activation in response to receptor stimulation.[52] Pontin has also been recently shown to play a role in the evolution of systemic sclerosis, a disease that may be associated with Raynaud phenomenon.[97] Further, pontin together with its partner reptin has been shown to interact with HSP90.[98] These data strongly indicate that multiple interactions with various chaperones along the biosynthetic pathway are responsible for the unique temperature-sensitive trafficking of human α_{2C}-AR.

5. CONCLUSIONS

α_{2C}-AR is a member of GPCR class with unique properties. This receptor has distinct subcellular localization in fibroblasts and vascular smooth muscle cells compared to neuronal or endocrine cells. This feature suggests that α_{2C}-AR has different trafficking determinants in different cell types. The mechanisms of α_{2C}-AR transport have been partly elucidated, particularly the identification of new molecular chaperones involved. The temperature-sensitive upregulation of α_{2C}-AR plasma membrane levels observed in fibroblasts and vascular smooth muscle cells is an important factor in the pathology of Raynaud phenomenon. Development of agents interfering with the trafficking process of α_{2C}-AR may eventually provide an effective therapy to this old disease.

ACKNOWLEDGMENTS

Supported by grants from National Institute of Arthritis and Musculoskeletal and Skin Diseases R03AR064008, National Institute of General Medicine P20-GM103514, and startup funds from Howard University.

REFERENCES

1. Ahlquist RP. The adrenergic receptor. *J Pharm Sci.* 1966;55:359–367.
2. Black JW, Stephenson JS. Pharmacology of a new adrenergic beta-receptor-blocking compound (Nethalide). *Lancet.* 1962;2:311–314.
3. Langer SZ. Presynaptic regulation of catecholamine release. *Biochem Pharmacol.* 1974;23:1793–1800.
4. Cambridge D, Davey MJ, Massingham R. Prazosin, a selective antagonist of post-synaptic alpha-adrenoceptors. *Br J Pharmacol.* 1977;59:514P–515P.
5. Grant JA, Scrutton MC. Novel alpha2-adrenoreceptors primarily responsible for inducing human platelet aggregation. *Nature.* 1979;277:659–661.
6. Rufolo RR, Nichols AJ, Stradel JM, Hieble JP. Structure and function of α-adrenoceptors. *Pharmacol Rev.* 1991;43:475–505.
7. Kobilka BK, Matsui H, Kobilka TS, et al. Cloning, sequencing, and expression of the gene coding for the human platelet α2-adrenergic receptor. *Science.* 1987;238:650–656.
8. Regan JW, Kobilka TS, Yang-Feng TL, Caron MG, Lefkowitz RJ, Kobilka BK. Cloning and expression of a human kidney cDNA for an alpha 2-adrenergic receptor subtype. *Proc Natl Acad Sci USA.* 1988;85:6301–6305.
9. Lomasney JW, Lorenz W, Allen LF, et al. Expansion of the alpha 2-adrenergic receptor family: cloning and characterization of a human alpha 2-adrenergic receptor subtype, the gene for which is located on chromosome 2. *Proc Natl Acad Sci USA.* 1990;87:5094–5098.
10. Lorenz W, Lomasney JW, Collins S, Regan JW, Caron MG, Lefkowitz RJ. Expression of three alpha 2-adrenergic receptor subtypes in rat tissues: implications for alpha 2 receptor classification. *Mol Pharmacol.* 1990;38:599–603.
11. Link R, Daunt D, Barsh G, Chruscinski A, Kobilka B. Cloning of two mouse genes encoding alpha 2-adrenergic receptor subtypes and identification of a single amino acid in the mouse alpha 2-C10 homolog responsible for an interspecies variation in antagonist binding. *Mol Pharmacol.* 1992;42:16–27.
12. Zhou F, Filipeanu CM, Duvernay MT, Wu G. Cell-surface targeting of alpha2-adrenergic receptors—inhibition by a transport deficient mutant through dimerization. *Cell Signal.* 2006;18:318–327.
13. Filipeanu CM, de Vries R, Danser AH, Kapusta DR. Modulation of α(2C) adrenergic receptor temperature-sensitive trafficking by HSP90. *Biochim Biophys Acta.* 1813;2011:346–357.
14. Blaxall HS, Hass NA, Bylund DB. Expression of alpha 2-adrenergic receptor genes in rat tissues. *Receptor.* 1994;4:191–199.
15. Eason MG, Liggett SB. Human alpha 2-adrenergic receptor subtype distribution: widespread and subtype-selective expression of alpha 2C10, alpha 2C4, and alpha 2C2 mRNA in multiple tissues. *Mol Pharmacol.* 1993;44:70–75.
16. Handy DE, Flordellis CS, Bogdanova NN, Bresnahan MR, Gavras H. Diverse tissue expression of rat alpha 2-adrenergic receptor genes. *Hypertension.* 1993;21:861–865.
17. Perälä M, Hirvonen H, Kalimo H, et al. Differential expression of two alpha 2-adrenergic receptor subtype mRNAs in human tissues. *Brain Res Mol Brain Res.* 1992;16:57–63.
18. Regard JB, Sato IT, Coughlin SR. Anatomical profiling of G protein-coupled receptor expression. *Cell.* 2008;135:561–571, Log average relative transcript abundance in mouse tissues measured by qPCR.

19. Björklund M, Sirviö J, Sallinen J, Scheinin M, Kobilka BK, Riekkinen Jr P. Alpha2C-adrenoceptor overexpression disrupts execution of spatial and non-spatial search patterns. *Neuroscience*. 1999;88:1187–1198.
20. Link RE, Desai K, Hein L, et al. Cardiovascular regulation in mice lacking alpha2-adrenergic receptor subtypes b and c. *Science*. 1996;273:803–805.
21. Scheinin M, Sallinen J, Haapalinna A. Evaluation of the alpha2C-adrenoceptor as a neuropsychiatric drug target studies in transgenic mouse models. *Life Sci*. 2001;68:2277–2285.
22. Zhang W, Ordway GA. The alpha2C-adrenoceptor modulates GABA release in mouse striatum. *Brain Res Mol Brain Res*. 2003;112:24–32.
23. Romana-Souza B, Nascimento AP, Brum PC, Monte-Alto-Costa A. Deletion of the α2A/α2C-adrenoceptors accelerates cutaneous wound healing in mice. *Int J Exp Pathol*. 2014;95:330–341.
24. Couto GK, Davel AP, Brum PC, Rossoni LV. Double disruption of α2A- and α2C-adrenoceptors induces endothelial dysfunction in mouse small arteries: role of nitric oxide synthase uncoupling. *Exp Physiol*. 2014;99:1427–1438.
25. Small KM, Forbes SL, Rahman FF, Bridges KM, Liggett SB. A four amino acid deletion polymorphism in the third intracellular loop of the human alpha 2C-adrenergic receptor confers impaired coupling to multiple effectors. *J Biol Chem*. 2000;275:23059–23064.
26. Small KM, McGraw DW, Liggett SB. Pharmacology and physiology of human adrenergic receptor polymorphisms. *Annu Rev Pharmacol Toxicol*. 2003;43:381–411.
27. Woodiwiss AJ, Badenhorst D, Sliwa K, et al. Beta1- and alpha2c-adrenoreceptor variants as predictors of clinical aspects of dilated cardiomyopathy in people of African ancestry. *Cardiovasc J Afr*. 2008;19:188–193.
28. Neumeister A, Drevets WC, Belfer I, et al. Effects of a alpha 2C-adrenoreceptor gene polymorphism on neural responses to facial expressions in depression. *Neuropsychopharmacology*. 2006;31:1750–1756.
29. Kim JO, Jeon YJ, Kim OJ, et al. Association between common genetic variants of α2A-, α2B- and α2C-adrenoceptors and the risk of silent brain infarction. *Mol Med Rep*. 2014;9:2459–2466.
30. Montgomery MD, Bylund DB. Lack of effect of the alpha2C-adrenoceptor Del322-325 polymorphism on inhibition of cyclic AMP production in HEK293 cells. *Br J Pharmacol*. 2010;159:820–830.
31. Li JL, Canham RM, Vongpatanasin W, Leonard D, Auchus RJ, Victor RG. Do allelic variants in alpha2A and alpha2C adrenergic receptors predispose to hypertension in blacks? *Hypertension*. 2006;47:1140–1146.
32. Regitz-Zagrosek V, Hocher B, Bettmann M, et al. Alpha2C-adrenoceptor polymorphism is associated with improved event-free survival in patients with dilated cardiomyopathy. *Eur Heart J*. 2006;27:454–459.
33. Brand FN, Larson MG, Kannel WB, McGuirk JM. The occurrence of Raynaud's phenomenon in a general population: the Framingham Study. *Vasc Med*. 1997;2:296–301.
34. Prete M, Fatone MC, Favoino E, Perosa F. Raynaud's phenomenon: from molecular pathogenesis to therapy. *Autoimmun Rev*. 2014;13:655–667.
35. Raynaud M. *Local Asphyxia and Symmetrical Gangrene of the Extremities*. Paris: L Leclerc; 1862.
36. Morrison SF. Carl Ludwig Distinguished Lectureship of the APS Neural Control and Autonomic Regulation Section: central neural pathways for thermoregulatory cold defense. *J Appl Physiol*. 2011;110:1137–1149.
37. Wood HM, Ernst ME. Renin-angiotensin system mediators and Raynaud's phenomenon. *Ann Pharmacother*. 2006;40:1998–2002.
38. Nguyen VA, Eisendle K, Gruber I, Hugl B, Reider D, Reider N. Effect of the dual endothelin receptor antagonist bosentan on Raynaud's phenomenon secondary to

systemic sclerosis: a double-blind prospective, randomized, placebo-controlled pilot study. *Rheumatology (Oxford)*. 2010;49:583–587.
39. Pope J, Fenlon D, Thompson A, et al. Ketanserin for Raynaud 'phenomenon in progressive systemic sclerosis. *Cochrane Database Syst Rev*. 2000;2:CD000954.
40. Flavahan NA, Lindblad LE, Verbeuren TJ, Shepherd JT, Vanhoutte PM. Cooling and alpha 1- and alpha 2-adrenergic responses in cutaneous veins: role of receptor reserve. *Am J Physiol*. 1985;249:H950–H955.
41. Freedman RR, Baer RP, Mayes MD. Blockade of vasospastic attacks by alpha 2-adrenergic but not alpha 1-adrenergic antagonists in idiopathic Raynaud's disease. *Circulation*. 1995;92:1448–1451.
42. Chotani MA, Flavahan S, Mitra S, Daunt D, Flavahan NA. Silent alpha(2C)-adrenergic receptors enable cold-induced vasoconstriction in cutaneous arteries. *Am J Physiol Heart Circ Physiol*. 2000;278:H1075–H1083.
43. Bailey SR, Mitra S, Flavahan S, Flavahan NA. Reactive oxygen species from smooth muscle mitochondria initiate cold-induced constriction of cutaneous arteries. *Am J Physiol Heart Circ Physiol*. 2005;289:H243–H250.
44. Jeyaraj SC, Chotani MA, Mitra S, Gregg HE, Flavahan NA, Morrison KJ. Cooling evokes redistribution of alpha2C-adrenoceptors from Golgi to plasma membrane in transfected human embryonic kidney 293 cells. *Mol Pharmacol*. 2001;60:1195–1200.
45. Angelotti T, Daunt D, Shcherbakova OG, Kobilka B, Hurt CM. Regulation of G-protein coupled receptor traffic by an evolutionary conserved hydrophobic signal. *Traffic*. 2010;11:560–578.
46. Daunt DA, Hurt C, Hein L, Kallio J, Feng F, Kobilka BK. Subtype-specific intracellular trafficking of alpha2-adrenergic receptors. *Mol Pharmacol*. 1997;51:711–720.
47. Duvernay MT, Filipeanu CM, Wu G. The regulatory mechanisms of export trafficking of G protein-coupled receptors. *Cell Signal*. 2005;17:1457–1465.
48. Dong C, Filipeanu CM, Duvernay MT, Wu G. Regulation of G protein-coupled receptor export trafficking. *Biochim Biophys Acta*. 2007;1768:853–870.
49. Duvernay MT, Zhou F, Wu G. A conserved motif for the transport of G protein-coupled receptors from the endoplasmic reticulum to the cell surface. *J Biol Chem*. 2004;279:30741–30750.
50. Dong C, Nichols CD, Guo J, Huang W, Lambert NA, Wu G. A triple arg motif mediates alpha2B-adrenergic receptor interaction with Sec24C/D and export. *Traffic*. 2012;13:857–868.
51. Dong C, Wu G. Regulation of anterograde transport of alpha2-adrenergic receptors by the N termini at multiple intracellular compartments. *J Biol Chem*. 2006;281:38543–38554.
52. Filipeanu CM, Pullikuth AK, Guidry JJ. Molecular determinants of the human α2C-adrenergic receptor temperature-sensitive intracellular traffic. *Mol Pharmacol*. 2015; 87:792–802.
53. Michelsen K, Yuan H, Schwappach B. Hide and run. Arginine-based endoplasmic-reticulum-sorting motifs in the assembly of heteromultimeric membrane proteins. *EMBO Rep*. 2005;6:717–722.
54. Cunningham MR, McIntosh KA, Pediani JD, et al. Novel role for proteinase-activated receptor 2 (PAR2) in membrane trafficking of proteinase-activated receptor 4 (PAR4). *J Biol Chem*. 2012;287:16656–16669.
55. Okamoto Y, Shikano S. Phosphorylation-dependent C-terminal binding of 14-3-3 proteins promotes cell surface expression of HIV co-receptor GPR15. *J Biol Chem*. 2011;286:7171–7181.
56. Stepanchick A, McKenna J, McGovern O, Huang Y, Breitwieser GE. Calcium sensing receptor mutations implicated in pancreatitis and idiopathic epilepsy syndrome disrupt an arginine-rich retention motif. *Cell Physiol Biochem*. 2010;26:363–374.

57. Laskey RA, Honda BM, Mills AD, Finch JT. Nucleosomes are assembled by an acidic protein that binds histones and transfers them to DNA. *Nature*. 1978;275:416–420.
58. Williams D, Devi LA. Escorts take the lead molecular chaperones as therapeutic targets. *Prog Mol Biol Transl Sci*. 2010;91:121–149.
59. Brady AE, Limbird LE. G protein-coupled receptor interacting proteins: emerging roles in localization and signal transduction. *Cell Signal*. 2002;14:297–309.
60. Cooray SN, Clark AJ. Melanocortin receptors and their accessory proteins. *Mol Cell Endocrinol*. 2011;331:215–221.
61. Dupré DJ, Hammad MM, Holland P, Wertman J. Role of chaperones in G protein coupled receptor signaling complex assembly. *Subcell Biochem*. 2012;63:23–42.
62. Filipeanu CM, Guidry JJ, Leonard ST, Winsauer PJ. Δ9-THC increases endogenous AHA1 expression in rat cerebellum and may modulate CB1 receptor function during chronic use. *J Neurochem*. 2011;118:1101–1112.
63. Ni M, Lee AS. ER chaperones in mammalian development and human diseases. *FEBS Lett*. 2007;581:3641–3651.
64. Caramelo JJ, Parodi AJ. Getting in and out from calnexin/calreticulin cycles. *J Biol Chem*. 2008;283:10221–10225.
65. Lanctot PM, Leclerc PC, Clément M, et al. Importance of N-glycosylation positioning for cell-surface expression, targeting, affinity and quality control of the human AT1 receptor. *Biochem J*. 2005;390:367–376.
66. Dudek J, Benedix J, Cappel S, et al. Functions and pathologies of BiP and its interaction partners. *Cell Mol Life Sci*. 2009;66(9):1556–1569.
67. Bermak JC, Li M, Bullock C, Zhou QY. Regulation of transport of the dopamine D1 receptor by a new membrane-associated ER protein. *Nat Cell Biol*. 2001;3:492–498.
68. Kuang YQ, Charette N, Frazer J, et al. Dopamine receptor-interacting protein 78 acts as a molecular chaperone for CCR5 chemokine receptor signaling complex organization. *PLoS One*. 2012;7:e40522.
69. Leclerc PC, Auger-Messier M, Lanctot PM, Escher E, Leduc R, Guillemette G. A polyaromatic caveolin-binding-like motif in the cytoplasmic tail of the type 1 receptor for angiotensin II plays an important role in receptor trafficking and signaling. *Endocrinology*. 2002;143:4702–4710.
70. Birdsall NJ. Can different receptors interact directly with each other? *Trends Neurosci*. 1982;5:137–138.
71. Hebert TE, Moffett S, Morello JP, et al. A peptide derived from a beta2-adrenergic receptor transmembrane domain inhibits both receptor dimerization and activation. *J Biol Chem*. 1996;271:16384–16392.
72. Cvejic S, Devi LA. Dimerization of the delta opioid receptor: implication for a role in receptor internalization. *J Biol Chem*. 1997;272:26959–26964.
73. Maggio R, Vogel Z, Wess J. Coexpression studies with mutant muscarinic/adrenergic receptors provide evidence for intermolecular “cross-talk” between G-protein-linked receptors. *Proc Natl Acad Sci USA*. 1993;90:3103–3107.
74. Hebert TE, Loisel TP, Adam L, Ethier N, Onge SS, Bouvier M. Functional rescue of a constitutively desensitized beta2AR through receptor dimerization. *Biochem J*. 1998;330:287–293.
75. White JH, Wise A, Main MJ, et al. Heterodimerization is required for the formation of a functional GABA(B) receptor. *Nature*. 1998;396:679–682.
76. Martin SC, Russek SJ, Farb DH. Molecular identification of the human GABABR2: cell surface expression and coupling to adenylyl cyclase in the absence of GABABR1. *Mol Cell Neurosci*. 1999;13:180–191.
77. Prinster SC, Holmqvist TG, Hall RA. Alpha2C-adrenergic receptors exhibit enhanced surface expression and signaling upon association with beta2-adrenergic receptors. *J Pharmacol Exp Ther*. 2006;318:974–981.

78. Saito H, Kubota M, Roberts RW, Chi Q, Matsunami H. RTP family members induce functional expression of mammalian odorant receptors. *Cell.* 2004;119:679–691.
79. Mainland J, Matsunami H. RAMP like proteins: RTP and REEP family of proteins. *Adv Exp Med Biol.* 2012;744:75–86.
80. Björk S, Hurt CM, Ho VK, Angelotti T. REEPs are membrane shaping adapter proteins that modulate specific g protein-coupled receptor trafficking by affecting ER cargo capacity. *PLoS One.* 2013;8:e76366.
81. Hurt CM, Björk S, Ho VK, Gilsbach R, Hein L, Angelotti T. REEP1 and REEP2 proteins are preferentially expressed in neuronal and neuronal-like exocytotic tissues. *Brain Res.* 2014;1545:12–22.
82. Hurt CM, Feng FY, Kobilka B. Cell-type specific targeting of the alpha 2c-adrenoceptor. Evidence for the organization of receptor microdomains during neuronal differentiation of PC12 cells. *J Biol Chem.* 2000;275:35424–35431.
83. Motawea HK, Jeyaraj SC, Eid AH, et al. Cyclic AMP-Rap1A signaling mediates cell surface translocation of microvascular smooth muscle α2C-adrenoceptors through the actin-binding protein filamin-2. *Am J Physiol Cell Physiol.* 2013;305:C829–C845.
84. Jeyaraj SC, Unger NT, Eid AH, et al. Cyclic AMP-Rap1A signaling activates RhoA to induce α(2c)-adrenoceptor translocation to the cell surface of microvascular smooth muscle cells. *Am J Physiol Cell Physiol.* 2012;303:C499–C511.
85. Eid AH, Chotani MA, Mitra S, Miller TJ, Flavahan NA. Cyclic AMP acts through Rap1 and JNK signaling to increase expression of cutaneous smooth muscle alpha2C-adrenoceptors. *Am J Physiol Heart Circ Physiol.* 2008;295:H266–H272.
86. Kim EY, Ridgway LD, Dryer SE. Interactions with filamin A stimulate surface expression of large-conductance Ca^{2+}-activated K+ channels in the absence of direct actin binding. *Mol Pharmacol.* 2007;72:622–630.
87. Smith L, Litman P, Kohli E, et al. RACK1 interacts with filamin-A to regulate plasma membrane levels of the cystic fibrosis transmembrane conductance regulator. *Am J Physiol Cell Physiol.* 2013;305:C111–C120.
88. Breitwieser GE. The calcium sensing receptor life cycle: trafficking, cell surface expression, and degradation. *Best Pract Res Clin Endocrinol Metab.* 2013;27:303–313.
89. Duvernay MT, Wang H, Dong C, Guidry JJ, Sackett DL, Wu G. Alpha2B-adrenergic receptor interaction with tubulin controls its transport from the endoplasmic reticulum to the cell surface. *J Biol Chem.* 2011;286:14080–14089.
90. Picard D. Heat-shock protein 90, a chaperone for folding and regulation. *Cell Mol Life Sci.* 2002;59:1640–1648.
91. Wandinger SK, Richter K, Buchner J. The Hsp90 chaperone machinery. *J Biol Chem.* 2008;283:18473–18477.
92. Taherian A, Krone PH, Ovsenek N. A comparison of Hsp90alpha and Hsp90beta interactions with cochaperones and substrates. *Biochem Cell Biol.* 2008;86:37–45.
93. Prodromou C. The 'active life' of Hsp90 complexes. *Biochim Biophys Acta.* 1823;2012:614–623.
94. Pai KS, Mahajan VB, Lau A, Cunningham DD. Thrombin receptor signaling to cytoskeleton requires Hsp90. *J Biol Chem.* 2001;276:32642–32647.
95. Bergmayr C, Thurner P, Keuerleber S, et al. Recruitment of a cytoplasmic chaperone relay by the A2A adenosine receptor. *J Biol Chem.* 2013;288:28831–28844.
96. Binda C, Génier S, Cartier A, et al. A G protein-coupled receptor and the intracellular synthase of its agonist functionally cooperate. *J Cell Biol.* 2014;204:377–393.
97. Kaji K, Fertig N, Medsger Jr TA, et al. Autoantibodies to RuvBL1 and RuvBL2: a novel systemic sclerosis-related antibody associated with diffuse cutaneous and skeletal muscle involvement. *Arthritis Care Res (Hoboken).* 2014;66:575–584.
98. Izumi N, Yamashita A, Hirano H, Ohno S. Heat shock protein 90 regulates phosphatidylinositol 3-kinase-related protein kinase family proteins together with the RUVBL1/2 and Tel2-containing co-factor complex. *Cancer Sci.* 2012;103:50–57.

CHAPTER TWELVE

N-Terminal Signal Peptides of G Protein-Coupled Receptors: Significance for Receptor Biosynthesis, Trafficking, and Signal Transduction

Claudia Rutz, Wolfgang Klein, Ralf Schülein[1]
Leibniz-Institut für Molekulare Pharmakologie (FMP), Berlin, Germany
[1]Corresponding author: e-mail address: schuelein@fmp-berlin.de

Contents

Abstract

Signal sequences play a key role during the first steps of the intracellular transport of G protein-coupled receptors (GPCRs). They are involved in targeting of the nascent chains to the membrane of the endoplasmic reticulum (ER) and initiate integration of the newly synthesized receptors into this compartment. Two classes of signal sequences are known: N-terminal signal peptides, which are usually cleaved-off following ER insertion and internal signal sequences, the so-called signal anchor sequences, which form part of the mature proteins. About 5–10% of the GPCRs contain N-terminal signal peptides; the vast majority possesses signal anchor sequences. The reason why only a subset of GPCRs require signal peptides for ER targeting/insertion was addressed in the past decade by a limited number of studies indicating that the presence of signal peptides facilitates N-tail translocation at the ER membrane. Interestingly, recent work showed that signal peptides of GPCRs do not only serve "classical" functions in the early secretory pathway. Uncleaved pseudo signal peptides may regulate receptor densities in the plasma membrane, receptor dimerization, and G protein coupling selectivity. On the other hand, even cleaved and released peptides may have post-ER functions.

Progress in Molecular Biology and Translational Science, Volume 132
ISSN 1877-1173
http://dx.doi.org/10.1016/bs.pmbts.2015.03.003

In this review, we summarize the current knowledge about cleavable signal peptides of GPCRs and address also the question whether these sequences may serve as future drug targets in pharmacology.

1. INTRODUCTION

The heptahelical G protein-coupled receptors (GPCRs) form a large protein family, play an important role in signal transduction, and are the most important drug targets. Like other integral membrane proteins, GPCRs must be delivered by intracellular transport mechanisms[1] to their correct subcellular location to function properly, usually to the plasma membrane. GPCRs use the secretory pathway[2,3] to reach the plasma membrane (Fig. 1A). The first step of the intracellular transport of GPCRs is their targeting to the

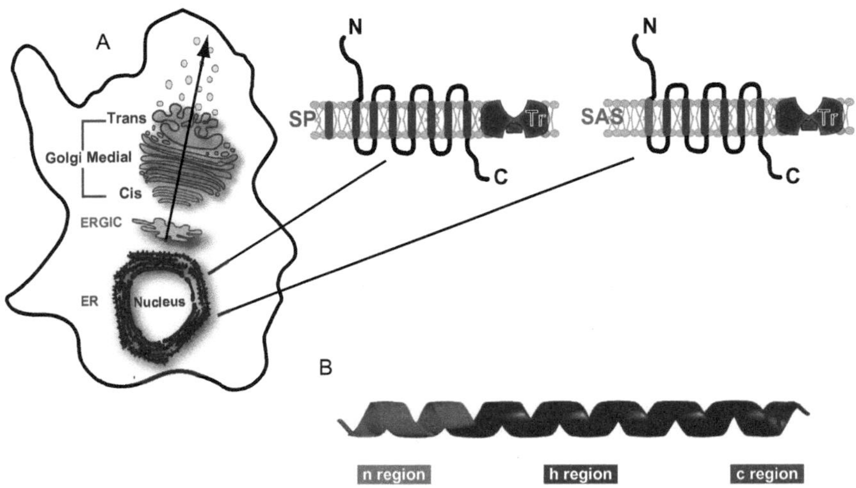

Figure 1 Intracellular transport of GPCRs. (A) Schematic depiction of the secretory pathway. Integration into the ER membrane is mediated by the translocon (Tr) and signal sequences of the receptors. GPCRs possess either signal peptides (SP, red), which are cleaved-off following ER integration or signal anchor sequences (SAS; usually TM1), which form part of the mature proteins. Receptors are then transported in the membrane of vesicles via the ER–Golgi intermediate compartment (ERGIC) and the individual compartments of the Golgi apparatus to the cell surface. (B) Basic architecture of signal peptides. At the N-terminus, signal peptides usually contain a short stretch of rather polar amino acid residues (n region). A longer hydrophobic core (h region) is followed by a C-terminal segment, which contains helix-breaking proline and glycine residues and small, uncharged residues at positions −1 and −3 of the cleavage site (c region). Signal peptides are thought to adopt an overall helical conformation. (See the color plate.)

endoplasmic reticulum (ER) membrane and their integration into the lipid bilayer of this compartment, which is mediated by the translocon complex. Both ER targeting and insertion are regulated by signal sequences of the nascent chains. Signal sequences fall into two classes: N-terminal signal peptides (also called cleaved signal sequences) are located at the extreme N-termini of the nascent chains and are cleaved-off during or after integration of the proteins into the ER membrane. The second type, the signal anchor sequences (usually the first transmembrane domains, TM1) form part of the mature proteins. Integral membrane proteins with an extracellular N-tail, such as GPCRs, possess either signal peptides or signal anchor sequences (Fig. 1A). Whereas the vast majority of the GPCRs (90–95%) contain signal anchor sequences, only a small subgroup (5–10%) harbors signal peptides. In contrast, membrane proteins with an intracellular N-tail invariantly have signal anchor sequences. Newly synthesized secretory proteins, which are not integrated into the ER membrane but are translocated across the bilayer, possess invariantly signal peptides. Membrane proteins with an extracellular N-tail are thus the only proteins, which may possess either type of signal sequence. Due to the large number of available sequences, the GPCR protein family was ideally suited to address the question why some proteins with an extracellular N-tail possess signal peptides, whereas the others do not.

2. STRUCTURE AND BASIC PROPERTIES OF SIGNAL PEPTIDES

N-terminal signal peptides of membrane proteins have a tripartite structure and share characteristic features with signal peptides of secretory proteins[4,5] (Fig. 1B): a hydrophobic α-helical core (h region), which is N-terminally flanked by rather polar amino acid residues (n region). The C-terminal side contains helix-breaking proline and glycine residues and small, uncharged residues at positions −1 and −3 of the cleavage site (c region). In eukaryotes, signal peptide length ranges from approximately 12 to 40 amino acid residues with an average length of about 23 residues. Due to their N-terminal location and their tripartite structure, many signal peptides can be identified at first sight from the primary sequences of the proteins. However, not all sequences follow the above rules unambiguously; and in these cases, prediction programs can be applied to identify signal peptides and to discriminate them from signal anchor sequences. SignalP,[6–9] for example, is widely used and was updated regularly to its current 4.1 version. It is striking that although signal peptides have a conserved architecture and secondary structure, they do not show any sequence homology. Even in the case of closely related proteins, signal peptide sequences vary substantially in length and sequence. Taking the basic functions of signal peptides into account, this sequence variability may lead to different ER targeting/insertion efficiencies and signal peptides may be consequently involved in regulating protein expression.

In contrast to signal peptides, signal anchor sequences are not cleaved-off and are inherent parts of the mature proteins. They do not have a cleavage site, and the hydrophobic region is substantially longer.[10] In the case of polytopic proteins like GPCRs, usually the TM1 exerts the signal anchor function although the more C-terminal TM's may have a similar potential when TM1 is deleted.[11]

3. SIGNAL PEPTIDE FUNCTIONS DURING THE EARLY SECRETORY PATHWAY

The functions of signal peptides during early protein biogenesis are very well studied for secretory proteins, which are translocated across the ER membrane.[12] The mechanisms for integral membrane proteins, which are integrated into the ER membrane, can be considered as variations of this process (see below). In the case of secretory proteins, signal peptides are recognized shortly after their synthesis at cytosolic ribosomes by the signal recognition particle (SRP), an RNA protein complex (Fig. 2). The formation of a complex among SRP, ribosome, and nascent chain (SRP-RNC complex) causes a delay of cytosolic translation (elongation arrest)[13] and mediates the movement of the built complex to the SRP receptor (SR) at the ER membrane. The SR mediates association of the SRP-RNC complex with the translocon and binding of the ribosome to its cytosolic site.[14,15] The main component of the translocon is the protein-conducting Sec61 channel, which is composed of three subunits: Sec61α (protein-conducting channel), Sec61β, and Sec61γ.[16] Disassembly of the SRP-RNC complex is enabled by a GTP-dependent interaction between the SRP and the SR. Thereafter, the signal peptides and the adjacent N-tail sequences enter the cytosolic side of Sec61 in a hairpin conformation. In these early steps of translocation, the Sec61-associated translocating chain associating membrane protein may be involved in signal peptide recognition.[17–20] Binding of the signal peptide destabilizes the closed conformation of the channel and mediates transition to the open state (translocon gating). Translocon gating is assisted by the translocon-associated protein (TRAP)[21] and the chaperone BiP, which also provides molecular

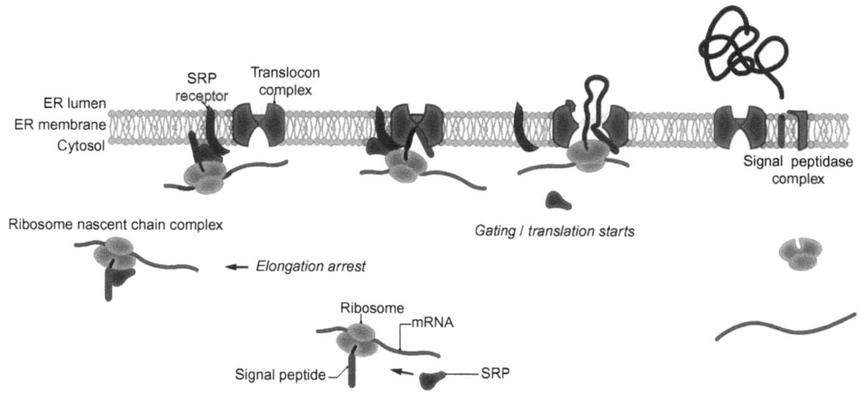

Figure 2 ER targeting/translocation of secretory proteins. The signal peptide (red) of the nascent chain associates with the SRP shortly after its synthesis at a cytosolic ribosome. Translation stops and the nascent chain/SRP/ribosome complex is targeted to the SRP receptor of the translocon complex of the ER membrane. The ribosomal tunnel exit gets in touch with the translocon and the signal peptide and the adjacent N-terminal region engage with the protein-conducting Sec61 channel in a hairpin conformation. After Sec61 gating, translation resumes and the nascent chain is translocated cotranslationally through the protein-conducting channel. The signal peptide leaves Sec61 at its lateral gate and the protein is released from the membrane at the luminal side by signal peptide cleavage through the signal peptidase complex. (See the color plate.)

ratchet functions (driving force) during translocation.[22–26] Nascent chains are translocated cotranslationally through Sec61 into the ER lumen while the signal peptide leaves the protein-conducting channel at its lateral gate between transmembrane helix 2–3 and 7–8.[27] N-glycosylation of the nascent protein is mediated by the translocon-associated oligosaccharyltransferase complex.[28] TRAP[29] and the ribosome-associated membrane protein 4[30,31] could also be involved. The signal peptide is finally cleaved-off by the signal peptidase complex of the ER membrane and the protein is liberated on the ER luminal side.[32] In mammalian cells, such a cotranslational translocation is the predominant mechanism. A recently described SRP-independent posttranslational mechanism seems to be restricted to smaller proteins (fewer than 100 amino acid residues).[33,34]

4. SIGNAL PEPTIDE FUNCTIONS OF GPCRs DURING THE ER INSERTION PROCESS

As mentioned above, the insertion mechanism of membrane proteins and thus that of GPCRs can be considered as a variation of the translocation mechanism of secretory proteins. Differences are due to the fact that membrane proteins have a defined orientation, which is established early at the translocon by charge differences along the transmembrane domains.[35–37] Recent data suggest that the TRAP complex may also be necessary for topogenesis, at least in the case of some proteins.[38] Membrane proteins are not translocated through Sec61. Instead, they leave the protein-conducting channel at its lateral gate to escape into the ER membrane. The crystal structure of the archaebacterial Sec61 ortholog SecY from *Methanocaldococcus jannaschii*[39] helped to study these processes in great detail.[40–45]

In contrast to secretory proteins, membrane proteins with an extracellular N-tail such as GPCRs contain either signal peptides or signal anchor sequences raising the question why only a subgroup of proteins requires additional cleavable signal peptides. By comparing the ER targeting mechanisms of GPCRs possessing signal peptides or signal anchor sequences, hypotheses could be derived.

If GPCRs possess signal peptides (Fig. 3), the initial steps are identical to that of secretory proteins. Translation starts at cytosolic ribosomes and is arrested by SRP binding once the signal peptides emerge from the ribosomal exit tunnel. The RNS complexes are subsequently targeted to the ER membrane, the nascent chains engage with the translocon, and following

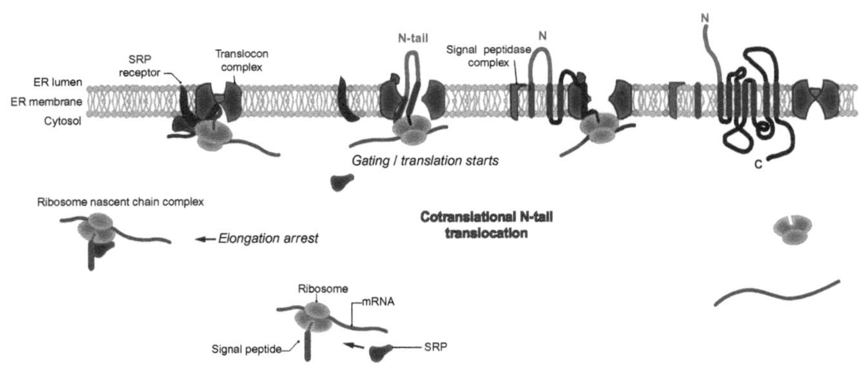

Figure 3 ER targeting/insertion of GPCRs possessing a signal peptide. The individual ER targeting/insertion steps of GPCRs containing a signal peptide represent a variation of the processes outlined for secretory proteins (see Fig. 2). As signal peptides are located at the extreme N-termini of the proteins, they prevent N-tail synthesis at cytosolic ribosomes by SRP binding and elongation arrest. Note that the N-tail can be translocated cotranslationally through Sec61 in this case following restart of translocation. Regarding membrane proteins, only the later extracellular domains are translocated into the ER lumen and the transmembrane domains leave Sec61 at its lateral gate, thereby anchoring the protein in the bilayer. The intracellular domains remain at the cytosolic side. (See the color plate.)

translocon gating, translation resumes. The N-tails are translocated through the protein-conducting channel and the newly synthesized transmembrane domains leave the channel laterally, while the extracellular domains are translocated into the ER lumen. Finally, the signal peptides are cleaved-off. Note that if signal peptides were present, the N-tails of the proteins could be translocated cotranslationally through Sec61 by ER-bound ribosomes like the secretory proteins described above. Because signal peptides are located at the extreme N-termini, they prevent N-tail synthesis at cytosolic ribosomes by SRP binding and elongation arrest (with the exception of the segment of the N-tails adjacent to the signal peptides, which are buried in the ribosome).

If GPCRs do not possess signal peptides, SRP binding and elongation arrest are mediated by signal anchor sequences and these domains are located C-terminally of the N-tails (Fig. 4). The N-tails are consequently synthesized at cytosolic ribosomes rather than at ER-bound ribosomes. They are exposed to the cytoplasm until the signal anchor sequences appear and must be translocated posttranslationally through Sec61.

Taking the considerations above into account, it was reasonable to speculate that cleavable signal peptides are necessary for those GPCRs, which are unable to translocate their N-tails posttranslationally either because they are very long or contain domains, which are rapidly folded or cannot be kept in

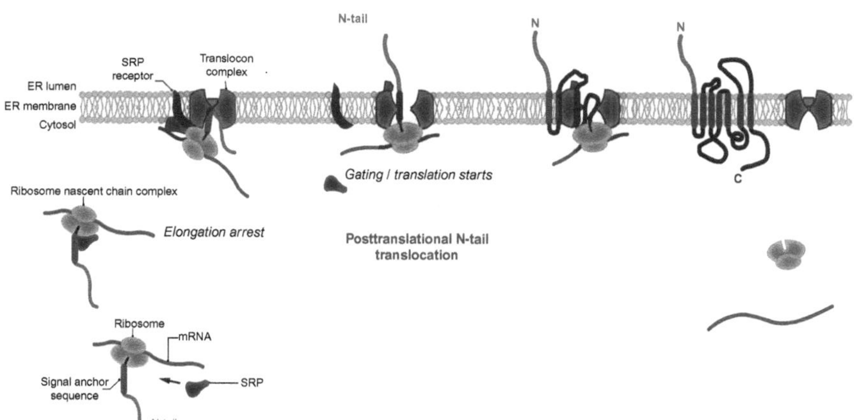

Figure 4 ER targeting/insertion of GPCRs possessing a signal anchor sequence. If GPCRs do not possess signal peptides, SRP binding and elongation arrest are mediated by signal anchor sequences. Note, that these domains are located C-terminally of the N-tails. The N-tails are consequently synthesized at cytosolic ribosomes until the signal anchor sequences appear and initiate the SRP-mediated elongation arrest. As a consequence, N-tails must be translocated posttranslationally through Sec61. (See the color plate.)

an unfolded state by chaperones.[46,47] On the other hand, signal peptides may also help to establish specific receptor expression levels at the cell surface by regulating ER targeting/insertion.

Statistical studies using prediction programs revealed that signal peptides are not distributed equally throughout the GPCR subfamilies. They are enriched in GPCRs possessing long N-tails.[2,46] In the latter work, signal peptides were found predominantly in the family 1c (glycoprotein hormone receptors; classification according to Ref. 48), family 2 (secretin receptor group), and family 3 (metabotropic glutamate receptor group). In the very large GPCR families 1a (rhodopsin family) and 1b (peptide receptors), signal peptides are rare. Putative signal peptides thus seem to occur in GPCR families where the N-tails contribute to the ligand-binding domains. Such N-tails are prone to form stably folded domains and this may consequently necessitate signal peptides. For the human endothelin B receptor (ET_BR), it was indeed shown that its signal peptide is a requirement for N-tail translocation at the ER membrane.[46] Here, a domain in the N-tail necessitates the signal peptide, most likely because it folds rapidly in the cytosol.[46] The human cannabinoid receptor 1 has a very long N-tail, which can also not be translocated efficiently.[47] Fusion of a signal peptide strongly facilitates this process,[47] consistent with the view that GPCRs, which have difficulties in N-tail translocation, benefit from signal peptides. The crucial role of the signal peptide for expression of the human vasoactive intestinal peptide and pituitary adenylyl cyclase activating peptide receptor 1[49] points in the same direction. Taken together, the available data show that signal peptides facilitate N-tail translocation, at least for some GPCRs.

However, not all GPCRs possessing signal peptides follow this pattern. One example is the rat corticotropin-releasing factor receptor type 1 (CRF_1R), where the signal peptide was not an absolute requirement for N-tail translocation and its deletion led to a fully functional receptor where TM1 fulfilled signal anchor sequence functions.[50] However, plasma membrane expression of the signal peptide deletion mutant of the CRF_1R was strongly reduced, suggesting that its signal peptide allows higher receptor densities at the cell surface.[50] At least for the CRF_1R, such a regulation may be of physiological significance (see below). In the examples mentioned above, signal peptide deletion mutants were nevertheless integrated into the ER membrane because the TM1 of the proteins could compensate for the loss of the signal peptides and function as signal anchor sequences. Recent work, however, demonstrated that this may not be true for all GPCRs: deletion of the signal peptide of the glucagon-like peptide-1 receptor prevented receptor synthesis completely, suggesting that the transmembrane domains of this receptor cannot function as signal anchor sequences.[51] The resulting non ER-bound, cytosolic forms may be rapidly subjected to proteolysis. It should be noted, however, that another interpretation of these results is possible. It was recently shown for the protease-activated receptor 1 (PAR1) that the sequence encoding the signal peptide stabilizes the mRNA of the receptor, most likely by stem loop formation.[52] An mRNA stabilizing effect of the sequence encoding the signal peptide of the glucagon-like peptide-1 receptor could also explain a complete biosynthesis defect if this sequences was removed.

Signal peptides play not only an important role for ER targeting/insertion of GPCRs, they are also involved in translocon gating. In the case of secretory proteins, it was shown that signal peptides control gating of Sec61 in a substrate-specific manner.[53] Whereas only minor differences were observed for the targeting functions of the different signal peptides, they exhibited substantially different translocon gating efficiencies. Moreover, they matched functionally to the domains located C-terminally of the signal peptides.[53] In other words, gating efficiency is not determined by the signal peptides alone, but by the combination of signal peptides and adjacent sequences. Consistent results were obtained for the signal peptides of GPCRs. For the endothelin B receptor (ET_BR), it was shown that the signal peptide alone is unable to open Sec61.[54] Efficient gating was only observed if the N-tail sequence of the ET_BR following the signal peptide was present (residues Glu^{28}-Trp^{54}), indicating that the signal peptide and the respective N-tail sequence form a functional unit.[54] Such a cooperative function is explicable considering that signal peptide and adjacent N-tail sequence enter Sec61 in a hairpin conformation and could consequently interact.

It is important to note that these results have consequences for the attempts to overexpress GPCRs. Here, fusion of "good" signal peptides is frequently used to increase GPCR expression, in particular for those GPCRs containing only signal anchor sequences (e.g., Refs.55–62). If signal peptides and adjacent sequences, however, form a functional unit, overexpression of GPCRs may not be achieved in all cases, in particular not in those where the signal peptides do not match the N-tail counterparts.

5. POST-ER FUNCTIONS OF GPCR SIGNAL PEPTIDES

It was long thought that the significance of signal peptides of GPCRs is limited to the ER targeting/insertion process discussed above. However, recent data showed that these sequences may have important post-ER functions, at least for some GPCRs. Additional properties may arise if the signal peptides were not cleaved-off and consequently become part of the mature receptors as so-called pseudo signal peptides. On the other hand, the cleaved and released signal peptides may also have functions.

The presence of a pseudo signal peptide was originally described for the corticotropin-releasing factor (CRF) receptors. Two subtypes of CRF receptors are known, the CRF_1R (already mentioned above) and the CRF_2R.[63,64] The CRF_1R is expressed predominantly in the central nervous system (CNS). In the anterior pituitary, it plays a central role in the regulation of the hypothalamic–pituitary–adrenal stress axis in mammals.[65] The CRF_2R has three splice variants, namely, the $CRF_{2(a)}R$, $CRF_{2(b)}R$, and $CRF_{2(c)}R$. The CRF_2R is not only expressed in the CNS but also in the periphery. It was reported to contribute in regulating the feeding behavior, stress recovery, and may also be involved in modulating anxiety-related behavior.[66,67] Both CRF receptor subtypes couple to the Gs/adenylyl cyclase system with cAMP as a second messenger. In the case of the CRF_1R, promiscuous coupling was described, involving proteins of the Gi, Go, and Gq families (e.g., Refs. 68–70).

According to the signal peptide prediction program SignalP,[6–9] both the CRF_1R and the $CRF_{2(a)}R$ possess N-terminal signal peptides (Fig. 5). Whereas the signal peptide of the CRF_1R is indeed cleaved,[50] that of the $CRF_{2(a)}R$ remains at the mature receptor and was described as the first pseudo signal peptide in the GPCR family.[71] Replacement of residue N13 of the pseudo signal peptide by hydrophobic or positively charged residues converts the sequence into a fully functional and cleaved signal peptide, demonstrating that conventional signal peptide functions are inhibited by a single amino acid residue.[71] Recently, the presence of the pseudo signal peptide could be confirmed by resolving the crystal structure of the N-tail of the $CRF_{2(a)}R$.[72]

The pseudo signal peptide of the $CRF_{2(a)}R$ forms a so far unique GPCR domain, and signal peptide exchange experiments were carried out with the conventional signal peptide of the CRF_1R to analyze the functional significance of this sequence in transfected cells (summary: Fig. 6). It was shown that the presence of the pseudo signal peptide strongly decreases receptor expression at the plasma membrane due to a stronger interaction of the receptor with the calnexin/calreticulin chaperone system in the ER.[73] Most receptors are consequently retained in the early secretory pathway, and only a limited amount is delivered to the vesicular transport through the Golgi apparatus. The CRF_1R with its conventional signal peptide is instead readily expressed at the cell surface. Interaction of the $CRF_{2(a)}R$ with the

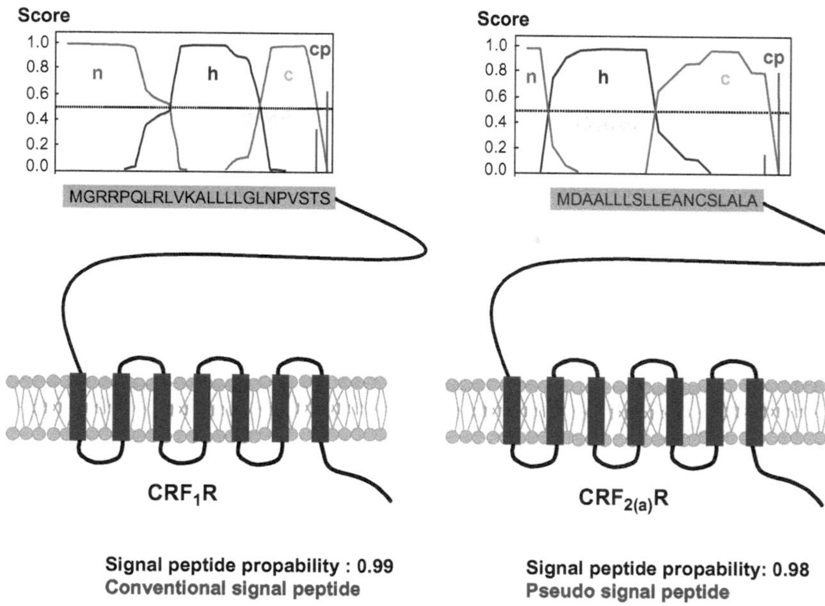

Figure 5 Depiction of the signal peptide sequences of the CRF_1R (left panel) and the $CRF_{2(a)}R$ receptor (right panel). The probabilities of the presence of n (green), h (black), and c (light blue) regions and the cleavage probabilities (cp, red) are indicated in a score ranging from 0 to 1. Note that a signal peptide was predicted in both cases. Whereas the CRF_1R indeed possesses a conventional and cleaved signal peptide, the $CRF_{2(a)}R$ contains a pseudo signal peptide. Predictions were carried out using the program SignalP.[6–9] (See the color plate.)

calnexin/calreticulin system in the ER is most likely mediated by an additional N-glycosylation site, which is present in the pseudo signal peptide. This bulky N-glycosylation could also be responsible for another effect of this domain, the prevention of $CRF_{2(a)}R$ homodimerization, which was described very recently.[74] Because of its pseudo signal peptide, the $CRF_{2(a)}R$ is expressed exclusively as a monomer, whereas the CRF_1R is able to form homodimers.[74] This has also an influence on signal transduction: the $CRF_{2(a)}R$ monomer couples exclusively to Gs, whereas the CRF_1R homodimer is able to activate both Gs and Gi.[73] The concentration response curve for cAMP formation is consequently monophasic for the $CRF_{2(a)}R$, but biphasic (bell-shaped) for the CRF_1R.[73] All properties mediated by the pseudo signal peptide could be completely transferred in signal peptide exchange experiments: a CRF_1R with a fused pseudo signal peptide is converted to a weakly expressed, monomeric receptor which couples only

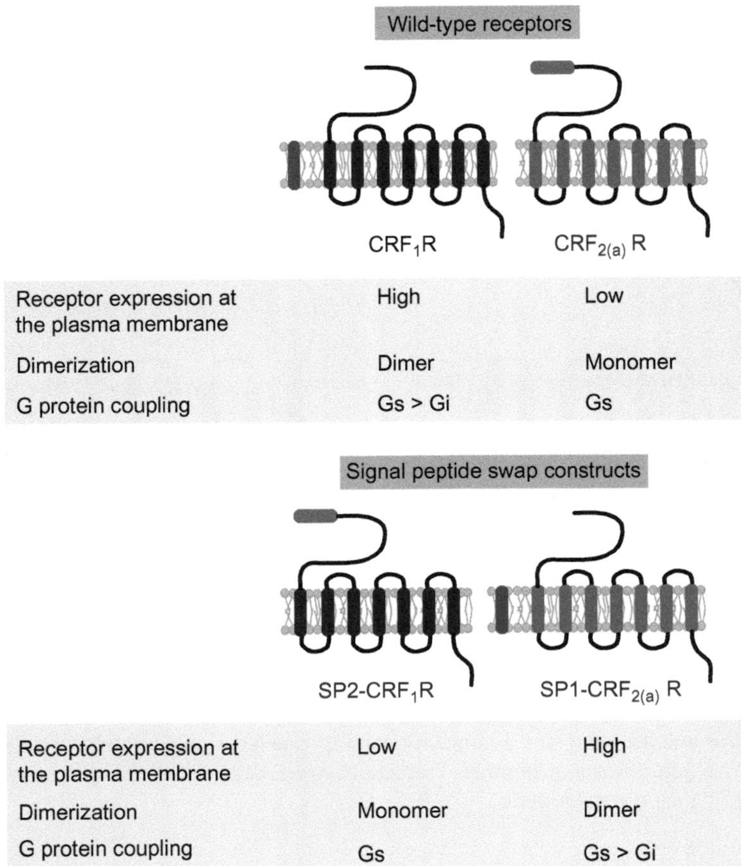

Figure 6 Functional significance of CRF receptor signal peptides: summary of the available results. Upper panel: The CRF_1R possesses a conventional cleaved signal peptide and is readily expressed in the plasma membrane. It forms dimers and couples to both Gs and Gi. The $CRF_{2(a)}R$ instead contains an uncleaved pseudo signal peptide and is expressed mainly intracellularly. It forms monomers and is only able to couple to Gs. Lower panel: All properties mediated by the pseudo signal peptide could be transferred in signal peptide swap experiments. In the case of construct SP2-CRF_1R, the signal peptide of the CRF_1R was replaced by the pseudo signal peptide of the $CRF_{2(a)}R$, whereas in construct SP1-$CRF_{2(a)}R$, the pseudo signal peptide of the $CRF_{2(a)}R$ was replaced by the signal peptide of the CRF_1R. (See the color plate.)

Gs; a $CRF_{2(a)}R$ containing the signal peptide of the CRF_1R becomes a highly expressed, dimeric receptor which couples both Gs and Gi (Fig. 6).[73,74]

It is not known whether the pseudo signal peptide of the $CRF_{2(a)}R$ has any physiological relevance. However, recently published data indicate that it may play a role in the stress response, in particular for the regulation of the cell surface expression of the $CRF_{2(a)}R$ in the serotonergic neurons of the dorsal raphe nucleus.[75] Both the CRF_1R and $CRF_{2(a)}R$ are expressed in these neurons; activation of the CRF_1R leads to a decrease of serotonin release and activation of the $CRF_{2(a)}R$ increases secretion of this neurotransmitter. In rats, the same distribution of the CRF receptors is found in these neurons *in vivo* as in transfected cells, i.e., a CRF_1R that is expressed in the plasma membrane in substantial amounts and a $CRF_{2(a)}R$ that is retained mainly in the cell's interior.[75] If the rats are exposed to social stress, distribution of the CRF receptors in the dorsal raphe neurons changes, depending on the reaction pattern of the animals. Rats coping with social stress show a mobilization of the $CRF_{2(a)}R$ to the plasma membrane, whereas the CRF_1R is removed from the cell surface by internalization; a similar CRF receptor redistribution was observed following swimming stress.[76] Rats reacting with anxiety and depression, however, preserve the original distribution.[75] Thus, one can speculate that coping with social stress is associated with releasing the trafficking restrictions mediated by the pseudo signal peptide of the $CRF_{2(a)}R$. For example, the ER environment may change under these conditions to a milieu that either favors cleavage of the pseudo signal peptide or leads to a decrease of calnexin/calreticulin binding. Other explanations, such as an increased transcription/translation of the $CRF_{2(a)}R$, or its decreased internalization is less likely, since these changes would not affect the restrictions provided by the pseudo signal peptide. Recent results at least showed that the ER environment of neurons might change substantially following psychological stress.[77,78]

It is currently not known whether the pseudo signal peptide of the $CRF_{2(a)}R$ is a unique domain within the GPCR protein family. However, the predicted signal peptide of the human α_{2C}-adrenoceptor seems to be uncleaved, too.[79] Interestingly, this sequence possesses also an N-glycosylation site, but it remains to be determined whether it forms a pseudo signal peptide similar to that of the $CRF_{2(a)}R$.

The pseudo signal peptide of the $CRF_{2(a)}R$ was studied intensively. In the case of cleaved signal peptides, it is an open question whether the peptides serve additional functions following their release from the mature GPCRs. Cleaved signal peptides are usually degraded, but in some cases, subsequent processing by an enzyme, namely, the signal peptide peptidase (SPP), was observed.[4,5] Membrane-embedded signal peptides of viral proteins, for example, have various functions during the infection cycle.[80–85] Processed signal peptides, however, may not only remain membrane-integrated to be functional. Some signal peptides are also retrotranslocated to the cytosol. Examples include the signal peptides of prolactin[86] and that of the mouse mammary tumor virus Rem protein.[87] The latter peptide is released into the cytosol by an SPP-independent mechanism and accumulates in nucleoli where it functions as a nuclear export factor for intron-containing transcripts of the virus.[87]

In the GPCR protein family, a functional significance of a cleaved signal peptide could be discussed at least in the case of PAR1. The N-terminal amino acid residues 1–41 of this receptor form a peptide called parstatin, which was originally thought to be released proteolytically from the mature receptor at the plasma membrane, following activation by the ligand thrombin. A synthetic parstatin peptide inhibits angiogenesis,[88,89] suppresses ocular neovascularization and inflammation,[90] and plays a role in cardioprotection and renal protection after ischemia and reperfusion injury.[91,92] It was recently demonstrated, however, that the PAR1 possesses a cleavable signal peptide.[52] Moreover, it was shown that the signal peptide alone could mediate the parstatin functions described above.[93] It may thus be speculated that the cleaved signal peptide of PAR1 is processed and retrotranslocated to the cytosol or to the nucleus similar to the above-mentioned viral signal peptides. It may then confer the described functions by as yet nondescribed mechanisms.

6. SIGNAL PEPTIDES OF GPCRs AS POTENTIAL DRUG TARGETS

Although signal peptides have a conserved secondary structure, they do not share any sequence homologies even between closely related proteins. For example, the mature forms of the two splice variants $CRF_{2(a)}R$ and $CRF_{2(b)}R$ are almost identical and cannot be differentiated by pharmacological means. Their signal peptides, however, are completely different in sequence and function: the $CRF_{2(a)}R$ contains the above-mentioned pseudo signal peptide, whereas the $CRF_{2(b)}R$ possesses a conventional and cleaved signal peptide (Fig. 7). In principle, signal peptides thus represent good novel drug targets. The idea is that inhibitors of signal peptide sequences may be used to block biosynthesis of specific GPCRs or other proteins.

Almost a decade ago, the first inhibitors of the translocon were published, which act in a signal sequence discriminatory manner.[94,95] Based on the fungal cyclodepsipeptide HUN-7293,[96,97] the derivative CAM741 was synthesized and shown to prevent cotranslational translocation of the

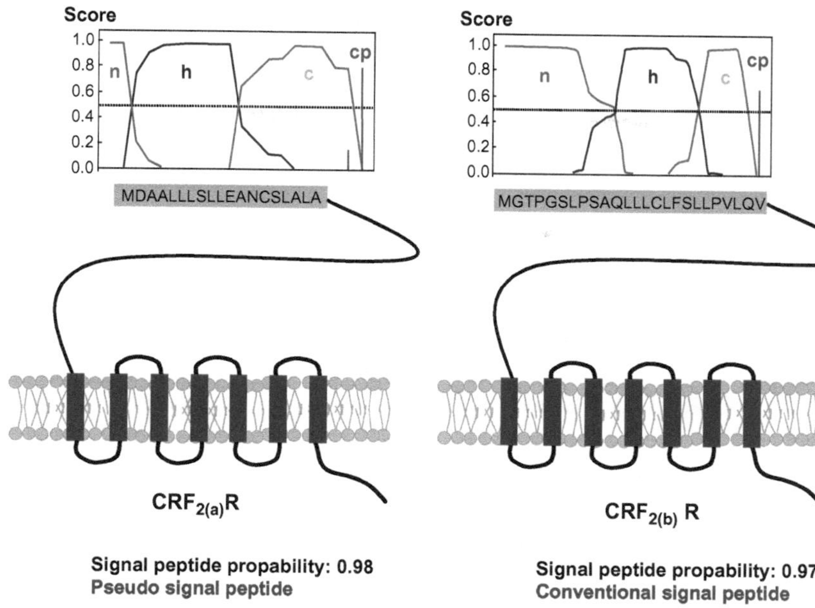

Figure 7 Depiction of the signal peptide sequences of the $CRF_{2(a)}R$ (left panel) and the $CRF_{2(b)}R$ (right panel). The probabilities of the presence of n (green), h (black), and c (light blue) regions and the cleavage probabilities (cp, red) are indicated in a score ranging from 0 to 1. Note that although the mature receptors are highly homologous, the signal peptides differ completely in sequence and function. Whereas the $CRF_{2(a)}R$ contains a pseudo signal peptide, the $CRF_{2(b)}R$ receptor possesses a conventional signal peptide similar to that of the CRF_1R. Predictions were carried out using the program SignalP.[6–9] (See the color plate.)

vascular cell adhesion molecule 1 (VCAM1) through Sec61.[94] Another substance, cotransin, represents a simplified derivative of HUN-7293 and also blocks cotranslational VCAM1 translocation.[95] It is noteworthy that all these cyclodepsipeptide inhibitors are selective rather than specific compounds meaning that they block the biosynthesis of a subset of proteins depending on properties of their signal sequences. It was shown recently that cotransin also inhibits biosynthesis of a GPCR, namely, the ET_BR with an IC_{50} value in the low micromolar range.[98] The detailed mechanism of action of these cyclodepsipeptides is not yet fully understood. It was shown that targeting of the nascent chains to the ER is not affected. However, productive interaction of the target signal sequences with Sec61 is prevented and the channel gating process is prohibited.[95] Since the cyclodepsipeptides act selectively, it was speculated that they interfere with binding of sensitive signal peptides to specific sites within Sec61.[99] A consensus sequence within the sensitive signal sequences could not be defined so far, although some critical residues were identified.[100]

The selective mechanism of action of these cyclodepsipeptides raises the question whether specific substances interfering with biosynthesis of only a single protein can be derived in the future. Such derivates may represent a novel active principle in GPCR pharmacology. In the case of cotransin, the synthesis of such derivatives may be facilitated by the established solid-phase synthesis protocol.[101,102]

7. CONCLUDING REMARKS

Recent work has shown that the idea of signal peptides playing only a role during the ER targeting/insertion processes is out of date. Pseudo signal peptides may have important post-ER functions for GPCRs and may even represent a new type of signal transduction regulation mechanism. It should be stressed that as of now, signal peptide prediction programs fail to identify pseudo signal peptides. Cleavage of predicted signal peptides should thus be analyzed experimentally for each GPCR by one of the described assays.[71] Rather neglected at the moment are the potential functions of cleaved and processed GPCR signal peptides. Here, novel types of GPCR-related regulation mechanisms could be identified in the future.

ACKNOWLEDGMENTS

We thank Janine Kirstein and Arthur Gibert for critical reading of the manuscript. Our research on the signal peptides of GPCRs has been funded by the Sonderforschungsbereich 449 and project 1116/2-1 of the Deutsche Forschungsgemeinschaft.

REFERENCES

1. Segev N, Alfonso A, Payne GS, et al. *Trafficking Inside Cells: Pathways, Mechanisms and Regulation*. Austin, TX: Landes Bioscience; 2010.
2. Wallin E, von Heijne G. Properties of N-terminal tails in G-protein coupled receptors: a statistical study. *Protein Eng*. 1995;8(7):693–698.
3. Higy M, Junne T, Spiess M. Topogenesis of membrane proteins at the endoplasmic reticulum. *Biochemistry*. 2004;43(40):12716–12722.
4. Martoglio B, Dobberstein B. Signal sequences: more than just greasy peptides. *Trends Cell Biol*. 1998;8(10):410–415.
5. Hegde RS, Bernstein HD. The surprising complexity of signal sequences. *Trends Biochem Sci*. 2006;31(10):563–571.
6. Nielsen H, Engelbrecht J, Brunak S, von Heijne G. Identification of prokaryotic and eukaryotic signal peptides and prediction of their cleavage sites. *Protein Eng*. 1997;10(1):1–6.

7. Bendtsen JD, Nielsen H, von Heijne G, Brunak S. Improved prediction of signal peptides: SignalP 3.0. *J Mol Biol.* 2004;340(4):783–795.
8. Petersen TN, Brunak S, von Heijne G, Nielsen H. SignalP 4.0: discriminating signal peptides from transmembrane regions. *Nat Methods.* 2011;8(10):785–786.
9. Emanuelsson O, Brunak S, von Heijne G, Nielsen H. Locating proteins in the cell using TargetP, SignalP and related tools. *Nat Protoc.* 2007;2(4):953–971.
10. Nilsson I, Whitley P, von Heijne G. The COOH-terminal ends of internal signal and signal-anchor sequences are positioned differently in the ER translocase. *J Cell Biol.* 1994;126(5):1127–1132.
11. Audigier Y, Friedlander M, Blobel G. Multiple topogenic sequences in bovine opsin. *Proc Natl Acad Sci USA.* 1987;84(16):5783–5787.
12. Dudek J, Pfeffer S, Lee PH, et al. Protein transport into the human endoplasmic reticulum. *J Mol Biol.* 2015;427(6):1159–1175.
13. Walter P, Blobel G. Translocation of proteins across the endoplasmic reticulum III. Signal recognition protein (SRP) causes signal sequence-dependent and site-specific arrest of chain elongation that is released by microsomal membranes. *J Cell Biol.* 1981;91(2): 557–561.
14. Walter P, Johnson AE. Signal sequence recognition and protein targeting to the endoplasmic reticulum membrane. *Annu Rev Cell Biol.* 1994;10(1):87–119.
15. Shan S-o, Walter P. Co-translational protein targeting by the signal recognition particle. *FEBS Lett.* 2005;579(4):921–926.
16. Denks K, Vogt A, Sachelaru I, Petriman N-A, Kudva R, Koch H-G. The Sec translocon mediated protein transport in prokaryotes and eukaryotes. *Mol Membr Biol.* 2014;31(2–3):58–84.
17. Gorlich D, Hartmann E, Prehn S, Rapoport TA. A protein of the endoplasmic reticulum involved early in polypeptide translocation. *Nature.* 1992;357(6373):47–52.
18. Görlich D, Rapoport TA. Protein translocation into proteoliposomes reconstituted from purified components of the endoplasmic reticulum membrane. *Cell.* 1993; 75(4):615–630.
19. Voigt S, Jungnickel B, Hartmann E, Rapoport TA. Signal sequence-dependent function of the TRAM protein during early phases of protein transport across the endoplasmic reticulum membrane. *J Cell Biol.* 1996;134(1):25–35.
20. Hegde RS, Voigt S, Rapoport TA, Lingappa VR. TRAM regulates the exposure of nascent secretory proteins to the cytosol during translocation into the endoplasmic reticulum. *Cell.* 1998;92(5):621–631.
21. Fons RD, Bogert BA, Hegde RS. Substrate-specific function of the translocon-associated protein complex during translocation across the ER membrane. *J Cell Biol.* 2003;160(4):529–539.
22. Dierks T, Volkmer J, Schlenstedt G, et al. A microsomal ATP-binding protein involved in efficient protein transport into the mammalian endoplasmic reticulum. *EMBO J.* 1996;15(24):6931–6942.
23. Schauble N, Lang S, Jung M, et al. BiP-mediated closing of the Sec61 channel limits Ca^{2+} leakage from the ER. *EMBO J.* 2012;31(15):3282–3296.
24. Tyedmers J, Lerner M, Wiedmann M, Volkmer J, Zimmermann R. Polypeptide-binding proteins mediate completion of co-translational protein translocation into the mammalian endoplasmic reticulum. *EMBO Rep.* 2003;4(5):505–510.
25. Nicchitta CV, Blobel G. Lumenal proteins of the mammalian endoplasmic reticulum are required to complete protein translocation. *Cell.* 1993;73(5):989–998.
26. Shaffer KL, Sharma A, Snapp EL, Hegde RS. Regulation of protein compartmentalization expands the diversity of protein function. *Dev Cell.* 2005;9(4):545–554.
27. Albers S-V, Szabó Z, Driessen AJM. Protein secretion in the archaea: multiple paths towards a unique cell surface. *Nat Rev Microbiol.* 2006;4(7):537–547.

28. Chavan M, Yan A, Lennarz WJ. Subunits of the translocon interact with components of the oligosaccharyl transferase complex. *J Biol Chem.* 2005;280(24):22917–22924.
29. Losfeld ME, Ng BG, Kircher M, et al. A new congenital disorder of glycosylation caused by a mutation in SSR4, the signal sequence receptor 4 protein of the TRAP complex. *Hum Mol Genet.* 2014;23(6):1602–1605.
30. Schroder K, Martoglio B, Hofmann M, et al. Control of glycosylation of MHC class II-associated invariant chain by translocon-associated RAMP4. *EMBO J.* 1999; 18(17):4804–4815.
31. Yamaguchi A, Hori O, Stern DM, Hartmann E, Ogawa S, Tohyama M. Stress-associated endoplasmic reticulum protein 1 (SERP1)/ribosome-associated membrane protein 4 (RAMP4) stabilizes membrane proteins during stress and facilitates subsequent glycosylation. *J Cell Biol.* 1999;147(6):1195–1204.
32. Evans EA, Gilmore R, Blobel G. Purification of microsomal signal peptidase as a complex. *Proc Natl Acad Sci USA.* 1986;83(3):581–585.
33. Shao S, Hegde RS. A calmodulin-dependent translocation pathway for small secretory proteins. *Cell.* 2011;147(7):1576–1588.
34. Johnson N, Vilardi F, Lang S, Leznicki P, Zimmermann R, High S. TRC40 can deliver short secretory proteins to the Sec61 translocon. *J Cell Sci.* 2012;125(Pt 15):3612–3620.
35. von Heijne G. Control of topology and mode of assembly of a polytopic membrane protein by positively charged residues. *Nature.* 1989;341(6241):456–458.
36. Hartmann E, Rapoport TA, Lodish HF. Predicting the orientation of eukaryotic membrane-spanning proteins. *Proc Natl Acad Sci U S A.* 1989;86(15):5786–5790.
37. Bogdanov M, Dowhan W, Vitrac H. Lipids and topological rules governing membrane protein assembly. *Biochim Biophys Acta.* 2014;1843(8):1475–1488.
38. Sommer N, Junne T, Kalies K-U, Spiess M, Hartmann E. TRAP assists membrane protein topogenesis at the mammalian ER membrane. *Biochim Biophys Acta.* 2013;1833(12):3104–3111.
39. Van den Berg B, Clemons WM, Collinson I, et al. X-ray structure of a protein-conducting channel. *Nature.* 2004;427(6969):36–44.
40. Hessa T, Kim H, Bihlmaier K, et al. Recognition of transmembrane helices by the endoplasmic reticulum translocon. *Nature.* 2005;433(7024):377–381.
41. Hessa T, Meindl-Beinker NM, Bernsel A, et al. Molecular code for transmembrane-helix recognition by the Sec61 translocon. *Nature.* 2007;450(7172):1026–1030.
42. Hessa T, Reithinger JH, von Heijne G, Kim H. Analysis of transmembrane helix integration in the endoplasmic reticulum in S. cerevisiae. *J Mol Biol.* 2009;386(5):1222–1228.
43. Reithinger JH, Yim C, Kim S, Lee H, Kim H. Structural and functional profiling of the lateral gate of the Sec61 translocon. *J Biol Chem.* 2014;289(22):15845–15855.
44. Gogala M, Becker T, Beatrix B, et al. Structures of the Sec61 complex engaged in nascent peptide translocation or membrane insertion. *Nature.* 2014;506(7486):107–110.
45. Demirci E, Junne T, Baday S, Bernèche S, Spiess M. Functional asymmetry within the Sec61p translocon. *Proc Natl Acad Sci USA.* 2013;110(47):18856–18861.
46. Köchl R, Alken M, Rutz C, et al. The signal peptide of the G protein-coupled human endothelin B receptor is necessary for translocation of the N-terminal tail across the endoplasmic reticulum membrane. *J Biol Chem.* 2002;277(18):16131–16138.
47. Andersson H, D'Antona AM, Kendall DA, Von Heijne G, Chin C-N. Membrane assembly of the cannabinoid receptor 1: impact of a long N-terminal tail. *Mol Pharmacol.* 2003;64(3):570–577.
48. Bockaert J, Pin JP. Molecular tinkering of G protein-coupled receptors: an evolutionary success. *EMBO J.* 1999;18(7):1723–1729.
49. Couvineau A, Rouyer-Fessard C, Laburthe M. Presence of a N-terminal signal peptide in class II G protein-coupled receptors: crucial role for expression of the human VPAC1 receptor. *Regul Pept.* 2004;123(1–3):181–185.

50. Alken M, Rutz C, Köchl R, et al. The signal peptide of the rat corticotropin-releasing factor receptor 1 promotes receptor expression but is not essential for establishing a functional receptor. *Biochem J.* 2005;390(pt 2):455–464.
51. Huang Y, Wilkinson GF, Willars GB. Role of the signal peptide in the synthesis and processing of the glucagon-like peptide-1 receptor. *Br J Pharmacol.* 2010;159(1):237–251.
52. Zampatis DE, Rutz C, Furkert J, et al. The protease-activated receptor 1 possesses a functional and cleavable signal peptide which is necessary for receptor expression. *FEBS Lett.* 2012;586(16):2351–2359.
53. Kim SJ, Mitra D, Salerno JR, Hegde RS. Signal sequences control gating of the protein translocation channel in a substrate-specific manner. *Dev Cell.* 2002;2(2):207–217.
54. Alken M, Schmidt A, Rutz C, et al. The sequence after the signal peptide of the G protein-coupled endothelin B receptor is required for efficient translocon gating at the endoplasmic reticulum membrane. *Mol Pharmacol.* 2009;75(4):801–811.
55. Guan XM, Kobilka TS, Kobilka BK. Enhancement of membrane insertion and function in a type IIIb membrane protein following introduction of a cleavable signal peptide. *J Biol Chem.* 1992;267(31):21995–21998.
56. Grisshammer R, Duckworth R, Henderson R. Expression of a rat neurotensin receptor in Escherichia coli. *Biochem J.* 1993;295(Pt 2):571–576.
57. Grünewald S, Haase W, Reiländer H, Michel H. Glycosylation, palmitoylation, and localization of the human D2S receptor in baculovirus-infected insect cells. *Biochemistry.* 1996;35(48):15149–15161.
58. Kempf J, Snook LA, Vonesch JL, Dahms TES, Pattus F, Massotte D. Expression of the human mu opioid receptor in a stable Sf9 cell line. *J Biotechnol.* 2002; 95(2):181–187.
59. Sugiyama K, Niki TP, Inokuchi K, Teranishi Y, Ueda M, Tanaka A. Heterologous expression of metabotropic glutamate receptor subtype 1 in Saccharomyces cerevisiae. *Appl Microbiol Biotechnol.* 2004;64(4):531–536.
60. Sugawara T, Ito K, Shiroishi M, et al. Fluorescence-based optimization of human bitter taste receptor expression in Saccharomyces cerevisiae. *Biochem Biophys Res Commun.* 2009;382(4):704–710.
61. Shepard BD, Natarajan N, Protzko RJ, Acres OW, Pluznick JL. A cleavable N-terminal signal peptide promotes widespread olfactory receptor surface expression in HEK293T cells. *PLoS One.* 2013;8(7):e68758.
62. Quitterer U, Pohl A, Langer A, Koller S, Abdalla S. A cleavable signal peptide enhances cell surface delivery and heterodimerization of cerulean-tagged angiotensin II AT1 and bradykinin B2 receptor. *Biochem Biophys Res Commun.* 2011;409(3):544–549.
63. Hauger RL, Risbrough V, Brauns O, Dautzenberg FM. Corticotropin releasing factor (CRF) receptor signaling in the central nervous system: new molecular targets. *CNS Neurol Disord Drug Targets.* 2006;5(4):453–479.
64. Hauger RL, Grigoriadis DE, Dallman MF, Plotsky PM, Vale WW, Dautzenberg FM. International Union of Pharmacology. XXXVI. Current status of the nomenclature for receptors for corticotropin-releasing factor and their ligands. *Pharmacol Rev.* 2003;55(1):21–26.
65. Denver RJ. Structural and functional evolution of vertebrate neuroendocrine stress systems. *Ann N Y Acad Sci.* 2009;1163:1–16.
66. Spina M, Merlo-Pich E, Chan RK, et al. Appetite-suppressing effects of urocortin, a CRF-related neuropeptide. *Science.* 1996;273(5281):1561–1564.
67. Coste SC, Kesterson RA, Heldwein KA, et al. Abnormal adaptations to stress and impaired cardiovascular function in mice lacking corticotropin-releasing hormone receptor-2. *Nat Genet.* 2000;24(4):403–409.
68. Grammatopoulos DK, Dai Y, Randeva HS, et al. A novel spliced variant of the type 1 corticotropin-releasing hormone receptor with a deletion in the seventh

transmembrane domain present in the human pregnant term myometrium and fetal membranes. *Mol Endocrinol.* 1999;13(12):2189–2202.
69. Wietfeld D, Heinrich N, Furkert J, et al. Regulation of the coupling to different G proteins of rat corticotropin-releasing factor receptor type 1 in human embryonic kidney 293 cells. *J Biol Chem.* 2004;279(37):38386–38394.
70. Gutknecht E, Van der Linden I, Van Kolen K, Verhoeven KFC, Vauquelin G, Dautzenberg FM. Molecular mechanisms of corticotropin-releasing factor receptor-induced calcium signaling. *Mol Pharmacol.* 2009;75(3):648–657.
71. Rutz C, Renner A, Alken M, et al. The corticotropin-releasing factor receptor type 2a contains an N-terminal pseudo signal peptide. *J Biol Chem.* 2006;281(34):24910–24921.
72. Pal K, Swaminathan K, Xu HE, Pioszak AA. Structural basis for hormone recognition by the human CRFR2{alpha} G protein-coupled receptor. *J Biol Chem.* 2010; 285(51):40351–40361.
73. Schulz K, Rutz C, Westendorf C, et al. The pseudo signal peptide of the corticotropin-releasing factor receptor type 2a decreases receptor expression and prevents Gi-mediated inhibition of adenylyl cyclase activity. *J Biol Chem.* 2010;285(43):32878–32887.
74. Teichmann A, Rutz C, Kreuchwig A, Krause G, Wiesner B, Schülein R. The pseudo signal peptide of the corticotropin-releasing factor receptor type 2A prevents receptor oligomerization. *J Biol Chem.* 2012;287(32):27265–27274.
75. Wood SK, Zhang X-Y, Reyes BAS, Lee CS, Van Bockstaele EJ, Valentino RJ. Cellular adaptations of dorsal raphe serotonin neurons associated with the development of active coping in response to social stress. *Biol Psychiatry.* 2013;73(11):1087–1094.
76. Waselus M, Nazzaro C, Valentino RJ, Van Bockstaele EJ. Stress-induced redistribution of corticotropin-releasing factor receptor subtypes in the dorsal raphe nucleus. *Biol Psychiatry.* 2009;66(1):76–83.
77. Pavlovsky AA, Boehning D, Li D, Zhang Y, Fan X, Green TA. Psychological stress, cocaine and natural reward each induce endoplasmic reticulum stress genes in rat brain. *Neuroscience.* 2013;246:160–169.
78. Huang G-B, Zhao T, Muna SS, et al. Effects of chronic social defeat stress on behaviour, endoplasmic reticulum proteins and choline acetyltransferase in adolescent mice. *Int J Neuropsychopharmacol.* 2013;16(7):1635–1647.
79. Jahnsen JA, Uhlen S. The predicted N-terminal signal sequence of the human alpha(2) C-adrenoceptor does not act as a functional cleavable signal peptide. *Eur J Pharmacol.* 2012;684(1–3):51–58.
80. York J, Romanowski V, Lu M, Nunberg JH. The signal peptide of the Junín arenavirus envelope glycoprotein is myristoylated and forms an essential subunit of the mature G1-G2 complex. *J Virol.* 2004;78(19):10783–10792.
81. Agnihothram SS, York J, Nunberg JH. Role of the stable signal peptide and cytoplasmic domain of G2 in regulating intracellular transport of the Junín virus envelope glycoprotein complex. *J Virol.* 2006;80(11):5189–5198.
82. York J, Nunberg JH. Role of the stable signal peptide of Junín arenavirus envelope glycoprotein in pH-dependent membrane fusion. *J Virol.* 2006;80(15):7775–7780.
83. Eichler R, Lenz O, Strecker T, Eickmann M, Klenk H-D, Garten W. Identification of Lassa virus glycoprotein signal peptide as a trans-acting maturation factor. *EMBO Rep.* 2003;4(11):1084–1088.
84. Froeschke M, Basler M, Groettrup M, Dobberstein B. Long-lived signal peptide of lymphocytic choriomeningitis virus glycoprotein pGP-C. *J Biol Chem.* 2003; 278(43):41914–41920.
85. Schrempf S, Froeschke M, Giroglou T, von Laer D, Dobberstein B. Signal peptide requirements for lymphocytic choriomeningitis virus glycoprotein C maturation and virus infectivity. *J Virol.* 2007;81(22):12515–12524.

86. Weihofen A, Lemberg MK, Ploegh HL, Bogyo M, Martoglio B. Release of signal peptide fragments into the cytosol requires cleavage in the transmembrane region by a protease activity that is specifically blocked by a novel cysteine protease inhibitor. *J Biol Chem.* 2000;275(40):30951–30956.
87. Dultz E, Hildenbeutel M, Martoglio B, Hochman J, Dobberstein B, Kapp K. The signal peptide of the mouse mammary tumor virus Rem protein is released from the endoplasmic reticulum membrane and accumulates in nucleoli. *J Biol Chem.* 2008;283(15):9966–9976.
88. Zania P, Gourni D, Aplin AC, et al. Parstatin, the cleaved peptide on proteinase-activated receptor 1 activation, is a potent inhibitor of angiogenesis. *J Pharmacol Exp Ther.* 2009;328(2):378–389.
89. Duncan MB, Kalluri R. Parstatin, a novel protease-activated receptor 1-derived inhibitor of angiogenesis. *Mol Interv.* 2009;9(4):168–170.
90. Huang H, Vasilakis P, Zhong X, et al. Parstatin suppresses ocular neovascularization and inflammation. *Invest Ophthalmol Vis Sci.* 2010;51(11):5825–5832.
91. Strande JL, Widlansky ME, Tsopanoglou NE, et al. Parstatin: a cryptic peptide involved in cardioprotection after ischaemia and reperfusion injury. *Cardiovasc Res.* 2009;83(2):325–334.
92. Diamantopoulos A, Kyriazis I, Geronatsiou K, et al. Parstatin prevents renal injury following ischemia/reperfusion and radiocontrast administration. *Am J Nephrol.* 2012;36(3):278–286.
93. Routhu KV, Tsopanoglou NE, Strande JL. Parstatin(1–26): the putative signal peptide of protease-activated receptor 1 confers potent protection from myocardial ischemia-reperfusion injury. *J Pharmacol Exp Ther.* 2010;332(3):898–905.
94. Besemer J, Harant H, Wang S, et al. Selective inhibition of cotranslational translocation of vascular cell adhesion molecule 1. *Nature.* 2005;436(7048):290–293.
95. Garrison JL, Kunkel EJ, Hegde RS, Taunton J. A substrate-specific inhibitor of protein translocation into the endoplasmic reticulum. *Nature.* 2005;436(7048):285–289.
96. Boger DL, Keim H, Oberhauser B, Schreiner EP, Foster CA. Total synthesis of HUN-7293. *J Am Chem Soc.* 1999;121(26):6197–6205.
97. Schreiner EP, Kern M, Steck A, Foster CA. Synthesis of ether analogues derived from HUN-7293 and evaluation as inhibitors of VCAM-1 expression. *Bioorg Med Chem Lett.* 2004;14(19):5003–5006.
98. Westendorf C, Schmidt A, Coin I, et al. Inhibition of biosynthesis of human endothelin B receptor by the cyclodepsipeptide cotransin. *J Biol Chem.* 2011;286(41):35588–35600.
99. Harant H, Wolff B, Schreiner EP, et al. Inhibition of vascular endothelial growth factor cotranslational translocation by the cyclopeptolide CAM741. *Mol Pharmacol.* 2007;71(6):1657–1665.
100. Harant H, Lettner N, Hofer L, Oberhauser B, de Vries JE, Lindley IJ. The translocation inhibitor CAM741 interferes with vascular cell adhesion molecule 1 signal peptide insertion at the translocon. *J Biol Chem.* 2006;281(41):30492–30502.
101. Coin I, Beyermann M, Bienert M. Solid-phase peptide synthesis: from standard procedures to the synthesis of difficult sequences. *Nat Protoc.* 2007;2(12):3247–3256.
102. Coin I, Beerbaum M, Schmieder P, Bienert M, Beyermann M. Solid-phase synthesis of a cyclodepsipeptide: cotransin. *Org Lett.* 2008;10(17):3857–3860.

CHAPTER THIRTEEN

Regulation of GPCR Anterograde Trafficking by Molecular Chaperones and Motifs

Brent Young*, Jaime Wertman†, Denis J. Dupré*,[1]
*Department of Pharmacology, Dalhousie University, Halifax, Nova Scotia, Canada
†Department of Microbiology and Immunology, Dalhousie University, Halifax, Nova Scotia, Canada
[1]Corresponding author: e-mail address: denis.dupre@dal.ca

Contents

Abstract

G protein-coupled receptors (GPCRs) make up a superfamily of integral membrane proteins that respond to a wide variety of extracellular stimuli, giving them an important role in cell function and survival. They have also proven to be valuable targets in the fight against various diseases. As such, GPCR signal regulation has received considerable attention over the last few decades. With the amplitude of signaling being determined in large part by receptor density at the plasma membrane, several endogenous mechanisms for modulating GPCR expression at the cell surface have come to light. It has been shown that cell surface expression is determined by both exocytic and endocytic processes. However, the body of knowledge surrounding GPCR trafficking from the endoplasmic reticulum to the plasma membrane, commonly known as anterograde

Progress in Molecular Biology and Translational Science, Volume 132
ISSN 1877-1173
http://dx.doi.org/10.1016/bs.pmbts.2015.02.012

trafficking, has considerable room for growth. We focus here on the current paradigms of anterograde GPCR trafficking. We will discuss the regulatory role of both the general and "nonclassical private" chaperone systems in GPCR trafficking as well as conserved motifs that serve as modulators of GPCR export from the endoplasmic reticulum and Golgi apparatus. Together, these topics summarize some of the known mechanisms by which the cell regulates anterograde GPCR trafficking.

1. INTRODUCTION

The vast majority of G protein-coupled receptor (GPCR) signaling occurs at the cell surface; thus, the number of receptors made available at the plasma membrane (PM) must be tightly regulated at the levels of export, internalization, and degradation.[1] The importance of GPCR trafficking is highlighted by the fact that some human disorders are associated with mutations in the trafficking pathway. Some forms of retinitis pigmentosa, nephrogenic diabetes, and ovarian dysgenesis are some of the disorders that are caused by abnormalities in GPCR folding and/or trafficking.[2] The endoplasmic reticulum (ER) is an organelle that plays a key role in this process. Here, proper folding, assembly, and quality control must occur before the receptor can continue along the anterograde transport pathway.[3] It is also in this organelle that GPCRs seem to associate closely with molecular chaperone proteins.

Broadly speaking, molecular chaperones are a group of proteins involved in the facilitation of the correct folding and maturation of polypeptides into their mature protein forms. GPCRs seem to require certain molecular chaperones to assist their folding and maturation, as knockdown or mutation of chaperones can impair proper expression of the GPCR at the cell surface. There are several chaperone groups that have been shown to couple with multiple GPCRs, such as the lectin-binding chaperones; however, there are also chaperones that regulate the maturation of specific GPCRs.[4] The following section aims to outline the important structures and motifs within the GPCRs that allow them to interact with chaperone proteins, and the involvement of both general and specific chaperones in GPCR maturation through the ER to the cell surface. Thus, it is the goal of this chapter to provide an overview of the mechanisms through which GPCRs transition from their synthesis through the Golgi to their expression at the cell surface.

2. GENERAL CHAPERONE SYSTEM INVOLVED IN GPCR TRAFFICKING

2.1 Maturation Within the ER

ER chaperone proteins are believed to be the most important regulators of GPCR trafficking. Indeed, they are the largest group of proteins to interact with GPCRs. Aside from their role in promoting receptor maturation via proper folding, chaperones also prevent misfolded proteins from exiting the ER.[5] Although we will discuss chaperones specific to certain GPCRs in Section 3, the body of knowledge surrounding the general chaperone system and GPCR trafficking is considerably larger.[4] Here, we will discuss the role of calnexin, calreticulin, and BiP as sensors of GPCR misfolding and modulators of GPCR trafficking.

Several studies have indicated that GPCRs interact with lectin, or carbohydrate-binding chaperones such as calnexin and calreticulin. Some examples include the angiotensin II type 1 receptor (AT1R),[6] arginine vasopressin receptor 2 (AVPR2),[7] arginine vasopressin receptor 3 (AVPR3),[8] luteinizing hormone receptor (LHR),[9] follicle-stimulating hormone receptor, thyrotropin receptor,[10] gonadotropin-releasing hormone receptor,[11] and the melanin-concentrating hormone receptor 1 (MCHR1).[12] In contrast with other studies, it has also been demonstrated that both homo- and heterodimers of the AT1R and the β_2-adrenergic receptor (β_2-AR) couple with calreticulin, but not calnexin during the maturation process.[13] The lectin chaperones play an important role in the quality control process of the GPCR life cycle. Oligosaccharyltransferases perform N-glycosylation of GPCRs, adding sugar chains to asparagine residues to the N-terminus of newly synthesized GPCRs. Following this, glucosides I and II cleave the two terminal glucose molecules from the immature GPCR, leaving a single glucose. Calnexin and calreticulin bind to these monoglycosylated proteins in the ER and prevent their aggregation or degradation.[14] Next, glucosidase II can remove the terminal glucose molecule from the polypeptide, allowing the properly folded GPCR to exit from the ER. If the GPCR is still improperly folded, it will be reglycosylated and will reenter the quality control cycle. The addition of carbohydrate molecules to asparagine residues in the N-terminal region and extracellular loops of the receptor was originally thought to be a requisite for the interaction between GPCRs and carbohydrate-binding chaperones. However, it has since been shown that N-linked glycosylation is not required to establish an interaction between calnexin and the AT1R.[6] Studies of the human δ-opioid receptor (DOR) have shown that this receptor undergoes quality control through both a calnexin-mediated glycosylation-dependent mechanism and a less stringent glycosylation-independent mechanism. Researchers suggested that this glycosylation-independent mechanism was an alternative quality control pathway and that its induction may be regulated by cell conditions.[15]

BiP (i.e., GRP 78) is an ER-resident chaperone that belongs to the heat-shock protein (Hsp)70 family. BiP is known to contain two domains, the peptide binding domain at the C-terminal and an ATPase domain at the amino-terminal.[16] With the help of Hsp40 cofactors, BiP binds the exposed hydrophobic regions of unfolded proteins and facilitates translocation of the newly synthesized peptide into the ER lumen.[13,17] It also mediates the reverse translocation of misfolded proteins to the ER-associated degradation (ERAD) pathway.[16] It has been shown that both mutant and normal wild-type (WT) GPCRs are degraded via the ERAD pathway.[18–20] BiP has been shown to interact with the AT1R,[6] rhodopsin,[21] the thyrotropin receptor,[10] and the LHR.[9] Interestingly, mutant forms of GPCRs seem to have an enhanced and/or prolonged interaction with both the carbohydrate-binding chaperones and BiP variants, likely due to their inability to move swiftly through the quality control process.[7,12] An example is that the adenosine A2A receptor can be precipitated out of the cytosol of cells with various components of the Hsp relay machinery, suggesting that these proteins are recruited to this receptor throughout its folding and maturation.[22]

2.2 Trafficking Beyond the ER

Once the GPCRs have left the ER, they move to the ER–Golgi intermediate compartment (ERGIC), which is a dynamic organelle in the trafficking of newly synthesized proteins. Trafficking within the ERGIC can move in the anterograde or retrograde direction, bringing the cargo to the *cis*-Golgi, the receiving face of the Golgi apparatus, or back to the ER for further folding and quality control.[23] Retrograde transport can also direct misfolded proteins back to the ERAD system, or to lysosomes for degradation.[24] Most of the trafficking within the ERGIC and the Golgi apparatus is dependent on the GPCR within the vesicle, along with Rab and Arf GTPases and the COPI coat proteins that surround the vesicles.[23]

Rab GTPases constitute a large family of small GTPases that exist in a GTP-bound active conformation and a GDP-bound inactive conformation. In general, Rab proteins are associated with membrane compartments of the cell, where they regulate trafficking along actin and tubulin networks. Rab1 seems to function in the trafficking of vesicles from the ER to the ERGIC or the ERGIC to the *cis*-Golgi, whereas Rab2 isoforms are implicated in the movement of vesicles toward the ER.[25,26] However, studies suggest that GPCRs may vary in their dependence on Rabs for proper trafficking; while overexpression of dominant-negative Rab1a and small-interfering RNA (siRNA) knockdown of Rab1b impairs forward trafficking of the AT1R and the β_2-AR, it does not affect the trafficking of α_{2B}-adrenergic receptor (α_{2B}-AR).[27] There is also interesting evidence to show that Rab dependence may be cell type dependent. For example, Rab1 is required for the ER to the plasma membrane transit of the cystic fibrosis transmembrane conductance regulator in HeLa and HEK293 cells, but not in BHK and CHO cells.[28]

As previously mentioned in this review, much less is known about GPCR trafficking within the Golgi apparatus, but it has been suggested that glycosylation status of the proteins is what regulates trafficking in this part of the cell. There is evidence that there are different glycosyltransferases at each cisternae, or flattened membrane, of the Golgi apparatus.[29] Studies examining δ-opioid receptor (DOR) in HEK293 cells suggested that O-glycosylation and N-linked oligosaccharide processing occur in distinct compartments of the Golgi.[30] In general, the cargo arrives at the Golgi apparatus via fusion with the *cis*-membrane.[31] It is known that Rab6 and Rab8 are two of the main factors regulating transport within the Golgi apparatus.[32] Expression of a guanine nucleotide binding defective (N125I) Rab6 protein was able to inhibit surface expression of the rhodopsin receptor in *Drosophila melanogaster*.[33] siRNA-mediated knockdown of Rab2 attenuated α_{2B}-AR-mediated ERK1/2 activation and β_2-AR-mediated cAMP production, while knockdown of Rab6 diminished signaling of β_2-AR, but not α_{2B}-AR, supporting the notion of GPCR-specific anterograde trafficking requirements in human cells.[34] Though there are conflicting theories concerning the exit of GPCRs from the Golgi apparatus, the substitution of adjacent tyrosine and serine amino acids within the N-terminus of the α_{2B}-AR resulted in the receptor being retained within the Golgi apparatus, suggesting the possibility of a Golgi apparatus–exit sequence in the N-terminus of this GPCR.[5,35]

3. RECEPTOR-SPECIFIC PROTEINS INVOLVED IN ANTEROGRADE TRAFFICKING

Thus far, this chapter has outlined the role of several molecular chaperones common to several GPCRs. However, it is also important to note that several GPCRs require specific chaperone proteins for their normal expression and, thus, function. While this section is not intended to provide a complete list of the chaperones and accessory proteins that interact with GPCRs, it serves to demonstrate that various GPCRs require the assistance of specialized "nonclassical private chaperones" for cell surface expression.[36]

3.1 DRiP78

A 78-kDa dopamine receptor-interacting protein (DRiP78) has been identified as an ER-resident chaperone.[37] This chaperone consists of two transmembrane spans, a dilysine ER-retention signal and a luminal zinc-finger domain that is essential for receptor binding.[37] As its name suggests, it interacts with the dopamine D1 receptor (D1R), and mutagenesis has revealed that this interaction is mediated by a D1R C-terminal $F(X)_3F(X)_3F$ export motif.[37] While overexpression of DRiP78 leads to ER retention and reduced cell surface expression of the D1R, sequestration of DRiP78 also slows export of the receptor. A similar pattern was observed with the M2 muscarinic receptor[37] and the adenosine A1 receptor.[38] Overexpression of DRiP78 was also shown to increase expression of the AT1R at the cell surface.[39] It has been speculated that aside from receptor binding, DRiP78 also facilitates the presentation of a C-terminal receptor sorting signal because only endogenous levels of DRiP78 promote proper levels of receptor expression at the cell surface.[37]

DRiP78 also interacts with C–X–C chemokine receptor type 4 (CXCR4) and C–C chemokine receptor type 5 (CCR5).[40] While DRiP78 was shown to interact with CCR5 and CXCR4 in their homodimeric forms, it does not interact with the CCR5/CXCR4 heterodimer.[40] Overexpression of DRiP78 was shown to decrease localization of CCR5 homodimer to the plasma membrane.[40] This suggests that receptor oligomerization has an important role in modulating the interaction between receptors and the trafficking machinery.

DRiP78 was also shown to interact with the Gγ subunit of the heterotrimeric G protein in the ER.[41] Here, it prevents degradation of the subunit prior to assembly with the Gβ subunit.[41] Given the fact that both Gβγ and the receptor interact with DRiP78, and DRiP78 expression is limited to the ER, it is suggested that DRiP78 mediates assembly of GPCRs and their respective heterotrimeric G proteins in the ER.[14]

3.2 Homer Proteins

Homer proteins possess an N-terminal enabled/VASP homology domain[42] that binds the proline-rich motif found in the C-terminus of group I metabotropic glutamate receptors (mGluRs) 1 and 5.[43] Through this interaction, Homer proteins modulate the expression and localization of mGluR1a and mGluR5.[43] Given the prominent role of group I mGluRs in the excitatory neurotransmission and neuronal plasticity,[44] Homer proteins have an important role in the function and development of the nervous system. Interestingly, various subtypes of the Homer protein have been shown to have different effects on mGluR trafficking. Where expression of Homer1a increases expression of mGluR1a at the cell surface,[45] expression of Homer1b promotes ER retention and, thus, decreases cell surface expression of the same receptor.[46] When Homer1a is coexpressed with mGluR5, there is no change in expression of the receptor at the cell surface.[45] However, when Homer1b is coexpressed with the same receptor, it leads to ER retention and a decrease in mGluR5 expression at the plasma membrane.[46] Interestingly, Homer1b-induced ER retention of mGluR5 can be reversed by expression of Homer1a, suggesting that the two Homer subtypes may compete for binding to mGluR5.[47]

3.3 Receptor Activity-Modifying Proteins

A class of single transmembrane spanning accessory proteins known as receptor activity-modifying proteins (RAMPs) have been shown to interact with the calcitonin and calcitonin-like receptors.[48] Aside from influencing the pharmacology of these receptors, RAMPs have also been shown to be essential for receptor transport to the cell surface. RAMP-mediated terminal glycosylation of the calcitonin and calcitonin-like receptors is a key regulator in this phenomenon.[48] This was also observed with the calcium-sensing receptor.[49] While the calcitonin and calcitonin-like receptors belong to the class B family of GPCRs,[50,51] the calcium-sensing receptor belongs to the class C family.[52] Therefore, it is speculated that RAMPs may have a more wide-ranging role in GPCR trafficking than previously thought.

3.4 NinaA

The neither inactivation nor afterpotential (*nina*) genes encode a family of cyclosporin binding proteins that are known as cyclophilins. A more prominent member of the *nina* group, *ninaA*, has been shown to encode a protein with a key role in rhodopsin trafficking.[53] The ninaA protein consists of an N-terminal signal sequence, a cyclophilin domain, a transmembrane domain, and a short cytoplasmic C-terminal tail.[54,55] As an integral membrane protein,[56] ninaA was shown to colocalize with *Drosophila* rhodopsin Rh1 in the ER and secretory vesicles.[53,57] Truncation of the ninaA transmembrane domain was shown to result in rhodopsin Rh1 accumulation in the ER.[53] Taken together, these data indicate that ninaA has a key role in export trafficking among specific rhodopsin receptors, and its function is dependent on ER membrane association.

3.5 Melanocortin2 Receptor Accessory Protein

The melanocortin2 receptor accessory protein (MRAP) is a small integral membrane protein that forms antiparallel homodimers.[58] In the presence of MRAP, the melanocortin2 receptor is glycosylated and expressed at the cell surface; however, in its absence, the receptor is retained within the ER.[58,59] Accordingly, melanocortin2 receptor expression levels are lower in the absence of MRAP.[58] Presumably, receptor expression levels are lower because improper processing of the receptor would have it targeted for degradation via the ER quality control machinery.[30]

Interestingly, MRAP was also the first GPCR accessory protein to be implicated in human disease.[59] While 25% of familial glucocorticoid deficiencies are related to mutations in the adrenocorticotropin receptor,[60–63] 20% are associated with mutations in MRAP.[64–66] Not only does this emphasize the importance of proper MC2 trafficking in human health, but it also emphasizes the role of MRAP in this phenomenon.

4. CONSERVED MOTIFS INVOLVED IN GPCR TRAFFICKING

4.1 C-Terminal Motifs

ER export serves as an important checkpoint for the proper folding and processing of GPCRs. For example, export of the DOR from the ER has been identified as the rate-limiting step in receptor maturation and cell surface expression.[30] Several studies have implicated the COPII vesicular coat as a gatekeeper in receptor trafficking, as ER export seems to be mediated solely by ER-derived COPII-coated vesicles.[67] These findings are supported by the existence of protein binding sites within the COPII machinery.[68] Specifically, the Sar1–Sec23/24 complex has been shown to bind ER export signals found within the C-terminus of several membrane spanning cargo proteins.[69]

When looking at GPCRs, the membrane-proximal region of the C-terminus appears to be a key determinant of ER export. Truncation of the membrane-proximal region leads to disruption in normal receptor trafficking with several GPCRs. This has been observed with rhodopsin,[70] the AVPR2,[71] the adenosine A1 receptor,[72] the α_{2B}-AR, the AT1R,[73] the MCHR1,[74] and the LHR.[75] This indicates that, by acting as a marker for proper translation and processing, the membrane-proximal region of the C-terminus serves as a modulator of GPCR trafficking.

Several conserved motifs have been identified as regulators of GPCR export from the ER. Indeed, a dileucine motif coupled with an upstream glutamate (E) residue [$E(X)_3LL$] in the C-terminus of the AVPR2 has been identified as a requisite for cell surface expression of the receptor. Mutation of the glutamate and/or leucine residues leads to a reduction in cell surface expression and receptor retention within the ER.[76]

While certain membrane-proximal C-terminal motifs have only been identified in a handful of receptors, mutation of these particular motifs has been shown to have a profound effect on the trafficking and signaling of these receptors. With the D1R, a triple phenylalanine motif [$F(X)_3F(X)_3F$] is essential for cell surface expression, and receptor trafficking to the cell surface is completely abolished when these three phenylalanine residues are mutated.[37] Likely as a result of being retained within the ER, these mutant D1Rs are also deficient in cAMP signaling.[37] With AVPR3, a dileucine surrounded by an upstream phenylalanine and asparagine and a downstream leucine [$FN(X)_2LL(X)_3L$] has been identified as an ER exit signal. Accordingly, this motif is required for cell surface localization of the receptor, and mutation of any one of these residues is enough to perturb cell surface expression.[8]

Several GPCRs have been shown to contain an ER export motif that consists of a phenylalanine and a dileucine spaced by six residues [$F(X)_6LL$].[35,77] Progressive truncation and alanine-scanning mutagenesis studies have indicated that this motif is required for export of the α_{2B}-AR, AT1R,[77] β_2-AR, and α_{1B}-AR.[35] The mechanism by which this motif regulates β_2-AR trafficking has since been attributed to Rab1 binding.[78]

It was shown that β_2-AR mutants that lack this motif are unable to bind Rab1 and, as a result, are unable to reach the cell surface.[78] With the α_{2B}-AR, substitution of Ile443 and Leu444 with a double phenylalanine disrupts receptor export from the ER.[77] This motif was also shown to act as a DRiP78 binding site for CCR5 and CXCR4, giving it direct implications for the trafficking of these receptors that are the coreceptors for the entry of the human immunodeficiency virus (HIV).[40] Duvernay *et al.* have also demonstrated that ER retention can be achieved by insertion or deletion of one or two amino acids between the phenylalanine and dileucine residues. This suggests that the specific spatial orientation of the phenylalanine and dileucine residues in relation to one another is essential for proper trafficking to occur.[77]

4.2 N-Terminal Motifs

When compared to the C-terminus, the GPCR N-terminal domain has received little attention with regard to its role in receptor trafficking. To date, the role of the N-terminus in anterograde trafficking remains controversial. Much like the C-terminus, however, the N-terminus has been shown to be essential for cell surface expression of certain receptors. Indeed, cleavage of the N-terminal domain disrupts cell surface expression of the α_{2B}-AR[35] and the endothelin B receptor.[79] To complicate matters, proteolytic cleavage of the α_{1D}-AR N-terminal domain increases expression of the receptor at the cell surface,[80] and elimination of the N-terminus has no effect on trafficking of the α_{1B}-AR.[81] These inconsistencies have made it difficult to assign a generalized regulatory role to the N-terminus in GPCR trafficking. This may suggest that the role of GPCR N-terminus may be receptor specific. However, there is a growing body of evidence to suggest the GPCR N-terminus modulates trafficking of several different adrenergic receptor subtypes at the Golgi.

Although it was once believed that protein export from the Golgi was constitutive, some studies have indicated that Golgi export is selective and regulated.[82,83] For example, a recent study demonstrated that the opsin E150K mutation is responsible for autosomal recessive retinitis pigmentosa. With this mutation, the receptor proceeds from the *cis*- to *medial*-Golgi compartments, but it is unable to exit from the *trans*-Golgi network.[84] While this falls short of naming a conserved N-terminal motif involved in GPCR export from the Golgi, there has been some success with the identification of an export motif specific to several α_2-ARs. An N-terminal motif consisting of a tyrosine and a serine is conserved among the membrane-proximal regions of the α_{2A}-AR, α_{2B}-AR, and α_{2C}-AR. Mutation of this motif leads to accumulation of the receptor within the Golgi.[35]

4.3 Other Motifs

In addition to motifs that promote GPCR export trafficking, there are several motifs that inhibit receptor trafficking to the cell surface. Indeed, Hermosilla *et al.* demonstrated that a fluorescently labeled AVPR2 mutant was incapable of reaching the cell surface.[85] This mutant consisted of the AVPR2 N-terminal domain, the first transmembrane domain, intracellular loop 1 (IL1), and intracellular loop 3 (IL3).[85] In addition to this, it was demonstrated that cell surface expression is comparable between the nonmutant AVPR2 and similar mutants lacking IL3. This suggested that IL3 alone inhibits cell surface expression of the AVPR2. An RXR retention motif in IL3 provides a regulatory mechanism for AVPR2 trafficking, and substitution of the arginine residues with lysine was shown to minimize intracellular retention of the fluorescently labeled receptor fragment.[85] Given the fact that the WT AVPR2 is capable of reaching the plasma membrane, it is speculated that the IL3 RXR retention signal is masked when proper folding and processing of the receptor occur. As such, this type of signal would provide an intrinsic quality control mechanism for receptor folding and assembly to ensure that only mature receptors are expressed at the plasma membrane.[5]

The first well-recognized, but now archetypal example of GPCR dimerization is an excellent example of the importance of receptor oligomerization in GPCR trafficking. The functional metabotropic γ-aminobutyric acid ($GABA_B$) receptor was found to encompass two subunits, $GABA_{BR1}$ and $GABA_{BR2}$.[86,87] When expressed alone, the $GABA_{BR1}$ subunit does not reach the plasma membrane, due to a carboxy-terminus ER-retention motif. In contrast, when the $GABA_{BR2}$ subunit is expressed individually, the subunit can reach the plasma membrane, but is unable to activate downstream effector pathways.[88,89] It was discovered that the ER-retention motif on $GABA_{BR1}$ is masked by heterodimerization of this subunit with $GABA_{BR2}$, allowing the effector-activating $GABA_{BR1}$ to be properly expressed at the plasma membrane.[86,87,90] This particular example was considered a unique circumstance, until images obtained via atomic force microscopy of rhodopsin homodimers in native membranes started to convince disbelievers.[91]

5. CONCLUSIONS

The importance of the strict control over GPCR maturation and expression is highlighted by the fact that approximately 30–40% of modern pharmaceuticals target this group of proteins.[92] Since the majority of agonist-induced GPCR signaling begins at the plasma membrane, researchers have identified the importance of understanding the regulatory mechanisms surrounding the maturation and trafficking of GPCRs from their synthesis to the cell surface. As we hope to have demonstrated, numerous studies have explored the role that molecular chaperones play in the anterograde trafficking of GPCRs. Both generalized molecular chaperones, common to many GPCRs, and receptor-specific chaperones have important roles in supporting the proper folding and maturation of these receptors, eventually allowing their expression at the plasma membrane. A plethora of techniques, including protein overexpression, knockdown, and mutation, have allowed researchers to examine these processes in detail. Lastly, these studies have also shown that receptors themselves may act as chaperone proteins, likely by one partner masking a retention motif on the other partner. Taken together, this chapter has emphasized the significant role that molecular chaperones have on GPCR maturation and expression, highlighting the importance of the continued investigation into their mechanism of action.

ACKNOWLEDGMENTS

B.Y. would like to acknowledge the Dalhousie Medical Research Foundation for funding via the Adopt-a-Researcher Program as well as Indspire for funding through the Health Careers Award program. J.W. acknowledges generous funding from the Killam Predoctoral Fellowship, the Dalhousie University President's Award, and the Scotia Scholar Award from the Nova Scotia Health Research Foundation. This work was supported by the Natural Sciences and Engineering Research Council of Canada to D.J.D. (NSERC Grant RGPIN-355310-2013).

REFERENCES

1. Duvernay MT, Filipeanu CM, Wu G. The regulatory mechanisms of export trafficking of G protein-coupled receptors. *Cell Signal.* 2005;17(12):1457–1465.
2. Conn PM, Ulloa-Aguirre A, Ito J, Janovick JA. G protein-coupled receptor trafficking in health and disease: lessons learned to prepare for therapeutic mutant rescue in vivo. *Pharmacol Rev.* 2007;59(3):225–250.
3. Ellgaard L, Helenius A. Quality control in the endoplasmic reticulum. *Nat Rev Mol Cell Biol.* 2003;4(3):181–191.
4. Kleizen B, Braakman I. Protein folding and quality control in the endoplasmic reticulum. *Curr Opin Cell Biol.* 2004;16(4):343–349.

5. Dong C, Filipeanu CM, Duvernay MT, Wu G. Regulation of G protein-coupled receptor export trafficking. *Biochim Biophys Acta*. 2007;1768(4):853–870.
6. Lanctot PM, Leclerc PC, Clement M, et al. Importance of N-glycosylation positioning for cell-surface expression, targeting, affinity and quality control of the human AT1 receptor. *Biochem J*. 2005;390(Pt 1):367–376.
7. Morello JP, Salahpour A, Petaja-Repo UE, et al. Association of calnexin with wild type and mutant AVPR2 that causes nephrogenic diabetes insipidus. *Biochemistry*. 2001;40(23):6766–6775.
8. Robert J, Clauser E, Petit PX, Ventura MA. A novel C-terminal motif is necessary for the export of the vasopressin V1b/V3 receptor to the plasma membrane. *J Biol Chem*. 2005;280(3):2300–2308.
9. Mizrachi D, Segaloff DL. Intracellularly located misfolded glycoprotein hormone receptors associate with different chaperone proteins than their cognate wild-type receptors. *Mol Endocrinol*. 2004;18(7):1768–1777.
10. Siffroi-Fernandez S, Giraud A, Lanet J, Franc JL. Association of the thyrotropin receptor with calnexin, calreticulin and BiP. Effects on the maturation of the receptor. *Eur J Biochem*. 2002;269(20):4930–4937.
11. Brothers SP, Janovick JA, Conn PM. Calnexin regulated gonadotropin-releasing hormone receptor plasma membrane expression. *J Mol Endocrinol*. 2006;37(3):479–488.
12. Fan J, Perry SJ, Gao Y, Schwarz DA, Maki RA. A point mutation in the human melanin concentrating hormone receptor 1 reveals an important domain for cellular trafficking. *Mol Endocrinol*. 2005;19(10):2579–2590.
13. Hammad MM, Dupre DJ. Chaperones contribute to G protein coupled receptor oligomerization, but do not participate in assembly of the G protein with the receptor signaling complex. *J Mol Signal*. 2010;5:16.
14. Dupre DJ, Hammad MM, Holland P, Wertman J. Role of chaperones in G protein coupled receptor signaling complex assembly. *Subcell Biochem*. 2012;63:23–42.
15. Lackman JJ, Markkanen PM, Hogue M, Bouvier M, Petaja-Repo UE. N-Glycan-dependent and -independent quality control of human delta opioid receptor N-terminal variants. *J Biol Chem*. 2014;289(25):17830–17842.
16. Hebert DN, Molinari M. In and out of the ER: protein folding, quality control, degradation, and related human diseases. *Physiol Rev*. 2007;87(4):1377–1408.
17. Vembar SS, Brodsky JL. One step at a time: endoplasmic reticulum-associated degradation. *Nat Rev Mol Cell Biol*. 2008;9(12):944–957.
18. Petaja-Repo UE, Hogue M, Laperriere A, Bhalla S, Walker P, Bouvier M. Newly synthesized human delta opioid receptors retained in the endoplasmic reticulum are retrotranslocated to the cytosol, deglycosylated, ubiquitinated, and degraded by the proteasome. *J Biol Chem*. 2001;276(6):4416–4423.
19. Lu M, Echeverri F, Moyer BD. Endoplasmic reticulum retention, degradation, and aggregation of olfactory G-protein coupled receptors. *Traffic*. 2003;4(6):416–433.
20. Robben JH, Knoers NV, Deen PM. Characterization of vasopressin V2 receptor mutants in nephrogenic diabetes insipidus in a polarized cell model. *Am J Physiol Renal Physiol*. 2005;289(2):F265–F272.
21. Anukanth A, Khorana HG. Structure and function in rhodopsin. Requirements of a specific structure for the intradiscal domain. *J Biol Chem*. 1994;269(31):19738–19744.
22. Bergmayr C, Thurner P, Keuerleber S, et al. Recruitment of a cytoplasmic chaperone relay by the A2A adenosine receptor. *J Biol Chem*. 2013;288(40):28831–28844.
23. Appenzeller-Herzog C, Hauri HP. The ER-Golgi intermediate compartment (ERGIC): in search of its identity and function. *J Cell Sci*. 2006;119(Pt 11):2173–2183.
24. Arvan P, Zhao X, Ramos-Castaneda J, Chang A. Secretory pathway quality control operating in Golgi, plasmalemmal, and endosomal systems. *Traffic*. 2002;3(11):771–780.

25. Tisdale EJ. Rab2 interacts directly with atypical protein kinase C (aPKC) iota/lambda and inhibits aPKCiota/lambda-dependent glyceraldehyde-3-phosphate dehydrogenase phosphorylation. *J Biol Chem.* 2003;278(52):52524–52530.
26. Moyer BD, Allan BB, Balch WE. Rab1 interaction with a GM130 effector complex regulates COPII vesicle cis–Golgi tethering. *Traffic.* 2001;2(4):268–276.
27. Wu G, Zhao G, He Y. Distinct pathways for the trafficking of angiotensin II and adrenergic receptors from the endoplasmic reticulum to the cell surface: Rab1-independent transport of a G protein-coupled receptor. *J Biol Chem.* 2003;278(47):47062–47069.
28. Yoo JS, Moyer BD, Bannykh S, Yoo HM, Riordan JR, Balch WE. Non-conventional trafficking of the cystic fibrosis transmembrane conductance regulator through the early secretory pathway. *J Biol Chem.* 2002;277(13):11401–11409.
29. Achour L, Labbe-Jullie C, Scott MG, Marullo S. An escort for GPCRs: implications for regulation of receptor density at the cell surface. *Trends Pharmacol Sci.* 2008;29(10):528–535.
30. Petaja-Repo UE, Hogue M, Laperriere A, Walker P, Bouvier M. Export from the endoplasmic reticulum represents the limiting step in the maturation and cell surface expression of the human delta opioid receptor. *J Biol Chem.* 2000;275(18):13727–13736.
31. Trucco A, Polishchuk RS, Martella O, et al. Secretory traffic triggers the formation of tubular continuities across Golgi sub-compartments. *Nat Cell Biol.* 2004;6(11): 1071–1081.
32. Jordens I, Marsman M, Kuijl C, Neefjes J. Rab proteins, connecting transport and vesicle fusion. *Traffic.* 2005;6(12):1070–1077.
33. Shetty KM, Kurada P, O'Tousa JE. Rab6 regulation of rhodopsin transport in Drosophila. *J Biol Chem.* 1998;273(32):20425–20430.
34. Dong C, Wu G. Regulation of anterograde transport of adrenergic and angiotensin II receptors by Rab2 and Rab6 GTPases. *Cell Signal.* 2007;19(11):2388–2399.
35. Dong C, Wu G. Regulation of anterograde transport of alpha2-adrenergic receptors by the N termini at multiple intracellular compartments. *J Biol Chem.* 2006;281(50): 38543–38554.
36. Millar RP, Newton CL. The year in G protein-coupled receptor research. *Mol Endocrinol.* 2010;24(1):261–274.
37. Bermak JC, Li M, Bullock C, Zhou QY. Regulation of transport of the dopamine D1 receptor by a new membrane-associated ER protein. *Nat Cell Biol.* 2001;3(5):492–498.
38. Malaga-Dieguez L, Yang Q, Bauer J, Pankevych H, Freissmuth M, Nanoff C. Pharmacochaperoning of the A1 adenosine receptor is contingent on the endoplasmic reticulum. *Mol Pharmacol.* 2010;77(6):940–952.
39. Leclerc PC, Auger-Messier M, Lanctot PM, Escher E, Leduc R, Guillemette G. A polyaromatic caveolin-binding-like motif in the cytoplasmic tail of the type 1 receptor for angiotensin II plays an important role in receptor trafficking and signaling. *Endocrinology.* 2002;143(12):4702–4710.
40. Kuang YQ, Charette N, Frazer J, et al. Dopamine receptor-interacting protein 78 acts as a molecular chaperone for CCR5 chemokine receptor signaling complex organization. *PLoS One.* 2012;7(7):e40522.
41. Dupre DJ, Robitaille M, Richer M, Ethier N, Mamarbachi AM, Hebert TE. Dopamine receptor-interacting protein 78 acts as a molecular chaperone for Ggamma subunits before assembly with Gbeta. *J Biol Chem.* 2007;282(18):13703–13715.
42. Brakeman PR, Lanahan AA, O'Brien R, et al. Homer: a protein that selectively binds metabotropic glutamate receptors. *Nature.* 1997;386(6622):284–288.
43. Tu JC, Xiao B, Yuan JP, et al. Homer binds a novel proline-rich motif and links group 1 metabotropic glutamate receptors with IP3 receptors. *Neuron.* 1998;21(4):717–726.
44. Luscher C, Huber KM. Group 1 mGluR-dependent synaptic long-term depression: mechanisms and implications for circuitry and disease. *Neuron.* 2010;65(4):445–459.

45. Ciruela F, Soloviev MM, McIlhinney RA. Co-expression of metabotropic glutamate receptor type 1alpha with homer-1a/Vesl-1S increases the cell surface expression of the receptor. *Biochem J*. 1999;341(Pt 3):795–803.
46. Roche KW, Tu JC, Petralia RS, Xiao B, Wenthold RJ, Worley PF. Homer 1b regulates the trafficking of group I metabotropic glutamate receptors. *J Biol Chem*. 1999; 274(36):25953–25957.
47. Ango F, Robbe D, Tu JC, et al. Homer-dependent cell surface expression of metabotropic glutamate receptor type 5 in neurons. *Mol Cell Neurosci*. 2002;20(2):323–329.
48. McLatchie LM, Fraser NJ, Main MJ, et al. RAMPs regulate the transport and ligand specificity of the calcitonin-receptor-like receptor. *Nature*. 1998;393(6683):333–339.
49. Bouschet T, Martin S, Henley JM. Receptor-activity-modifying proteins are required for forward trafficking of the calcium-sensing receptor to the plasma membrane. *J Cell Sci*. 2005;118(Pt 20):4709–4720.
50. Gorn AH, Lin HY, Yamin M, et al. Cloning, characterization, and expression of a human calcitonin receptor from an ovarian carcinoma cell line. *J Clin Invest*. 1992;90(5):1726–1735.
51. Muff R B, Hunziker W, Fischer JA, Born W. A human orphan calcitonin receptor-like structure. *Biochem Biophys Res Commun*. 1995;206(1):341–347.
52. Aida K, Koishi S, Tawata M, Onaya T. Molecular cloning of a putative Ca(2+)-sensing receptor cDNA from human kidney. *Biochem Biophys Res Commun*. 1995; 214(2):524–529.
53. Colley NJ, Baker EK, Stamnes MA, Zuker CS. The cyclophilin homolog ninaA is required in the secretory pathway. *Cell*. 1991;67(2):255–263.
54. Schneuwly S, Shortridge RD, Larrivee DC, Ono T, Ozaki M, Pak WL. Drosophila ninaA gene encodes an eye-specific cyclophilin (cyclosporine A binding protein). *Proc Natl Acad Sci USA*. 1989;86(14):5390–5394.
55. Shieh BH, Stamnes MA, Seavello S, Harris GL, Zuker CS. The ninaA gene required for visual transduction in Drosophila encodes a homologue of cyclosporin A-binding protein. *Nature*. 1989;338(6210):67–70.
56. Stamnes MA, Shieh BH, Chuman L, Harris GL, Zuker CS. The cyclophilin homolog ninaA is a tissue-specific integral membrane protein required for the proper synthesis of a subset of Drosophila rhodopsins. *Cell*. 1991;65(2):219–227.
57. Baker EK, Colley NJ, Zuker CS. The cyclophilin homolog NinaA functions as a chaperone, forming a stable complex in vivo with its protein target rhodopsin. *EMBO J*. 1994;13(20):4886–4895.
58. Sebag JA, Hinkle PM. Melanocortin-2 receptor accessory protein MRAP forms antiparallel homodimers. *Proc Natl Acad Sci USA*. 2007;104(51):20244–20249.
59. Metherell LA, Chapple JP, Cooray S, et al. Mutations in MRAP, encoding a new interacting partner of the ACTH receptor, cause familial glucocorticoid deficiency type 2. *Nat Genet*. 2005;37(2):166–170.
60. Clark AJ, McLoughlin L, Grossman A. Familial glucocorticoid deficiency associated with point mutation in the adrenocorticotropin receptor. *Lancet*. 1993;341(8843): 461–462.
61. Naville D, Barjhoux L, Jaillard C, et al. Demonstration by transfection studies that mutations in the adrenocorticotropin receptor gene are one cause of the hereditary syndrome of glucocorticoid deficiency. *J Clin Endocrinol Metab*. 1996;81(4):1442–1448.
62. Tsigos C, Arai K, Hung W, Chrousos GP. Hereditary isolated glucocorticoid deficiency is associated with abnormalities of the adrenocorticotropin receptor gene. *J Clin Invest*. 1993;92(5):2458–2461.
63. Weber A, Toppari J, Harvey RD, et al. Adrenocorticotropin receptor gene mutations in familial glucocorticoid deficiency: relationships with clinical features in four families. *J Clin Endocrinol Metab*. 1995;80(1):65–71.

64. Clark AJ, Metherell LA, Cheetham ME, Huebner A. Inherited ACTH insensitivity illuminates the mechanisms of ACTH action. *Trends Endocrinol Metab*. 2005;16(10):451–457.
65. Chan LF, Clark AJ, Metherell LA. Familial glucocorticoid deficiency: advances in the molecular understanding of ACTH action. *Horm Res*. 2008;69(2):75–82.
66. Cooray SN, Chan L, Metherell L, Storr H, Clark AJ. Adrenocorticotropin resistance syndromes. *Endocr Dev*. 2008;13:99–116.
67. Barlowe C. Signals for COPII-dependent export from the ER: what's the ticket out? *Trends Cell Biol*. 2003;13(6):295–300.
68. Aridor M, Weissman J, Bannykh S, Nuoffer C, Balch WE. Cargo selection by the COPII budding machinery during export from the ER. *J Cell Biol*. 1998;141(1):61–70.
69. Tang BL, Wang Y, Ong YS, Hong W. COPII and exit from the endoplasmic reticulum. *Biochim Biophys Acta*. 2005;1744(3):293–303.
70. Heymann JA, Subramaniam S. Expression, stability, and membrane integration of truncation mutants of bovine rhodopsin. *Proc Natl Acad Sci USA*. 1997;94(10):4966–4971.
71. Oksche A, Dehe M, Schulein R, Wiesner B, Rosenthal W. Folding and cell surface expression of the vasopressin V2 receptor: requirement of the intracellular C-terminus. *FEBS Lett*. 1998;424(1–2):57–62.
72. Pankevych H, Korkhov V, Freissmuth M, Nanoff C. Truncation of the A1 adenosine receptor reveals distinct roles of the membrane-proximal carboxyl terminus in receptor folding and G protein coupling. *J Biol Chem*. 2003;278(32):30283–30293.
73. Gaborik Z, Mihalik B, Jayadev S, Jagadeesh G, Catt KJ, Hunyady L. Requirement of membrane-proximal amino acids in the carboxyl-terminal tail for expression of the rat AT1a angiotensin receptor. *FEBS Lett*. 1998;428(3):147–151.
74. Tetsuka M, Saito Y, Imai K, Doi H, Maruyama K. The basic residues in the membrane-proximal C-terminal tail of the rat melanin-concentrating hormone receptor 1 are required for receptor function. *Endocrinology*. 2004;145(8):3712–3723.
75. Rodriguez MC, Xie YB, Wang H, Collison K, Segaloff DL. Effects of truncations of the cytoplasmic tail of the luteinizing hormone/chorionic gonadotropin receptor on receptor-mediated hormone internalization. *Mol Endocrinol*. 1992;6(3):327–336.
76. Schulein R, Hermosilla R, Oksche A, et al. A dileucine sequence and an upstream glutamate residue in the intracellular carboxyl terminus of the vasopressin V2 receptor are essential for cell surface transport in COS.M6 cells. *Mol Pharmacol*. 1998;54(3):525–535.
77. Duvernay MT, Zhou F, Wu G. A conserved motif for the transport of G protein-coupled receptors from the endoplasmic reticulum to the cell surface. *J Biol Chem*. 2004;279(29):30741–30750.
78. Hammad MM, Kuang YQ, Morse A, Dupre DJ. Rab1 interacts directly with the beta2-adrenergic receptor to regulate receptor anterograde trafficking. *Biol Chem*. 2012;393(6):541–546.
79. Grantcharova E, Furkert J, Reusch HP, et al. The extracellular N terminus of the endothelin B (ETB) receptor is cleaved by a metalloprotease in an agonist-dependent process. *J Biol Chem*. 2002;277(46):43933–43941.
80. Andersson H, D'Antona AM, Kendall DA, Von Heijne G, Chin CN. Membrane assembly of the cannabinoid receptor 1: impact of a long N-terminal tail. *Mol Pharmacol*. 2003;64(3):570–577.
81. Hague C, Chen Z, Pupo AS, Schulte NA, Toews ML, Minneman KP. The N terminus of the human alpha1D-adrenergic receptor prevents cell surface expression. *J Pharmacol Exp Ther*. 2004;309(1):388–397.
82. Nishimura N, Plutner H, Hahn K, Balch WE. The delta subunit of AP-3 is required for efficient transport of VSV-G from the trans-Golgi network to the cell surface. *Proc Natl Acad Sci USA*. 2002;99(10):6755–6760.

83. Stockklausner C, Klocker N. Surface expression of inward rectifier potassium channels is controlled by selective Golgi export. *J Biol Chem.* 2003;278(19):17000–17005.
84. Zhu L, Imanishi Y, Filipek S, et al. Autosomal recessive retinitis pigmentosa and E150K mutation in the opsin gene. *J Biol Chem.* 2006;281(31):22289–22298.
85. Hermosilla R, Schulein R. Sorting functions of the individual cytoplasmic domains of the G protein-coupled vasopressin V(2) receptor in Madin Darby canine kidney epithelial cells. *Mol Pharmacol.* 2001;60(5):1031–1039.
86. Jones KA, Borowsky B, Tamm JA, et al. GABA(B) receptors function as a heteromeric assembly of the subunits GABA(B)R1 and GABA(B)R2. *Nature.* 1998; 396(6712):674–679.
87. White JH, Wise A, Main MJ, et al. Heterodimerization is required for the formation of a functional GABA(B) receptor. *Nature.* 1998;396(6712):679–682.
88. Ng GY, Clark J, Coulombe N, et al. Identification of a GABAB receptor subunit, gb2, required for functional GABAB receptor activity. *J Biol Chem.* 1999; 274(12):7607–7610.
89. Robbins MJ, Calver AR, Filippov AK, et al. GABA(B2) is essential for g-protein coupling of the GABA(B) receptor heterodimer. *J Neurosci.* 2001;21(20):8043–8052.
90. Sullivan R, Chateauneuf A, Coulombe N, et al. Coexpression of full-length gamma-aminobutyric acid(B) (GABA(B)) receptors with truncated receptors and metabotropic glutamate receptor 4 supports the GABA(B) heterodimer as the functional receptor. *J Pharmacol Exp Ther.* 2000;293(2):460–467.
91. Fotiadis D, Liang Y, Filipek S, Saperstein DA, Engel A, Palczewski K. Atomic-force microscopy: rhodopsin dimers in native disc membranes. *Nature.* 2003; 421(6919):127–128.
92. Filmore D. It's a GPCR world. *Mod Drug Discov.* 2004;7(11):24–28.

INDEX

Note: Page numbers followed by "*f*" indicate figures and "*t*" indicate tables.

D

E

F

G

N

O

P

Q

R

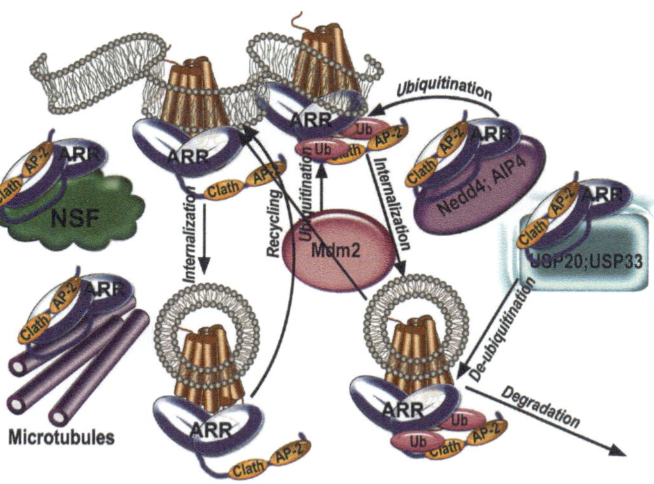

Vsevolod V. Gurevich and Eugenia V. Gurevich, Figure 1 Arrestins play many roles in GPCR trafficking. Arrestins (ARR) bind active phosphorylated GPCRs (shown as a seven-helix bundle). Receptor binding induces the release of the arrestin C-tail, which carries binding sites for clathrin (Clath) and adaptor protein-2 (AP2). The interactions of these sites with clathrin and AP2 promote receptor internalization via coated pits. Arrestins also recruit ubiquitin ligases Mdfm2, Nedd4, and AIP4 to the complex, which favors ubiquitination of both non-visual arrestins and at least some GPCRs. Arrestins also recruit certain deubiquitination enzymes (USP20 and USP33 are shown), facilitating receptor deubiquitination. The role of arrestin interactions with microtubules, centrosome, and *N*-ethylmaleimide-sensitive factor (NSF) in trafficking of GPCRs and/or other proteins remains to be elucidated.

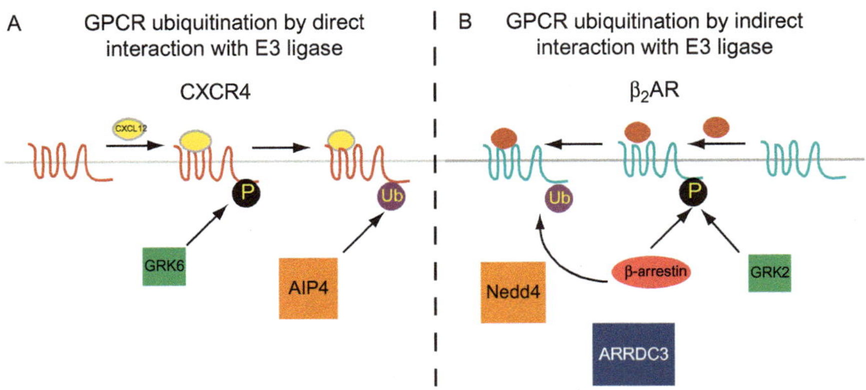

Justine E. Kennedy and Adriano Marchese, Figure 1 Mechanisms of GPCR ubiquitination at the plasma membrane by E3 ubiquitin ligases. E3 ubiquitin ligases belonging to the HECT domain Nedd4-like family can interact with GPCRs directly (A) or indirectly (B). (A) Upon binding to its cognate ligand CXCL12 (yellow oval), CXCR4 is phosphorylated by GRK6 on serine residues 324 and 325 at the plasma membrane. This is followed by recruitment of AIP4, a Nedd4-like E3 ubiquitin ligase, to the receptor, resulting in ubiquitination of nearby lysine residues. The WW domains of AIP4 mediate the interaction with CXCR4 via a noncanonical mechanism with phosphorylated serine residues 324 and 325. (B) Upon binding to the selective β-agonist isoproterenol (red oval) β_2AR is rapidly phosphorylated, likely by GRK2, leading to recruitment of β-arrestin-2. β-Arrestin-2 interacts with Nedd4, a Nedd4-like E3 ubiquitin ligase, and serves to bridge the interaction between β_2AR and Nedd4, thereby enabling ubiquitination of the receptor. The α-arrestin protein ARRDC3 has also been implicated in this event, but it may in fact act at a later step in the endocytic pathway.

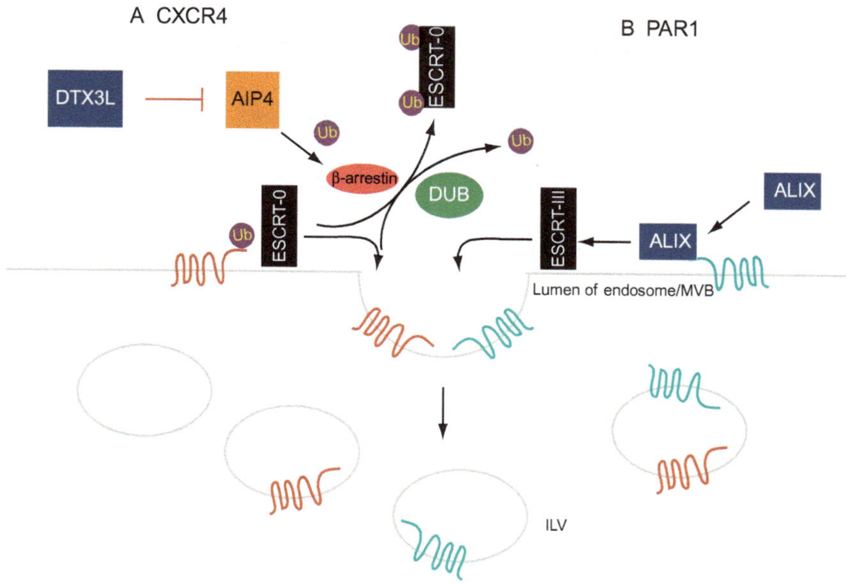

Justine E. Kennedy and Adriano Marchese, Figure 2 Mechanisms of GPCR sorting into intraluminal vesicles (ILVs) of endosomes or multivesicular bodies (MVBs). GPCRs targeted for lysosomal degradation can be found on the limiting membrane of endosomes or MVBs and on the membrane of ILVs. A mature MVB fuses with lysosomes where degradation of its contents occurs. The ILVs are formed by a complex process that involves the ESCRT (endosomal sorting complex required for transport) pathway. ESCRTs also deliver ubiquitinated GPCRs or other ubiquitinated transmembrane proteins into the invaginating regions of the limiting membrane that eventually give rise to the ILVs. (A) For GPCRs such as CXCR4, the attached ubiquitin moieties interact with ubiquitin-binding ESCRTs such as ESCRT-0. ESCRT-I, -II, and –III are also required and act in a coordinated manner to deliver CXCR4 into ILVs, but for clarity are not shown in the figure. ESCRT-0 recognizes and recruits ubiquitinated CXCR4 into the pathway. This step is regulated by ubiquitination of ESCRT-0. β-Arrestin-1 likely serves as an adaptor for ESCRT-0 ubiquitination by the Nedd4-like E3 ubiquitin ligase AIP4. Ubiquitination of ESCRT-0 is believed to inhibit its sorting activity, possibly to allow ubiquitinated CXCR4 to interact with the other ubiquitin-binding ESCRTs. AIP4 ligase activity is inhibited by the RING-finger E3 ubiquitin ligase DTX3L through a mechanism that is not completely understood, but this is thought to reduce ESCRT-0 ubiquitination and facilitate CXCR4 sorting into ILVs. (B) For some GPCRs, for which PAR1 serves as an example, delivery into ILVs does not require receptor ubiquitination and nor are the ubiquitin-binding ESCRTs (ESCRT-0, -I, and –II) required. Instead, the adaptor protein ALIX is required to target PAR1 into ILVs. ALIX interacts with a $YP(X)_3L$ motif present in the second intracellular loop of PAR1 and it also interacts with ESCRT-III, thereby delivering PAR1 into ILVs.

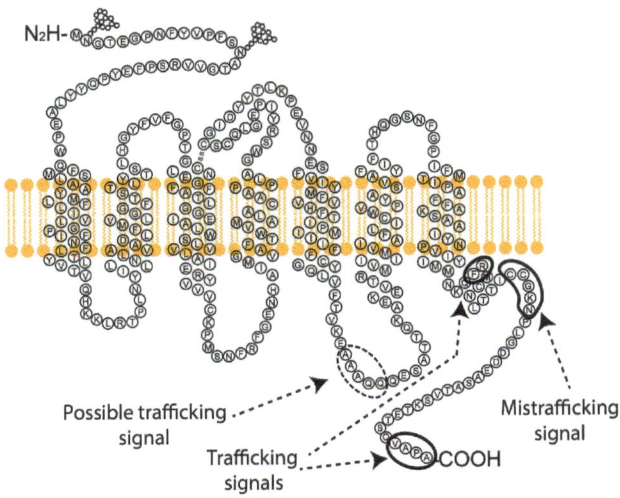

Ina Nemet *et al.*, Figure 1 Rhodopsin trafficking and mistrafficking signals. VAPA (VXPX) trafficking motif is located at the very end of the C-terminus of rhodopsin. Defects in this signal lead to rhodopsin mislocalization. Another putative trafficking signal is FR, which has not been validated *in vivo*. AAAQQ is a possible trafficking signal because it shares homology with the trafficking signal responsible for ciliary targeting of other GPCRs (SSTR3 and 5HTR6). CCGKN is a mistrafficking signal. Disruption of the VXPX motif leads to unmasking of this mistrafficking signal and causes rhodopsin mislocalization.

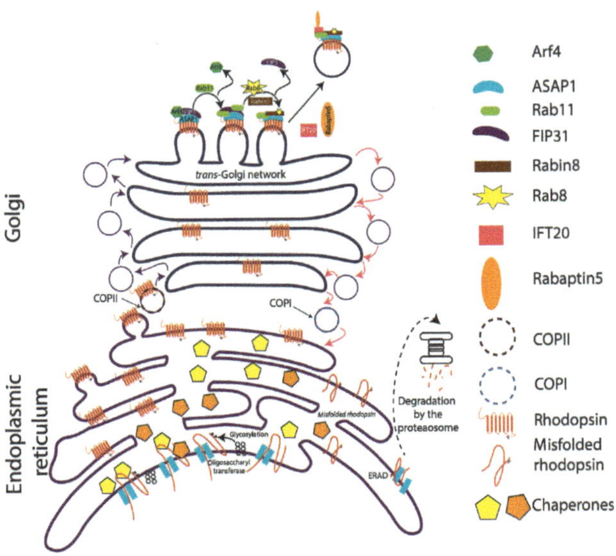

Ina Nemet *et al.*, Figure 2 Biogenesis of rhodopsin in the endoplasmic reticulum and formation of rhodopsin transport carriers in the Golgi. After folding in the ER, rhodopsin is transported to the Golgi where rhodopsin interacts with Arf4, FIP3, ASAP1, Rab8, and Rabbin8. Then, rhodopsin transport carriers bud off from the *trans*-Golgi network. IFT20 and Rabaptin5 probably bind to the carrier, but the timing of the binding is unclear.

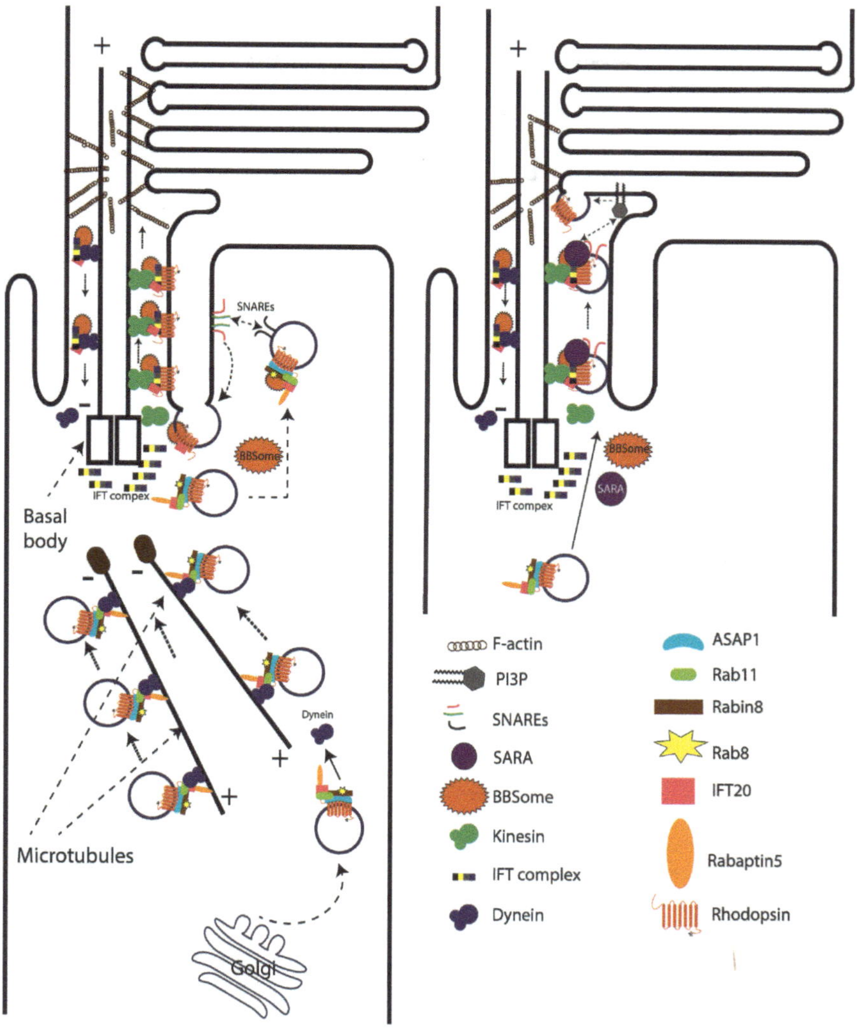

Ina Nemet *et al.*, Figure 3 Vectorial trafficking of rhodopsin from the Golgi to the distal end of the connecting cilium. Dynein and microtubules may mediate the vectorial trafficking of rhodopsin transport carriers from the Golgi to the basal body. The BBSome, IFT particles, and kinesin may mediate the subsequent trafficking of rhodopsin to the region of new disk synthesis. Two possible mechanisms for rhodopsin delivery are proposed: the first is the fusion of rhodopsin transport carriers with the IS periciliary plasma membrane and then trafficking of rhodopsin-containing rafts to the nascent disks (left panel); the second is the trafficking of rhodopsin transport carriers directly to the nascent disks (right panel).

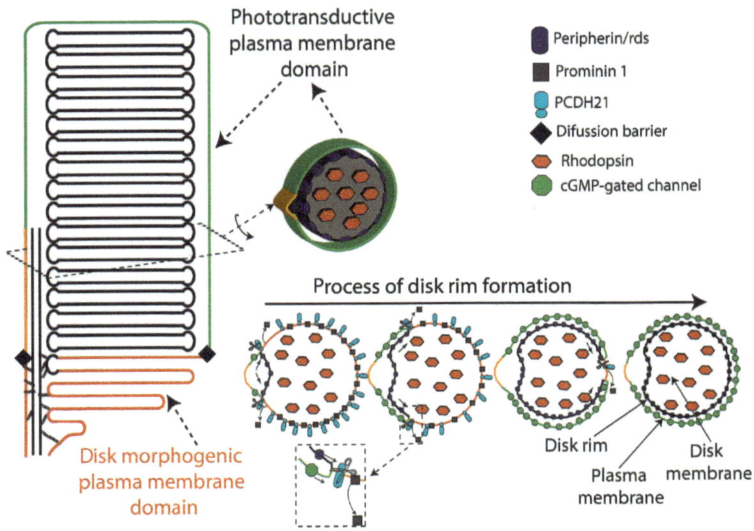

Ina Nemet *et al.*, Figure 4 Sequestration of rhodopsin into disk membranes. The OS consists of two major compartments: disk and plasma membrane. OS plasma membrane is further divided into two domains: phototransductive plasma membrane domain and disk morphogenic plasma membrane domain. After delivery of rhodopsin to the disk morphogenic plasma membrane, rhodopsin is incorporated into the mature disks in the process of disk rim formation. Rim formation includes the entrance of peripherin/rds to the evaginations, the proteolytic cleavage of PCDH21, and the removal of prominin1 by an unknown mechanism.

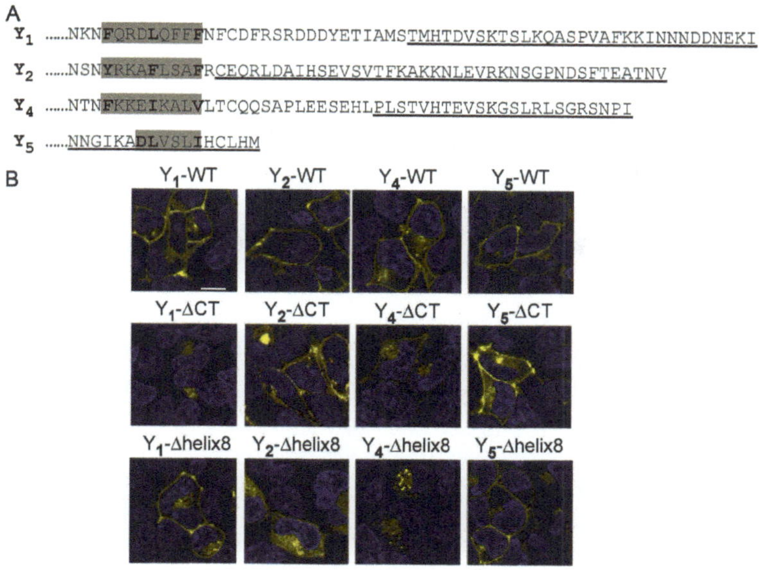

Karin Mörl and Annette G. Beck-Sickinger, Figure 1 C-terminal sequences regulate cell surface expression of Y receptors: (A) C-terminal sequences of human Y receptors. C-terminal deletion (ΔCT) is depicted by underlined sequences. Helix eight motifs are shaded in gray. (B) Representative fluorescence microscopy images of Y receptors and mutants, C-terminally fused to EYFP (yellow), transiently expressed in HEK293 cells as described in Walther *et al.*[53] Cell nuclei were visualized with Hoechst33342 (blue). ΔCT, deletion of C-terminal sequences underlined in (A); Δhelix8, deletion of helix eight motifs shaded in (A); scale bar: 10 μM.

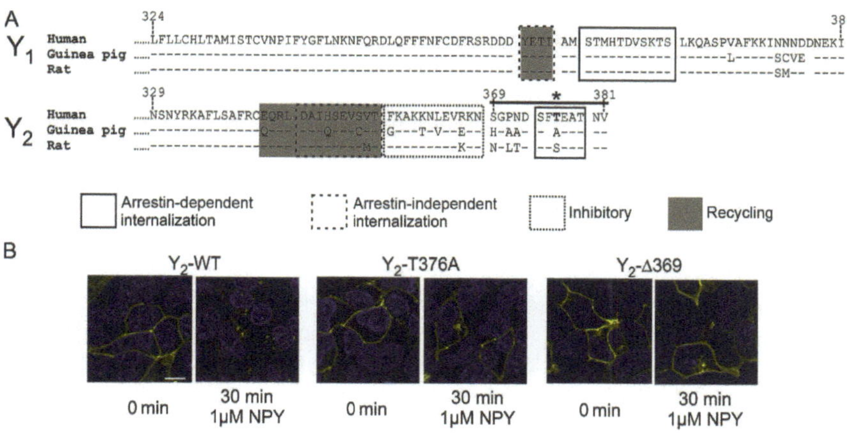

Karin Mörl and Annette G. Beck-Sickinger, Figure 2 Localization of identified trafficking motifs in C-terminal Y_1 and Y_2 receptor sequences: (A) alignment of human, guinea pig, and rat Y_1 and Y_2 sequences. Sequence motifs identified in human Y_2 receptors are labeled as depicted in the legend. The tyrosine mutated in Y_2-T376A is marked by an asterisk. The sequence truncated in Y_2-Δ369 is labeled by a bar above it. (B) Representative fluorescence microscopy images of human Y_2 receptor wild-type (Y_2-WT), the mutant Y_2-T376A, and the truncated receptor Y_2-Δ369, C-terminally fused to EYFP (yellow) and transiently expressed in HEK293 cells as described in Walther *et al.*,[80] before (0 min, left) and after 30 min (right) stimulation with 1 μM NPY. Cell nuclei were visualized with Hoechst33342 (blue). Scale bar: 10 μM.

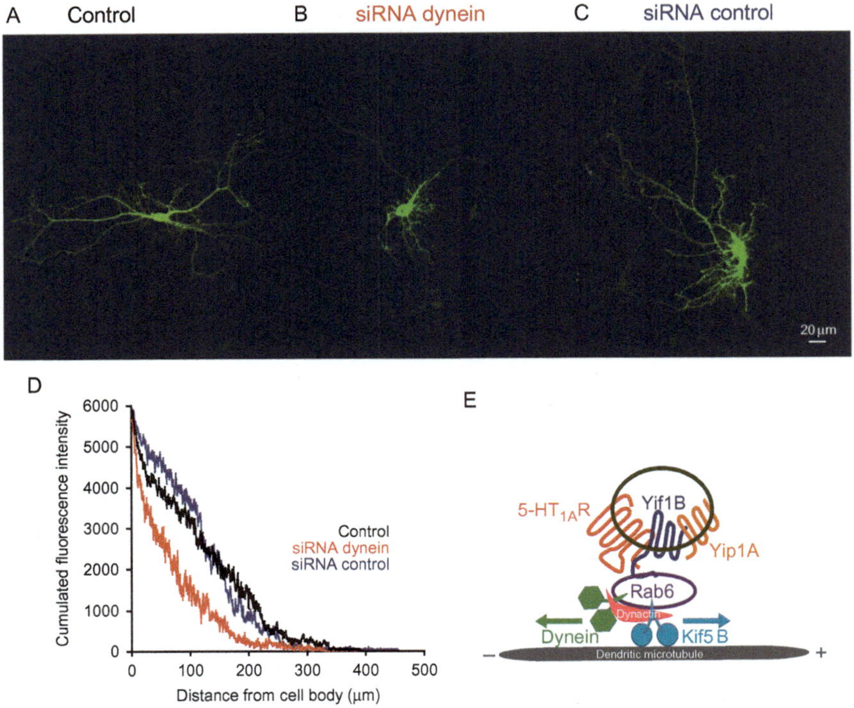

Michèle Darmon *et al.*, Figure 2 Dynein depletion disturbs 5-HT_{1A}R targeting toward the distal part of dendrites. Immunofluorescence of neurons transfected with the 5-HT_{1A}R–eGFP alone (A), cotransfected with siRNA control (B), or siRNA against endogenous dynein (C). Immunolabeling was performed with anti-GFP antibody to enhance the GFP signal on rat hippocampal neurons transfected at DIV7 and visualized 24 h after transfection. In this condition, the dendritic tree and axon were not perturbed by dynein depletion (as visualized by tubulin labeling, not shown). (D) The graph represents the cumulated fluorescence intensities of the 5-HT_{1A}R (cumulated fluorescence intensity, arbitrary unit) along the longest dendrite of monitored neurons (μm), in control condition (black) compared to siRNA control (blue) or to siRNA dynein (red). (E) Schematic representation of the Yif1B-scaffolding complex involved in 5-HT1AR trafficking toward the dendrites. Yif1B-dependent transport would involve Yif1B as the scaffold protein assembling the 5-HT1AR, Yip1A, and Rab6 in the same vesicles trafficking along the dendritic microtubules using two opposite molecular motors, the conventional kinesin Kif5B and the dynein for their bidirectional movements. The dynactin subunit p150 would enable the switch between the molecular motors and also the bidirectional transport observed in the dendrites. Rab6 might play the role of an intermediate protein for the interaction of these molecular motors with the Yif1B–scaffolding complex. Scale bar: 20 μm.

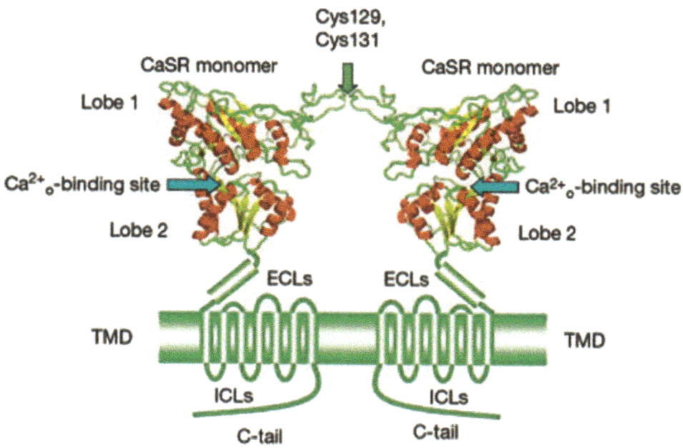

Kausik Ray, Figure 1 Schematic representation of the dimeric structure of the CaSR based on the known 3D crystal structure of the ECD of mGluR1. The dimer of the ECD is linked by two intermolecular disulfide linkages between cysteine 129-cysteine 129 and cysteine 131-cysteine 131, with each monomer assuming a Venus flytrap-like (VFT) conformation by lobe 1 and lobe 2. Each monomer is assumed to have at least one binding site for calcium in the crevice between the two lobes. Additional binding sites for calcium are likely present elsewhere in the ECD and TMD crevices. Calcimimetics, in contrast, bind to a site within the TMD, with the amino group of the drug in the linker between the two hydrophobic ends anchored to Glu837. The ICL and the long C-tail contain binding sites and determinants for binding of several intracellular proteins. ECL, extracellular loop; TMD, transmembrane domain; ICL, intracellular loop; C-tail, carboxyl-terminal tail. *Reproduced with copyright permission from Encyclopedia of Biological Chemistry; license content authors—W.A. Cramer, E. Yamashita, D. Baniulis, J. Whitelegge, S.S. Hasan.*

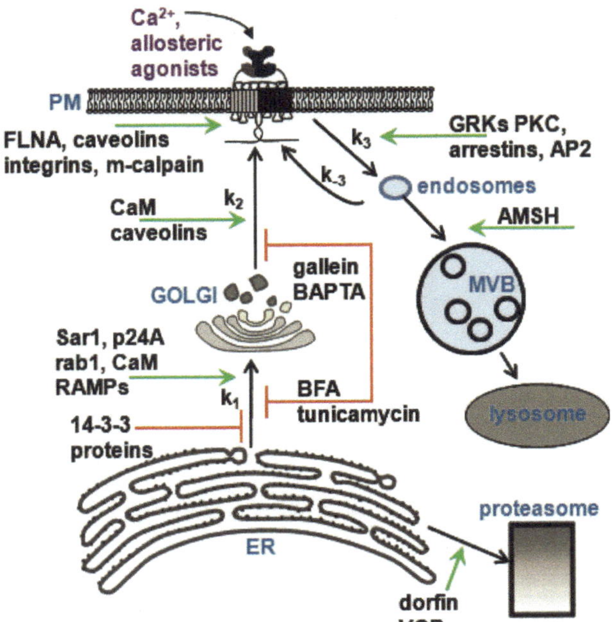

Kausik Ray, Figure 2 CaSR life cycle depicting major intracellular compartments CaSR traverses during its life cycle. Proteins and chemical compounds which impede trafficking are indicated with red lines, and protein interactions that promote trafficking of CaSR are indicated with green arrows. AP-2, adaptor protein-2; AMSH, associated molecule with the SH3 domain of STAM; BAPTA, 1,2-bis(90-aminophenoxy)ethane-*N,N,N′,N′*-tetraacetic acid; BFA, brefeldin A; CAM, calmodulin; ER, endoplasmic reticulum; FLNA, filamin A; GRKs, G-protein-coupled receptor kinases; MVB, multivesicular body; PM, plasma membrane; RAMPs, receptor-activity-modifying proteins; VCP/p97, valosin-containing protein. *Reproduced with copyright permission from Elsevier Press: Best Practice & Research Clinical Endocrinology & Metabolism 2013; 27:303–313; Author—Gerda Breitweiser; Title—The calcium sensing receptor life cycle; trafficking, cell surface expression, and degradation. Review article.*

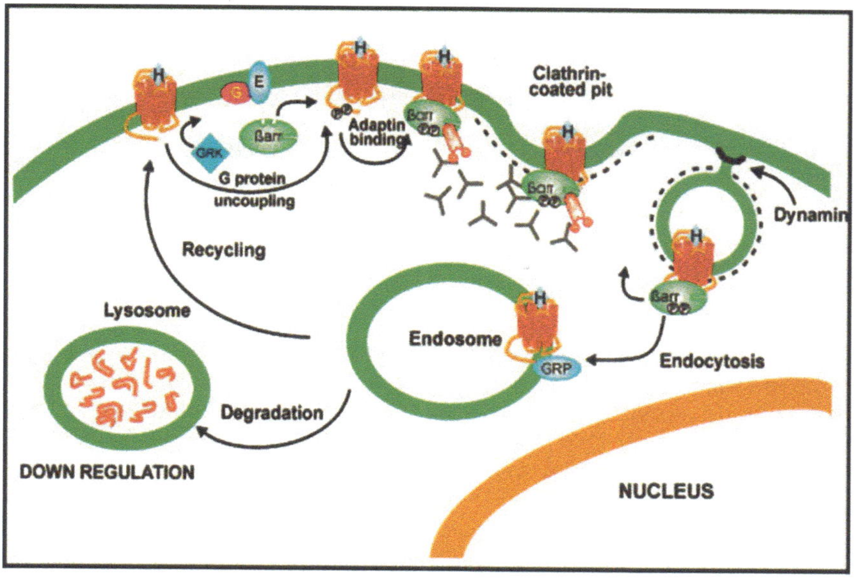

Kausik Ray, Figure 3 Schematic representation of GPCR desensitization and recycling following agonist activation. βarr, β-arrestin; E, effector enzyme; G, heterotrimeric G protein; GRK, G-protein-coupled receptor kinase; GRP, G-protein-coupled receptor phosphatase; H, hormone; P, phosphate group. See text for details. *Reproduced with copyright permission from Elsevier Press: Authors—Dhami GK, Ferguson SS. Book—Physiology and Therapeutics; Volume 111, Issue 1, July 2006, Pages 260–271; Regulation of metabotropic glutamate receptor signaling, desensitization and endocytosis.*

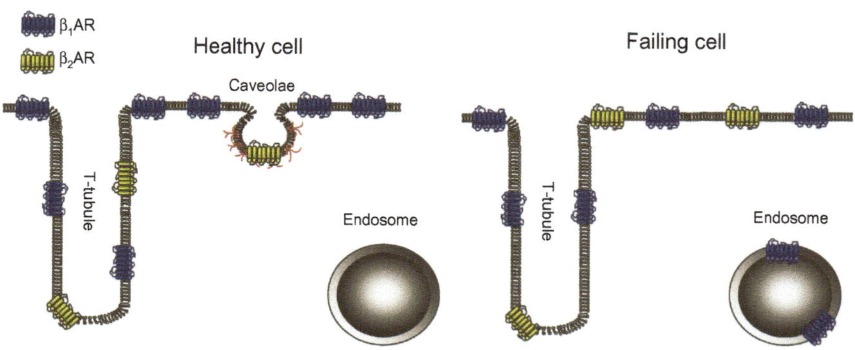

Qin Fu and Yang K. Xiang, Figure 1 Distribution of βAR in heart failure. In normal cardiac myocytes, β_1AR is distributed between the T-tubules and the crest membrane and β_2AR in T-tubules and caveolae membrane. In failing cardiac myocytes, β_1AR is relocated to endosomes, whereas β_2AR is not confined in caveolae due to reduction of numbers of caveolae.

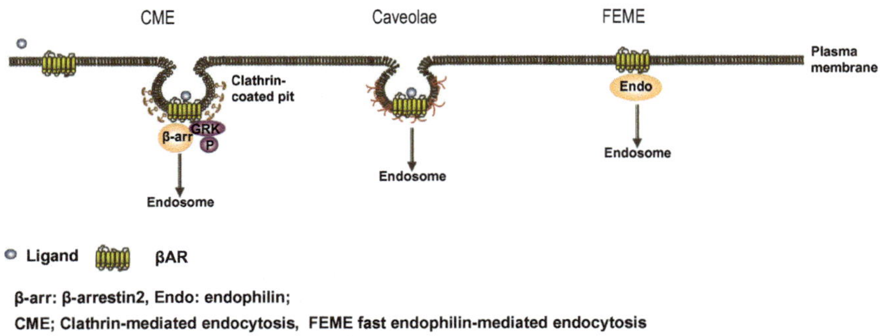

Qin Fu and Yang K. Xiang, Figure 2 βAR endocytosis machinery. Three major endocytic machineries involved βAR endocytosis. β_1AR utilizes clathrin-dependent and -independent (FEME and caveolae-mediated) endocytosis. In contrast, β_2AR undergoes endocytosis primarily through clathrin-dependent pathway.

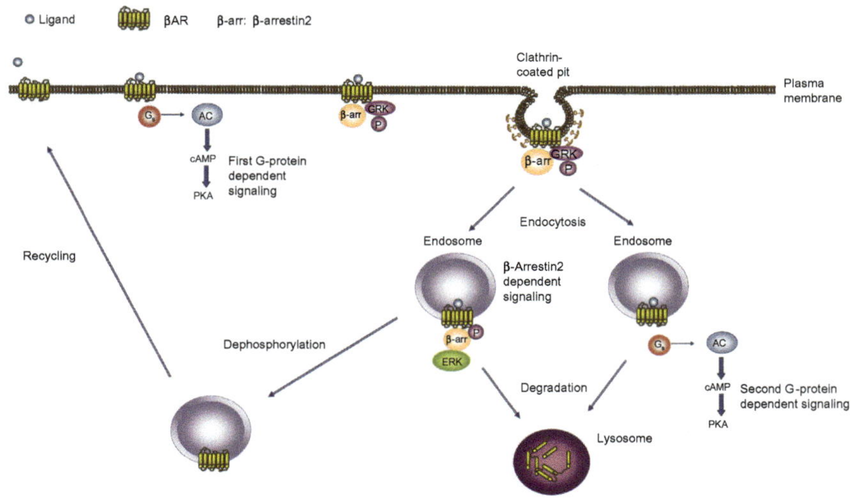

Qin Fu and Yang K. Xiang, Figure 3 Endosome βAR signaling. Agonist stimulation promotes βAR endocytosis, and the receptor undergoes subsequent sorting for either recycling or degradation. After agonist stimulation, βAR can signal to G proteins at the cell surface and at the endosome after agonist-induced endocytosis. In addition, β_1AR can signal through β-arrestin pathways.

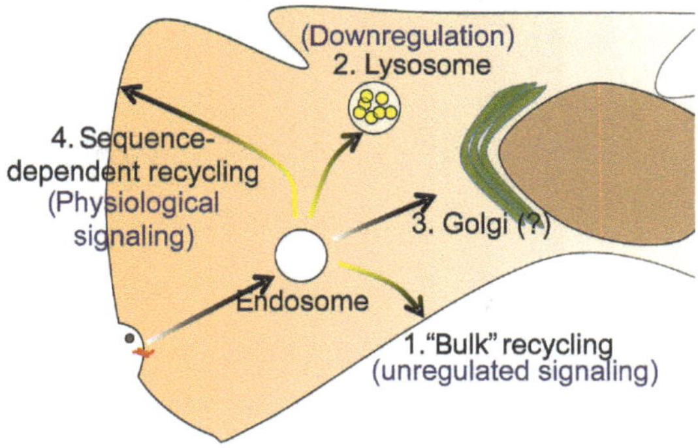

Shanna L. Bowman and Manojkumar A. Puthenveedu, Figure 1 Potential pathways for endocytosed GPCRs from the endosome. Sequence-dependent recycling is the physiological scenario for recovery of signaling, while sorting to the lysosome leads to degradation and downregulation. Bulk recycling pathway, taken normally by nutrient receptors but not GPCRs, potentially leads to unregulated signaling for GPCRs. While some proteins can travel to the Golgi apparatus, this is not established as a common route for GPCRs.

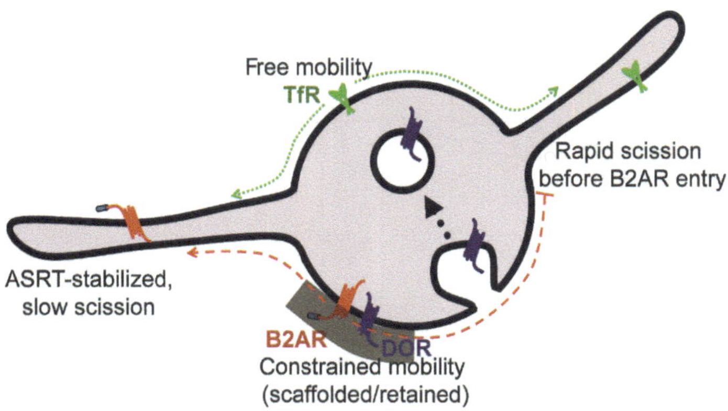

Shanna L. Bowman and Manojkumar A. Puthenveedu, Figure 2 A kinetic model for sorting in the endosome. The mobility of sequence-dependent recycling proteins or degraded proteins on the endosome is constrained by an unknown mechanism. This slow mobility reduces their entry into bulk recycling tubules which form and undergo scission rapidly. Bulk recycling proteins have unconstrained mobility on the endosome, which allows them to enter the short-lived bulk recycling tubules. ASRT domains are stabilized by the actin cytoskeleton, which provides sequence-dependent recycling proteins enough time to diffuse into these tubules. Degraded proteins are captured by the ESCRT machinery and packaged into intraluminal vesicles.

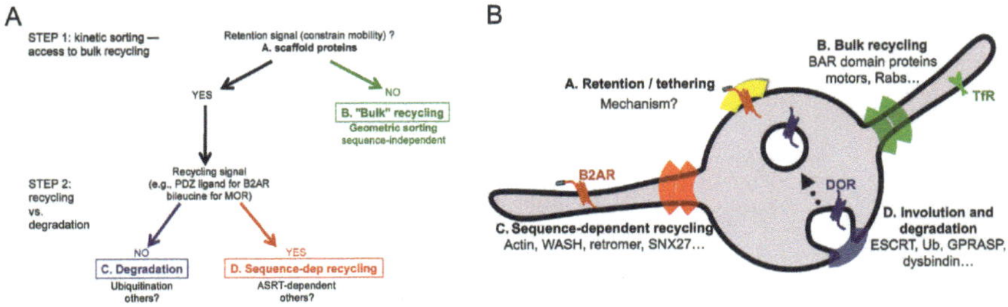

Shanna L. Bowman and Manojkumar A. Puthenveedu, Figure 3 A hierarchical model for GPCR sorting in the endosome. (A) The first step in sorting is an unknown mechanism that tethers GPCRs on the body of the endosome, excluding them from entering bulk recycling tubules. This could be a consequence of decreased mobility, as in the kinetic sorting model described in Fig. 2. Bulk recycling proteins are not tethered by this mechanism, allowing them to recycle by geometric sorting. Tethered proteins are either recycled, if they have a recycling sequence, or degraded, if they contain one of multiple signals including ubiquitination. (B) The proposed protein complexes on the endosome that mediate these hierarchical steps in sorting are shown.

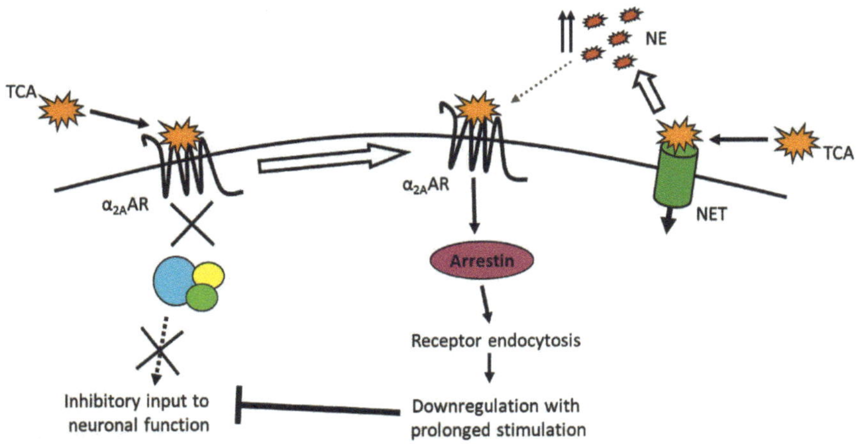

Christopher Cottingham *et al.*, Figure 2 Working model for a TCA-induced therapeutic mechanism relying on α_{2A}AR downregulation. Exposure to a TCA leads to acute α_{2A}AR endocytosis in an arrestin-dependent fashion. With chronic exposure, this endocytic response transitions into a stable downregulation of receptor expression. This direct TCA-induced mechanism is necessary given that NE levels achieved by reuptake inhibition are insufficient to drive receptor downregulation. Downregulation of α_{2A}ARs would be expected to reduce overall signaling activity by this receptor, activity which represents generally inhibitory input to neuronal function.

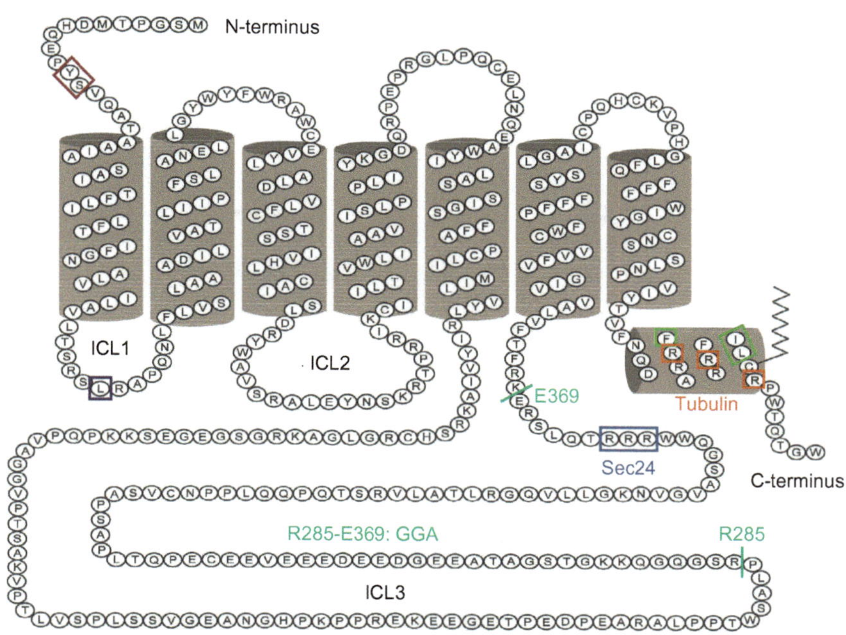

Guangyu Wu *et al.*, Figure 1 Specific motifs required for α_{2B}-AR export trafficking (see text for detail).

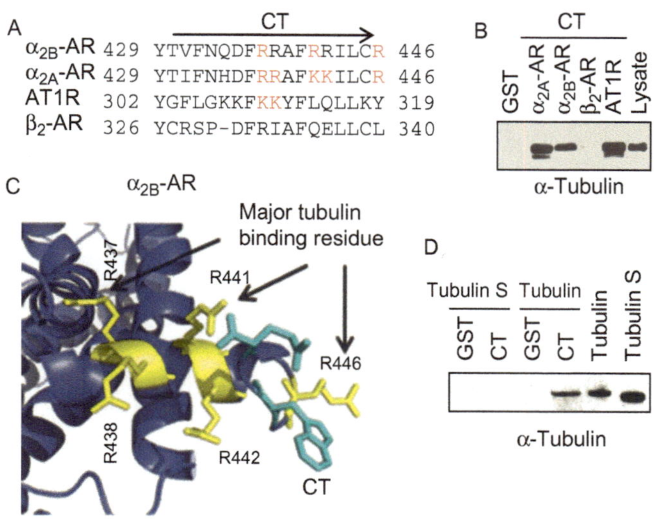

Guangyu Wu *et al.*, Figure 4 Interaction of the C-termini of GPCRs with tubulin. (A) Alignment of the C-termini (CT) of four GPCRs. (B) Interaction of the C-termini with purified tubulin measured in GST fusion protein pull-down assays. (C) Homology modeling of α_{2B}-AR based on the crystal structure of β_2-AR. (D) Tubulin S lacking the C-terminus does not bind to the α_{2B}-AR C-terminus. *The data are adapted from Refs. 63,64.*

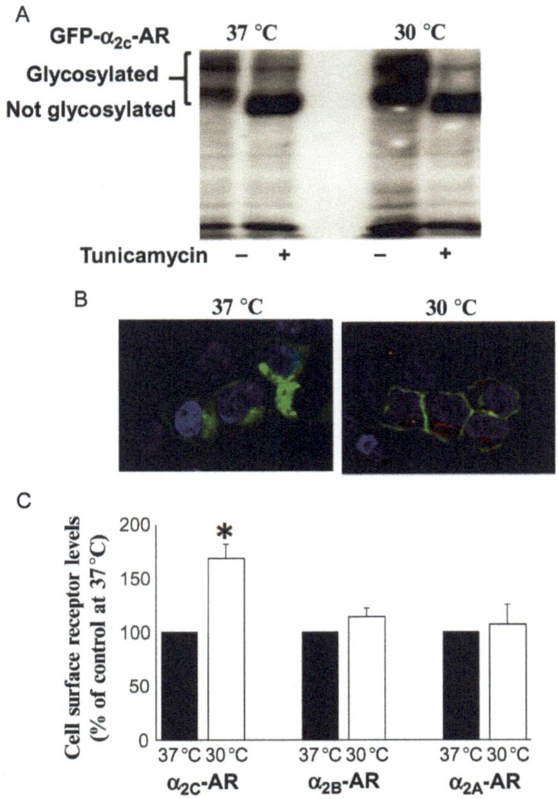

Catalin M. Filipeanu, Figure 1 Subcellular localization of α_{2C}-AR. (A) The levels of N-glycosylation of α_{2C}-AR at 37 °C (left) and 30 °C (right). HEK293T cells were transfected with 5 μg of GFP-tagged α_{2C}-AR for 48 h at 37 °C. Subsequently, the cells were serum starved and incubated at 30 °C for 18 h in the presence or absence of tunicamycin, an inhibitor of N-glycosylation (10 μg/ml). α_{2C}-AR species were detected in cell lysates by Western blot using a specific GFP antibody. (B) Examination of α_{2C}-AR subcellular localization demonstrating that in HEK293 cells transfected with GFP-tagged α_{2C}-AR at 37 °C, the majority of the receptor is localized in the cytosol (left), whereas at 30 °C the receptor is clearly present at the cell surface (right). (C) The cell surface levels of the three α_2-ARs in transfected HEK293 cells at 37 °C (black columns) and at 30 °C (white columns) were measured by radioligand binding using [^{3}H]RX821002 as described.[13] Only α_{2C}-AR plasma membrane levels were significantly upregulated at reduced temperature. $n > 6$ in each case, $*p < 0.05$.

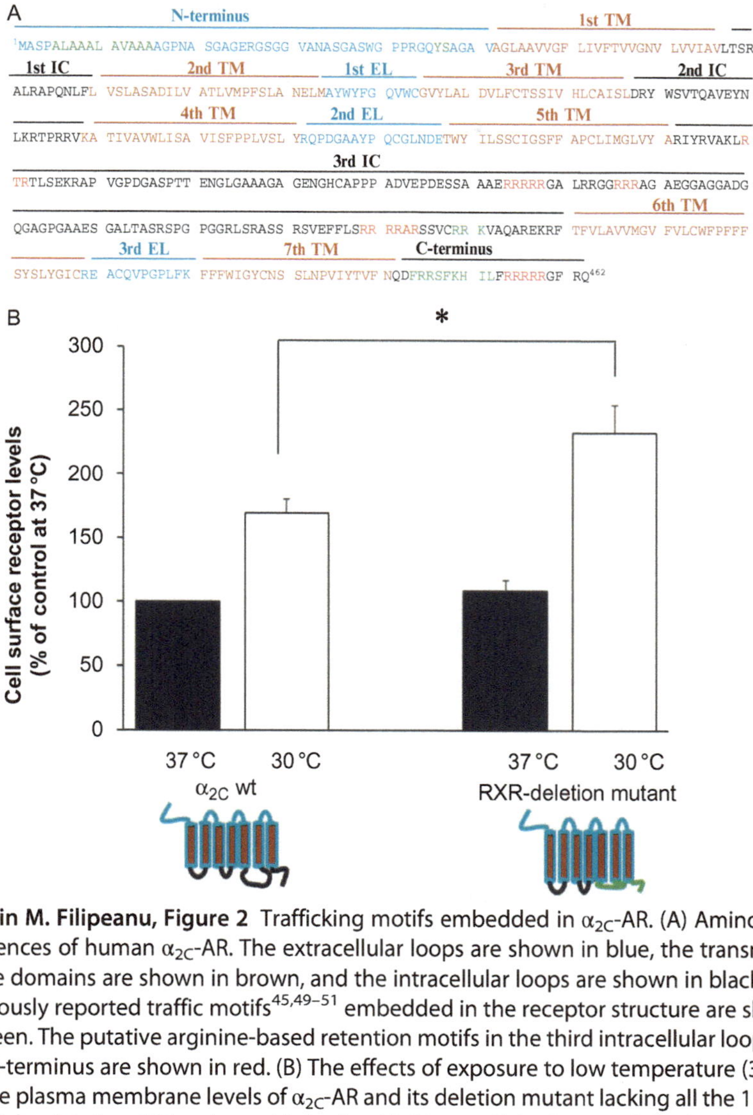

Catalin M. Filipeanu, Figure 2 Trafficking motifs embedded in α_{2C}-AR. (A) Amino-acid sequences of human α_{2C}-AR. The extracellular loops are shown in blue, the transmembrane domains are shown in brown, and the intracellular loops are shown in black. The previously reported traffic motifs[45,49–51] embedded in the receptor structure are shown in green. The putative arginine-based retention motifs in the third intracellular loop and the C-terminus are shown in red. (B) The effects of exposure to low temperature (30 °C) on the plasma membrane levels of α_{2C}-AR and its deletion mutant lacking all the 11 RXR motifs by deleting 125 amino acids in the third intracellular loop and the last 9 amino acids in the C-terminus. The experiments were performed in transfected HEK293 cells in similar manner as described.[13] $n > 6$ in each case, $*p < 0.05$. *Adapted from Ref. 52.*

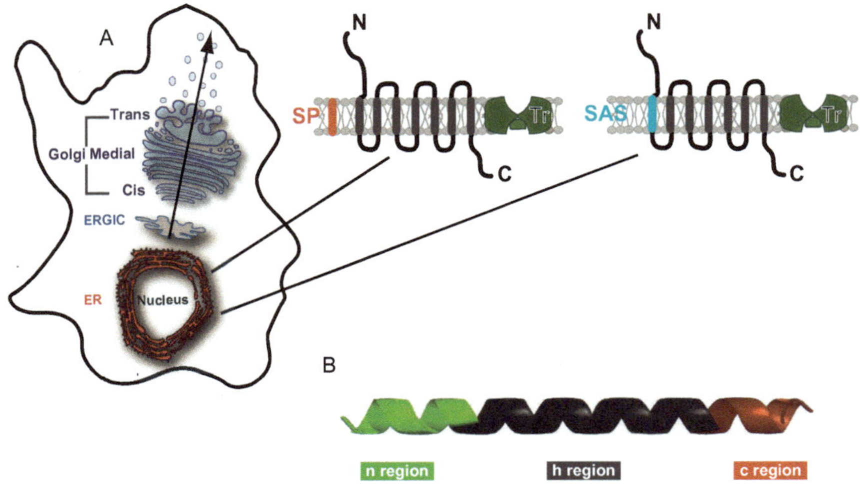

Claudia Rutz *et al.*, Figure 1 Intracellular transport of GPCRs. (A) Schematic depiction of the secretory pathway. Integration into the ER membrane is mediated by the translocon (Tr) and signal sequences of the receptors. GPCRs possess either signal peptides (SP, red), which are cleaved-off following ER integration or signal anchor sequences (SAS; usually TM1), which form part of the mature proteins. Receptors are then transported in the membrane of vesicles via the ER–Golgi intermediate compartment (ERGIC) and the individual compartments of the Golgi apparatus to the cell surface. (B) Basic architecture of signal peptides. At the N-terminus, signal peptides usually contain a short stretch of rather polar amino acid residues (n region). A longer hydrophobic core (h region) is followed by a C-terminal segment, which contains helix-breaking proline and glycine residues and small, uncharged residues at positions −1 and −3 of the cleavage site (c region). Signal peptides are thought to adopt an overall helical conformation.

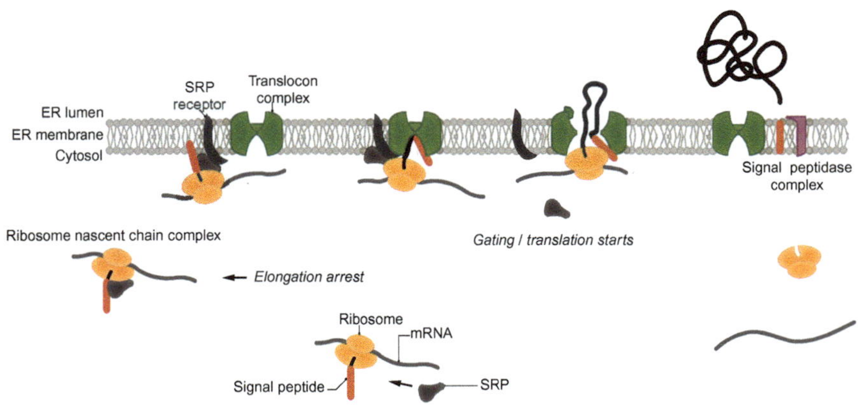

Claudia Rutz *et al.*, Figure 2 ER targeting/translocation of secretory proteins. The signal peptide (red) of the nascent chain associates with the SRP shortly after its synthesis at a cytosolic ribosome. Translation stops and the nascent chain/SRP/ribosome complex is targeted to the SRP receptor of the translocon complex of the ER membrane. The ribosomal tunnel exit gets in touch with the translocon and the signal peptide and the adjacent N-terminal region engage with the protein-conducting Sec61 channel in a hairpin conformation. After Sec61 gating, translation resumes and the nascent chain is translocated cotranslationally through the protein-conducting channel. The signal peptide leaves Sec61 at its lateral gate and the protein is released from the membrane at the luminal side by signal peptide cleavage through the signal peptidase complex.

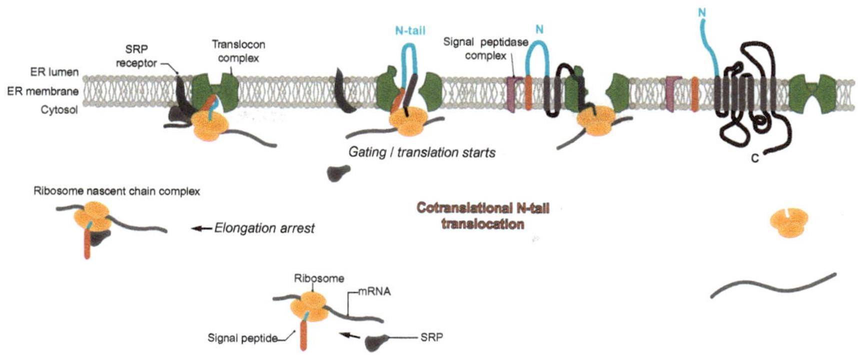

Claudia Rutz *et al.*, Figure 3 ER targeting/insertion of GPCRs possessing a signal peptide. The individual ER targeting/insertion steps of GPCRs containing a signal peptide represent a variation of the processes outlined for secretory proteins (see Fig. 2). As signal peptides are located at the extreme N-termini of the proteins, they prevent N-tail synthesis at cytosolic ribosomes by SRP binding and elongation arrest. Note that the N-tail can be translocated cotranslationally through Sec61 in this case following restart of translocation. Regarding membrane proteins, only the later extracellular domains are translocated into the ER lumen and the transmembrane domains leave Sec61 at its lateral gate, thereby anchoring the protein in the bilayer. The intracellular domains remain at the cytosolic side.

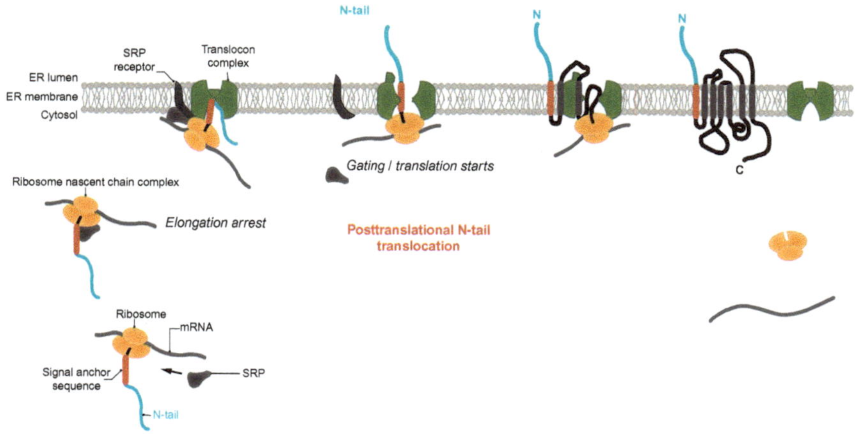

Claudia Rutz *et al.*, Figure 4 ER targeting/insertion of GPCRs possessing a signal anchor sequence. If GPCRs do not possess signal peptides, SRP binding and elongation arrest are mediated by signal anchor sequences. Note, that these domains are located C-terminally of the N-tails. The N-tails are consequently synthesized at cytosolic ribosomes until the signal anchor sequences appear and initiate the SRP-mediated elongation arrest. As a consequence, N-tails must be translocated posttranslationally through Sec61.

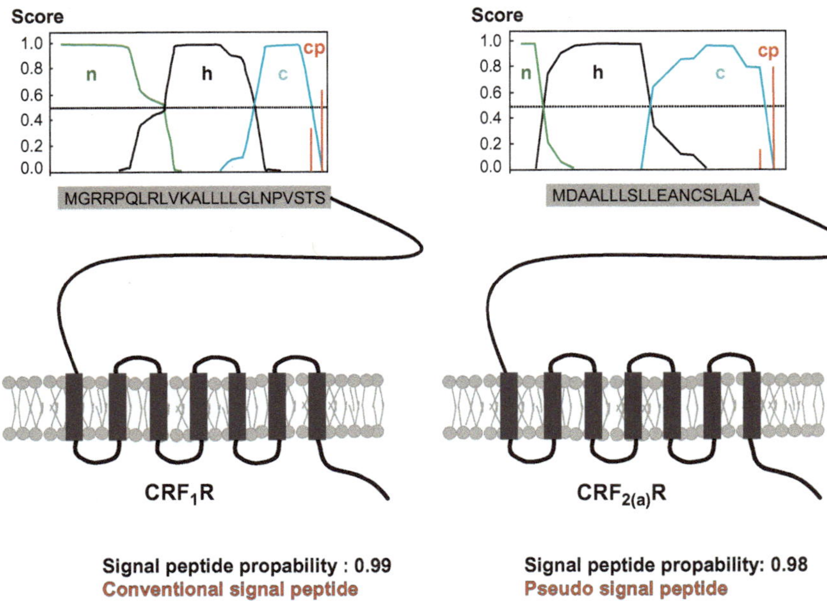

Claudia Rutz *et al.*, Figure 5 Depiction of the signal peptide sequences of the CRF_1R (left panel) and the $CRF_{2(a)}R$ receptor (right panel). The probabilities of the presence of n (green), h (black), and c (light blue) regions and the cleavage probabilities (cp, red) are indicated in a score ranging from 0 to 1. Note that a signal peptide was predicted in both cases. Whereas the CRF_1R indeed possesses a conventional and cleaved signal peptide, the $CRF_{2(a)}R$ contains a pseudo signal peptide. Predictions were carried out using the program SignalP.[6–9]

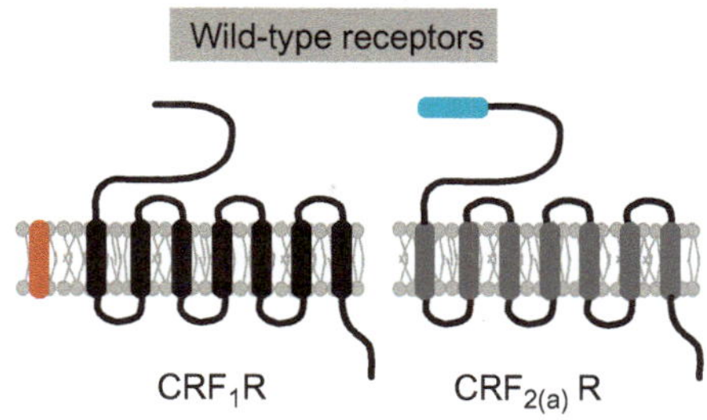

	CRF_1R	$CRF_{2(a)}R$
Receptor expression at the plasma membrane	High	Low
Dimerization	Dimer	Monomer
G protein coupling	Gs > Gi	Gs

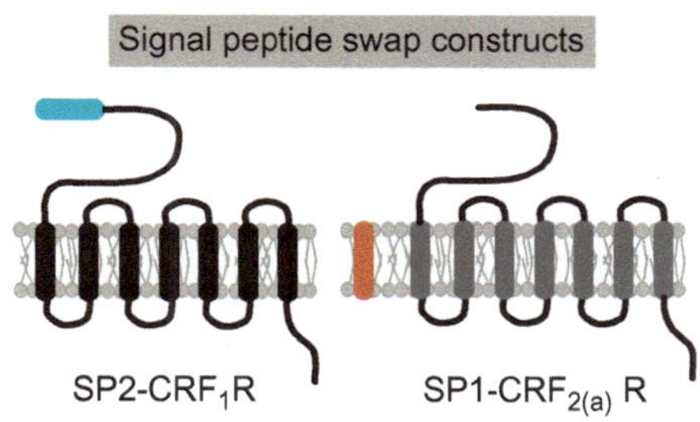

	SP2-CRF_1R	SP1-$CRF_{2(a)}R$
Receptor expression at the plasma membrane	Low	High
Dimerization	Monomer	Dimer
G protein coupling	Gs	Gs > Gi

Claudia Rutz *et al.*, Figure 6 Functional significance of CRF receptor signal peptides: summary of the available results. Upper panel: The CRF_1R possesses a conventional cleaved signal peptide and is readily expressed in the plasma membrane. It forms dimers and couples to both Gs and Gi. The $CRF_{2(a)}R$ instead contains an uncleaved pseudo signal peptide and is expressed mainly intracellularly. It forms monomers and is only able to couple to Gs. Lower panel: All properties mediated by the pseudo signal peptide could be transferred in signal peptide swap experiments. In the case of construct SP2-CRF_1R, the signal peptide of the CRF_1R was replaced by the pseudo signal peptide of the $CRF_{2(a)}R$, whereas in construct SP1-$CRF_{2(a)}R$, the pseudo signal peptide of the $CRF_{2(a)}R$ was replaced by the signal peptide of the CRF_1R.

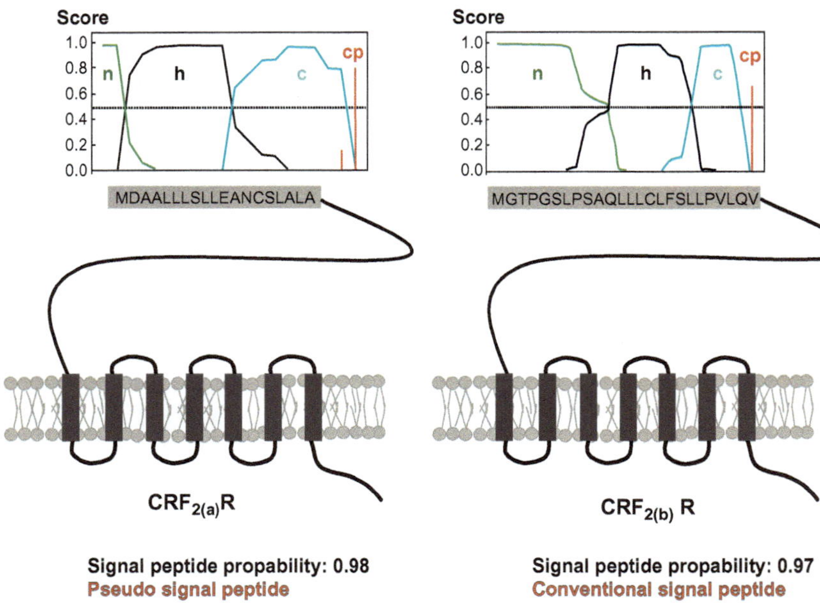

Claudia Rutz *et al.*, Figure 7 Depiction of the signal peptide sequences of the $CRF_{2(a)}R$ (left panel) and the $CRF_{2(b)}R$ (right panel). The probabilities of the presence of n (green), h (black), and c (light blue) regions and the cleavage probabilities (cp, red) are indicated in a score ranging from 0 to 1. Note that although the mature receptors are highly homologous, the signal peptides differ completely in sequence and function. Whereas the $CRF_{2(a)}R$ contains a pseudo signal peptide, the $CRF_{2(b)}R$ receptor possesses a conventional signal peptide similar to that of the CRF_1R. Predictions were carried out using the program SignalP.[6–9]

FSC
www.fsc.org
MIX
Papier | Fördert
gute Waldnutzung
FSC® C083411

Kolibri 360 GmbH
Ferdinand-Jühlke-Straße 7
99095 Erfurt, Deutschland
produktsicherheit@kolibri360.de